GENERAL EQUILIBRIUM ANALYSIS OF PRODUCTION AND INCREASING RETURNS

Series on Mathematical Economics and Game Theory

Published

Vol. 1: Theory of Regular Economies
by Ryo Nagata

Vol. 2: Theory of Conjectural Variations
by C. Figuières, A. Jean-Marie, N. Quérou & M. Tidball

Vol. 3: Cooperative Extensions of the Bayesian Game
by Tatsuro Ichiishi & Akira Yamazaki

Vol. 4: General Equilibrium Analysis of Production and Increasing Returns
by Takashi Suzuki

Forthcoming

Vol. 5: Fixed Points and Economic Equilibria
by Ken Urai

Series on Mathematical Economics and Game Theory

Vol. 4

GENERAL EQUILIBRIUM ANALYSIS OF PRODUCTION AND INCREASING RETURNS

Takashi Suzuki

Meiji-Gakuin University, Japan

NEW JERSEY • LONDON • SINGAPORE • BEIJING • SHANGHAI • HONG KONG • TAIPEI • CHENNAI

Published by

World Scientific Publishing Co. Pte. Ltd.
5 Toh Tuck Link, Singapore 596224
USA office: 27 Warren Street, Suite 401-402, Hackensack, NJ 07601
UK office: 57 Shelton Street, Covent Garden, London WC2H 9HE

British Library Cataloguing-in-Publication Data
A catalogue record for this book is available from the British Library.

GENERAL EQUILIBRIUM ANALYSIS OF PRODUCTION AND INCREASING RETURNS
Series on Mathematical Economics and Game Theory — Vol. 4

ISBN-13 978-981-283-331-0
ISBN-10 981-283-331-5

Typeset by Stallion Press
Email: enquiries@stallionpress.com

Printed in Singapore by Mainland Press Pte Ltd.

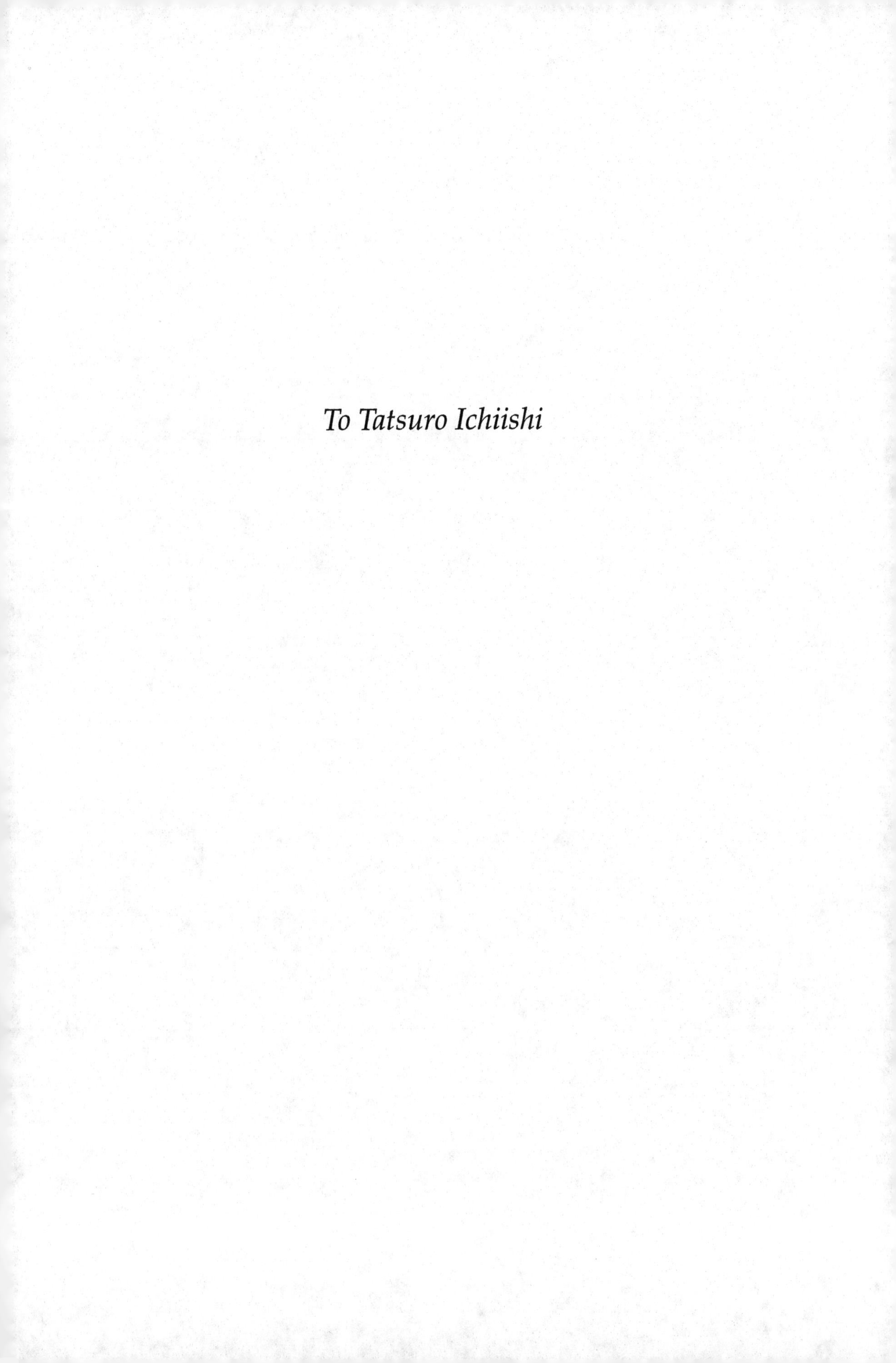

To Tatsuro Ichiishi

PREFACE

The economies of the modern advanced countries are characterized by the enormous extension of the markets, the rapid development of the technologies, and the further monopolization of the markets by small numbers of big companies. These characters of the markets are mutually related and, at least the last two of them belong to the Marshallian tradition of the economic thought. The modern general equilibrium theory which has essentially grown out of the Walrasian tradition so far, therefore, is not sufficient for considering the problems arising from the markets with these characters. In order to tackle these problems successfully, I believe that one has to go back to Marshall and construct the theory which unifies the ideas of the two great masters of the equilibrium analysis. The following chapters try to be a first step toward such a theory. I want to show in this book that the main body of the Marshallian concepts including the increasing returns and the monopolistic competition does indeed fit into the scheme of the modern general equilibrium theory. At the same time, I hope that this book will be a text book of the general equilibrium theory for graduate students. Therefore, its purpose will be achieved if it becomes a good companion for the students who are ambitious to contribute to any further developments of the theory.

From my own student time and in writing this book, I have been indebted to many people. I thank, in particular, Professors Kenjiro Ara, Marcus Berliant, Pierre Dehez, Jacques Dreze, Jean-Michel Grandmont, Kazuya Kamiya, Tohru Maruyama, Lionel McKenzie, Takashi Negishi, Kazuo Nishimura, Mitsunori Noguchi, Sin-ichi Takekuma, Akira Yamazaki, and Makoto Yano.

I am most thankful to Professor Tatsuro Ichiishi for inviting me to write this book. It is no doubt that the book would not be written if his invitation and encouragement did not exist. The whole volume of this book is dedicated to him.

I take this opportunity to express my gratitude to my wife Nobuko, since it is extremely doubtful whether the book could be completed without her help and support. I also want to thank my friend Mr. Susumu Sakuma who has cheered me up by his music. Finally, I express my gratitude to my cat *Kijiko* (1991 ~ 2008) and my dog *Tetsu* (1997 ~) for giving both of us the joy of life.

Takashi Suzuki
August, 2008 Tokyo

CONTENTS

INTRODUCTION

Chapter **1**

1.1. FROM THE 19TH CENTURY TO THE 1940s

The mathematical analysis of market equilibria was started by Cournot (1838). He constructed the first mathematical model of an economic market in which a downward sloping total demand curve was given and there existed finite number of firms whose technologies were given by the cost functions. In such an economy, he established the concepts of monopolistic equilibrium and oligopolistic equilibrium which now we call after his name.

Although limited within partial equilibrium analysis, these contributions are truly original and we cannot help being surprised by clarity and modernity of his analysis. Indeed, his profound insight had reached to almost the same perspective as that of modern theorists. He characterized perfectly competitive equilibrium as the limit of a sequence of imperfectly competitive equilibria with increasing number of producers. In other words, he had realized an effect of large number of economic agents in the market. This style of analysis was followed by Edgeworth (1881) and Debreu and Scarf (1963) in the study of the core of an exchange economy which will be discussed in Section 2.6 of this book.

Furthermore, in the Cournot's model, the total demand function $F(p)$ was not put into the model simply "by hands", but he gave an intuitive argument which justifies the function $F(p)$ to be continuous. He wrote:

> "... We will assume that the function $F(p)$, which expresses the law of demand or of the market, is a continuous function, i.e., a function which does not pass suddenly from one value to another, but which takes in passing all intermediate values. It might be otherwise if the number of consumers were limited....
>
> ... But the wider the market extends, and the more the combinations of needs, of fortunes, or even caprices, and varied among consumers, the closer the function $F(p)$ will come to varying with price in a continuous manner ... (1838, English translated version 1927, Section 22, pp. 49–50)."

What Cournot was explaining in the above statements is nothing but the "smoothing effect of aggregation" which we will discuss in Section 3.5 of this monograph.

As is well known, the market demand function was derived from the individual demand functions via utility maximization behaviors by Menger (1871), Jevons (1871), Walras (1874, 1877) and Marshall (1890).

Among these "marginal revolutionarists", the credit of establishing the general equilibrium model goes to Walras. He defined the general competitive equilibrium as a solution of an equations system; the system consists of the first-order conditions of utility maximization problems of the consumers, the first-order conditions of profit maximization problems of the firms, the budget equations of consumers, and the market conditions of the supply and the demand for each commodities.[1]

In the process of constructing his system of equations, Walras recognized that the only the $\ell - 1$ relative prices of the ℓ commodities were determined by the equations, and that the budget equations, when they were summed up over all households, made up of an identity equation which we now call after his name. Consequently, equations of the system were not all independent, and the number of equations was equal to the number of unknowns of the model. It seems that Walras was satisfied with this fact and he thought that it showed the consistency of his model. As we will see in Section 1.2, the question of the existence of economically meaningful solutions of the Walras system was left as a legacy to mathematical theorists of the next century.

Instead, Walras discussed the stability of equilibrium. His stability theory called as the *tâtonnements* is as follows.

The price adjustment process is considered in some definitive order of markets. The first market equilibrium is achieved by the change of price of that commodity which increases when the demand exceeds the supply and vice versa. Then the second market is adjusted in the same way. However, since the price of the second market affects the first market, the equilibrium of the first market will be disturbed. Therefore, when the price adjustment of the final market was completed, only this market will be in equilibrium. Walras assumed that the impact of the price change is the largest for its own market and alleged that the price of each market will be more close to the equilibrium price than what was before the adjustment process, and if one repeated this process, the market price systems will converge to the equilibrium price system.

The welfare economics of general equilibrium markets was studied by Edgeworth (1881) and Pareto (1909). Pareto formulated the maximum social welfare as a solution of the maximization problem of the social welfare function, and proposed the criterion of the social optimum which now we call after his name.

---

[1]We neglect markets of the intermediate (capital) commodities in his model.

Edgeworth considered a market model in which there exist two commodities and two traders (consumers) each of which has its own utility function and the initial endowment. He represented the model by a diagram which is now called by his name and used as a basic tool by modern theorists. He assumed that (a) the traders would not make a trade if there was another that would be more beneficial for both and (b) neither would make a trade that would make him worse off than in the absence of trade.

He called the set of allocations satisfying the conditions (a) and (b) the contract curve and observed that the competitive equilibrium allocation was contained in this set. He went on to suppose that the market which contained two types of traders rather than two individuals, and each type has the same number of traders with the same utility function and the same initial endowment. Edgeworth generalized the contract curve to this situation in which he assumed that no trade among any traders of any type would be completed if there existed a group of traders which could make another trade among themselves possible, using only their own initial endowments in such a way that nobody in the group was worse off and at least one member of the group would be better off than the initially proposed trade. It goes without saying that the set of allocations satisfying this condition is the core of the market game in the modern terminology.

Using the diagram neatly, Edgeworth concluded that as the number of individuals of each type became large, the core shrank to the competitive equilibrium allocation.

In the terminology of modern game theory, Cournot considered the competitive equilibrium as a limit of non-cooperative solutions of market games with increasing number of producers, and Edgeworth considered it as a limit of cooperative solutions of market games with increasing number of consumers.

Among many contributions of Marshall (1890, 8th edition 1920),[2] we are especially interested in his idea of external economies and increasing returns.

Marshall divided increasing returns to two types on account of their sources. His definitions are as follows.

> "We may divide the economies arising from an increase in the scale of production of any kind of goods, into two classes — firstly, those dependent on the general development of the industry; and secondly, those dependent on the resources of the individual houses of business engaged in it, on their organization and the efficiency of their management. We may call the former *external economies*, and the latter *internal economies* (1920, Chap. IX, p. 221)."

[2] Pigou said "It's all in Marshall."

Of course, as a theorist of the 19th century, he agreed that the limitations of production factors such as land naturally led to the decreasing returns. Whatever external or internal, he considered that increasing returns are something which are related to the efficiency of the human skill and technology.

> "... We say broadly that while the part which nature plays in production shows a tendency to diminishing return, the part which man plays shows a tendency to increasing return. The *law of increasing return* may be worded thus: an increase of labour and capital leads generally to improved organization, which increases the efficiency of the work of labour and capital (1920, Chap. XIII, p. 265)."

Therefore it seems fair to assume that (at least) most parts of "internal economies" in Marshall's sense correspond to non-convex production sets in the modern terminology, and "external economies" to (positive) external effects between individual firms when one tries to express his idea of economies of scale within a framework of modern equilibrium theories.

In any case, Marshall proposed these "economies of scale" as theoretical concepts, not simply as observations of actual facts. This means that he had assumed that the increasing returns whatever they were internal or external were consistent with his system as a whole, or more specifically, he assumed that the increasing returns were compatible with perfectly competitive equilibria.

At least, Marshall had recognized the inconsistency between the internal economies and the competitive (or price taking) behavior of individual firms. A part of the reason which supports his intuition seems that his equilibrium concept of perfect competition is something which is defined over long period of time so that free entries and exits are allowed, and distinguished from the temporary equilibrium. Invoking the famous metaphor of trees of the forest, he gave an image of his equilibrium concept in which each firm lives its own life time in the market.

> "But here we may read a lesson from the young trees of the forest as they struggle upwards through the benumbing shade of their older rivals. Many succumb on the way, and a few only survive; those few become stronger with every year, they get a larger share of light and air with every increase of their height, and at last in their turn they tower above their neighbours, and seem as though they would grow on forever, and forever become stronger as they grow. But they do not. One tree will last longer in full vigour and attain a greater size than another; but sooner or later age tells on them all. Though the taller ones have a better access to light and air than their rivals,

> they gradually lose vitality; and one after another they give place to others, which, though of less material strength, have on their side the vigour of youth (1920, Chap. XIII, p. 263)."[3]

Unfortunately, by the intuitive and verbatim nature of his exposition, the "Marshallian increasing returns (or equivalently decreasing marginal costs)" invited furious debates among theorists of the younger generations. Indeed one of the most critical opponent, Knight wrote in his reply (1925) to Graham (1925) who supported the external increasing returns that they were "empty economic boxes".

In his 1970 paper of the external increasing returns, Chipman wrote:

> "(The compatibility of increasing returns with perfectly competitive equilibrium) was once a lively subject of debate. The debate appears to have petered out in the 1930s, with nobody the apparent winner. That this was the outcome seems evident from later writings of some of the participants. Thus, Sir Dennis Robertson presented in 1957 on account which was substantially unaltered from his contribution to the 1930 Symposium on Increasing Returns, supporting the compatibility of increasing returns with perfect competition. On the other hand, Sir Roy Harrod in 1967 was able to state flatly, without any qualification as to whether economies were internal or external, that: "Increasing returns can, of course only occur if competition is less than perfect." In the contemporary international trade literature, some authors maintain that perfect competition can prevail under conditions of increasing returns, provided the economies of scale are external to individual firms; whereas others deny the compatibility of economies of scale with perfect competition under any circumstances, and with equal confidence[4] ... (1970, pp. 347–349)".

Although they did not reach any definitive conclusion as to whether the increasing returns were compatible with the perfect competition or not, the

[3]In many places of *The Principles*, Marshall emphasized the analogy between economics and biology. His attitude toward biology is sometimes compared to that of Walras toward mechanics.

[4]Lipsey states on pp. 511–512 (p. 277 of the reprinted version): "It is, of course, well known that exhausted economies of scale are incompatible with the existence of perfect competition, but it is equally well known that unexhausted economies of scale are compatible with the existence of imperfect competition as long as long-run marginal cost is declining faster [sic] than the marginal revenue (quoted by Chipman)."

controversies, from our point of view, had brought two important consequences as byproducts.

The first one is an idea due to Young (1928) that the (external) increasing returns were the force which drove the economy to grow. At first Young (1913) was skeptical about the increasing returns. But in the 1928 paper, he observed the division of labor resulting in the increasing returns was limited by the extent of the market, and conversely the extent of the market is, in turn, enlarged by the division of labor. As a consequence, he wrote "the division of labor depends in large part upon the division of labor," and he added that "this is more than mere tautology." According to him, the increasing returns are the source of a theory of growth. In fact, from the modern point of view we see clearly in the above statement the birth of the endogenous growth theory.

The second is the theory of imperfect competition which, by Newman (1960)s words, "rose from the ashes."

In one of the most influential paper (1926) which attacked the external increasing returns, Sraffa pointed out that the very concept of an industrial supply function was illegitimate, since it did not take into account of the interdependence of industries. He rejected the concept of competitive equilibrium itself, and proposed an alternative equilibrium concept, namely the monopolistically competitive equilibrium.

Two influential monographs appeared along this line of research, Robinson (1933) and Chamberlin (1933). An essential property of these theories is that firms in the market do not take prices as given when they make their production decision, which means that, in the context of the partial equilibrium analysis, each firm faces a downward sloping subjective demand curve. Robinson (1933) pursued the study of the welfare properties of imperfectly competitive equilibrium using the consumer surplus. Chamberlin (1933) introduced the commodity differentiation and free entry of the firms into the model, and clarified the concept of monopolistic competition.

The purpose of the present monograph is to shed some light on these Marshallian tradition of the economic thought, increasing returns and monopolistic competition from the perspective which has grown out of the Walrasian tradition of general equilibrium theory (see the following sections.). The goal of the book is Chap. 5 in which we will demonstrate the existence of a competitive equilibrium with external increasing returns in the dynamic infinite time horizon economy whose equilibrium consumption and production paths grow without bound. This model is considered to be a realization of the Young's view of increasing returns.

In Chap. 6, we will present and demonstrate a monopolistically competitive equilibrium of a market in which each firms perceives the (subjective) downward sloping demand curve and its technology exhibits a kind of (internal) increasing returns coming from large setup costs.

1.2. AFTER THE 1950s; EXISTENCE OF EQUILIBRIUM AND CORE LIMIT THEOREM

Mathematical foundations of the modern general equilibrium theory were established by the results of Arrow and Debreu (1954), McKenzie (1954), and Nikaido (1956) which proved the existence of the competitive equilibrium of the Walrasian market with finite number of commodities and economic agents. In these papers, all basic concepts such as the commodity space, the prices, the agents' characteristics, and so on are represented by the set theoretic concepts, and since then, the general equilibrium theory has been a "geometry of agents and commodities". Indeed, theorems on the competitive equilibrium and its welfare properties are stated and proved in terms of the topology and convex analysis, and the key results used there were the fixed point theorems and the separation hyperplane theorems (see Section 2.5).

By these mathematical techniques, the question of existence of equilibrium left by Walras has been answered affirmatively and definitively, probably the problem has been solved more generally than Walras himself has expected.

The success continued to solve the Edgeworth's conjecture on the core limit theorem by Debreu and Scarf (1963). They generalized Edgeworth's result for two commodities and two consumers to the case of arbitrarily finite number of commodities and types of consumers with strictly concave utility functions. More specifically, they considered a sequence of economies, as Edgeworth did, in which the number of consumers becomes large, proportionally for each type, and proved that the core allocations shrink to the competitive equilibrium in an appropriate sense. In a crucial step of the proof, the separation hyperplane theorem was used (see Theorem 6.1 of Chap. 2).

In view of these results, the power of mathematical method used in the economic analysis has been evident; the economic concepts and propositions have been clarified and the open questions and conjectures held by theorists in the 19th century have been solved definitively. In later chapters of this book, we will present the several results which have been obtained by the modern mathematical technique, and we will pursue this stream of researches further.

1.3. THE LOCAL UNIQUENESS AND STABILITY OF EQUILIBRIA

Once the existence of equilibrium has been established, we are interested in its stability and uniqueness. Walras' *tâtonnements* was formulated by Samuelson (1947) as a system of differential equations. This work was followed by Arrow and Hurwicz (1958) and Arrow, Block and Hurwicz (1959).

The uniqueness of equilibrium was first studied by Wald (1936) and he gave alternative sufficient conditions for the uniqueness; the weak axiom of revealed preference holds for the market excess demand functions, or all commodities are gross substitutes. Both of the conditions are quite strong, but Wald's results stimulated almost all later studies of this issue. Samuelson (1948, 1953–1954) discussed the problem in the context of factor price equalization in the world trade. Gale and Nikaido (1965) pointed out an error in Samuelson's argument and they generalized the condition of gross substitutes of Wald. A quite general condition was given by Dierker (1972), using a technique of differential topology.

The purpose of all researches given above has been to get some sufficient conditions for the (global) uniqueness of equilibrium. What these studies have made clear is that the equilibrium which is globally unique is very special and appears under very strong conditions.

Debreu (1970) discussed the problem from a drastically different angle. He studied the local uniqueness of equilibria, namely the equilibria which are discrete and locally stable, which means they change continuously when the agents' characteristics are perturbed continuously. An economy in which the competitive equilibria are discrete and locally stable is now called the regular economy, otherwise it is called a singular economy. The regularity is an important property in economic analysis, for most applications of theoretical model are performed by the method of comparative statistics, which perturbs the economic parameters slightly, and examines the associating changes of equilibria. Hence, it will lose any predictive power for singular economies. The regularity of equilibria is a theoretical basis of the comparative statistics.

Debreu did not ask any sufficient conditions under which economies become regular. Rather, he asked how generally economies are regular, and his answer was that "almost all" economies are regular. We explain this point more precisely, since this result has brought a drastic change of economic analysis methodologically and philosophically.

Recall that in the study of the core limit theorem, Edgeworth, Debreu, and Scarf have already considered a *sequence* of economies. But here, Debreu considers a *topological space* of economies. He showed that under suitable differentiability assumptions, the set of singular economies is a closed subset with the Lebesgue measure zero of the whole space of economies.

Before this paper appeared, economists had been interested in an *individual economy* and examined its economic properties. Debreu opened up another possibility to be explored, namely the mathematical structures of the *space of economies*.

Methodologically, his paper introduced the differential topology to economic analysis. Sard's theorem has been added to theorists' tool box (see Appendix E). We will review this theory in Section 2.8.

1.4. MARKETS WITH A CONTINUUM OF TRADERS

In a competitive equilibrium, economic agents behave as market prices are given to them by definition. This price-taking behavior has been usually justified in such a way that there exist a very large number of traders in the market and the influence of each individual on prices is negligible. However, this picture does not fit into the traditional economic models with finite number of traders which have been discussed so far.

Aumann (1964) introduced a model of an exchange economy with a continuum of consumers in order to present the idea of perfect competition rigorously from the mathematical point of view, and showed in such an economy, the set of core allocations coincides with the set of competitive allocations. Edgeworth's conjecture was realized as the core equivalence theorem (Section 3.4).

The actual markets certainly contain only finitely many traders. What is the economic meaning of the market with a continuum of traders? Aumann explained it by an analogy of physics. He wrote:

> "Actually, it is no stranger than a continuum of prices or of strategies or a continuum of "particles" in fluid mechanics. In all these cases, the continuum can be considered an approximation to the "true" situation in which there is a large but finite number of particles (or traders or possible prices) ... There is perhaps a certain psychological difference between a fluid with a continuum of particles and a market with a continuum of traders. Though we are intellectually convinced that a fluid contains only finitely many particles, to the naked eyes it still looks quite continuous. The economic structure of a shopping center, on the other hand, does not look like continuous at all. But, for the economic policy maker in Washington, or for any professional macro-economist, there is no such difference. He works with figures that are summarized for geographic regions, different industries, and so on; the individual consumer (or merchant) is as anonymous to him as the individual molecule is to the physicists ... (1964, p. 41)."

Hildenbrand (1974) gave more statistical explanations. As will be shown in Chap. 3, the economic model with a continuum of traders induces an atomless or continuous distribution over the set of agent's characteristics such as the income distribution.

Then he wrote:

> "To view the distribution of agent's characteristics of a *finite* set A of agents as atomless distribution means, strictly speaking, that the "actual" distribution is considered as a distribution of a *sample* of size $\#A$ drawn from a "hypothetical" population.[5] This statistical point of view is based on the well-known fact that the sample distributions converge with increasing sample size to the hypothetical distribution (1974, p. 110)."

The studies of general equilibrium models with finitely many traders has relied heavily on the convexity assumptions on the agents' characteristics. For instance, the existence of the competitive equilibria (Theorem 5.1) is obtained under the assumption that every consumer has a convex preference defined on a convex consumption set.

Aumann (1966) showed that in an exchange economy with a continuum of consumers, the existence of equilibrium could be obtained without the convexity assumption on preferences. Mathematically speaking, this is a consequence of the Liapunov's theorem (Theorem H1) which says that the integral of a measurable correspondence over an atomless measure space is convex valued. Therefore, even if individual demand correspondences are not convex valued, the market demand correspondence, which is defined as the integral of the individual demand correspondence over the atomless measure space of consumers, is convex valued, and we can apply Kakutani's theorem.

In subsequent studies, Mas-Colell (1977) proved the existence of equilibria without the convexity assumption on the consumption sets in an exchange economy with indivisible commodities, or the commodities which can be consumed in the unit of integers. Hence, the consumption set of this case is $\mathbb{R}_+ \times \mathbb{N}_+^{\ell-1}$, where $\mathbb{R}_+$ is the coordinate of a divisible commodity and the other $\ell - 1$ commodities are indivisible, which are obviously not convex set. Yamazaki (1978) generalized this result to the consumption sets which are just closed and bounded from below.

When one does not assume that the consumption sets are convex, the individual demand will generally exhibit discontinuous behaviors for continuous changes of the price system. The key observation of Mas-Colell and Yamazaki is that the discontinuous behaviors of the individual traders will be smoothed out when the distribution of agents' characteristics such as the income distribution is sufficiently dispersed. This phenomena is exactly what Cournot had already expected over 150 years ago. We will explain the "smoothing effect of the aggregation" in Section 3.5 and discuss the problems which will arise when one consider the production in Section 3.6.

---

[5] $\#A$ means the number of elements of the set A.

1.5. INCREASING RETURNS AND MONOPOLISTIC COMPETITION

The idea of external increasing returns of Marshall was made clear by Chipman (1970). He called it "parametric economies of scale", since in his formulation, each firm is supposed to take a scale parameter in its production function as given and believe that it operates under constant returns to scale, but actually the parameter is affected by the total amount of the input level of the industry as a whole, and consequently, the objective production function of the firm exhibits the increasing returns to scale. We give a simple example to illustrate the idea.

Suppose that there exist ν identical firms in an industry, each of which has the same production function $y = kz$, where y is output and z is input, and the coefficient k is a parameter. The firm takes k as given when it makes the production decision, so that its subjective production function is constant returns to scale, but actually k is assumed to be determined endogenously at the level of the total input of the industry, $\sum_{j=1}^{\nu} z_j$, where z_j is the input level of the firm j. That is, assuming that all firms use the same amount of the input z, $k = \nu z$ holds in equilibrium. Then, the firm's objective production function is $y = \nu z^2$, which exhibits the increasing returns and the resulting equilibrium concept is a competitive equilibrium with production externalities.

The idea of the parametric economies of scale originally came from Edgeworth. In the midst of the desperately confusing debates on the compatibility between competitive equilibria and the increasing returns (decreasing costs) which we saw in Section 1.1 (see also Notes for this section), he was looking at the truth. We quote Chipman:

> "The essential idea put forward by Edgeworth (1905, pp. 66–68; Papers, III, pp. 140–141) was that marginal cost was a function of a particular firm's output, and also of aggregate industrial output; and that it might be rising with respect to the former and falling with respect to the latter. According to this conception, rising marginal cost curves for the individual firms would shift downwards with a rise in industrial output, leading to a falling supply curve for the industry ... (1965, p. 739)."

For the first time in the history of this subject, Chipman (1970) showed that the external increasing returns were indeed compatible with competitive equilibrium and even Pareto optimality in one consumer economy. In Chap. 4, we will generalize his result to the case of several consumers. This generalization will play an essential role in Chap. 5.

Much efforts have been devoted to generalize the classical general equilibrium models of production economies to models including the (internal) increasing returns to scale technologies. This increasing returns, or the non-convex production sets are of course incompatible with the competitive behavior of firms. Instead, the firms are assumed to operate under some pricing rule, for instance, the pricing rule which satisfies the first-order condition of profit maximization. Mathematically speaking, the pricing rule in this case is what assigns to each firms a normal vector for each efficient production plan. This pricing rule is called the marginal cost pricing (MCP) rule.

Under the MCP rule, a firm with increasing returns to scale technology possibly earns the negative profit in the equilibrium. Such losses are assumed to be covered by lump sum taxes collected from the household in the economy. Therefore, the firms with increasing returns technologies in this theory can be thought of as privately owned public utilities, which are regulated. Mantel (1979) and Beato (1982) proved the existence of the MCP equilibrium.

On the other hand, the study of general equilibrium model of imperfect competition has been relatively poor, because of its fundamental difficulties, one conceptual and the other technical. As we have been seen, most of the results on general equilibrium theory have relied on the convexity assumptions. Theretofore, the theory has covered only the firms with constant or decreasing returns to scale technologies with a few exceptions such as MCP theory explained above. However, as a matter of fact, the monopolistically competitive firms with constant or decreasing returns to scale technologies are possibly rare, so that the economic meaning of the models with such firms seems to be restricted. Moreover, from the technical point of view, even if one assumes that the monopolistically competitive firms have convex technologies, the conditions on the fixed-point map to which Kakutani's theorem can be effectively applied are not generally guaranteed, as Roberts and Sonnenschein (1977) showed.

Nevertheless, we have two remarkable achievements: those by Negishi (1961) and Gabszewicz and Vial (1972).

Negishi introduced the downward sloping subjective demand function for each firms which passes through the equilibrium point, meaning that the firms observe the prices consistently with the equilibrium. Gabszewicz and Vial (1972) constructed a model in which the firms' behaviors are monopolistically competitive in the sense of Cournot and they make their production decisions using the true market demand function.

Both of the papers proved the existence of equilibria under the assumption that both the production functions and the profit functions are concave, which were necessary by the reason explained above.

In Chap. 6, we will generalize the Negishi type model of monopolistic competition to the case in which the firms are allowed to have a non-convex production sets which comes from large setup costs. We will prove the existence of equilibrium by using techniques which have been developed by Dehez and Dreze (1988a) in the study of equilibrium with the increasing returns and the pricing rule.

1.6. MARKETS WITH INFINITELY MANY COMMODITIES

An economic model of infinite time horizon first appeared in the context of the optimal growth theory by Ramsey (1928). His problem was to find the level of savings which would maximize a utility sum over future time for a population. Von Neumann (1937, 1945) presented the general equilibrium model of growth, in which there were no demand functions, only productions of linear activities. The economy is in equilibrium at each time period, and the equilibrium configurations were the same from period to period (stationary equilibrium). In order to prove the existence of equilibrium, he generalized a saddle point theorem of a bilinear form which was used to show the existence of equilibrium in two person zero-sum games. He deduced the saddle point theorem from a fixed-point theorem, and it was the first time that the fixed-point theorem appeared in the economic analysis.

The need of infinite-dimensional commodity spaces for general equilibrium analysis was pointed out clearly by Debreu (1959). In Note 2 of Chap. 2, he wrote:

> "The assumption of a finite number of dates has the great mathematical convenience of enabling one to stay within a finite-dimensional commodity spaces. There are, however, conceptual difficulties in postulating a predetermined instant beyond which all economic activity either ceases or is outside the scope of the analysis. It is therefore worth noticing that many results of the following chapters can be extended to infinite-dimensional commodity spaces. In general, the *commodity space* would be assumed to be a linear space L over the reals and instead of a price vector p, one would consider a linear form v on L defining for every action $a \in L$ its *value* $v(a)$. In this framework could also be studied cases where the date, the location, the quality of commodities are treated as continuous variables (1959, pp. 35–36)."

This program has been first carried out by Peleg and Yaari (1969) and Bewley (1970), both of which proved the existence of equilibrium for infinite time horizon economies. The commodity space of Peleg and Yaari is the space of all sequences endowed with the product topology, or the space $\mathbb{R}^\infty$. In order to prove the existence, they applied a Debreu-Scarf type core limit theorem. The commodity space of Bewley's paper is the space of all bounded sequence (or the space of all essentially bounded measurable functions, see Appendix H), which is denoted by ℓ^∞

(or L^∞ for the function space). He considered a sequence of equilibria of finite, say ℓ-dimensional sub-economies and showed that the limit of the finite-dimensional equilibria as $\ell \to \infty$ is indeed an equilibrium of the original infinite-dimensional economy.

Mas-Colell (1975) and Jones (1984) worked with the space of Borel measures on a compact set K, denoted by $ca(K, \mathcal{B}(K))$, where the intended economic meaning of the set K is the set of commodity characteristics. The commodity vector $\boldsymbol{x}$ is postulated as a measure on K and the value (distribution) for a Borel set $B \subset K, \boldsymbol{x}(B)$ represents the quantity of characteristics contained in the set B. This commodity space is considered to represent the commodity differentiation in a most general form.

Mas-Colell (1986) proved the existence for a general topological vector space including ℓ^∞ and $ca(K, \mathcal{B}(K))$ in an exchange economy, and Zame (1987) generalized his result to a production economy.

These accomplishments of the past will culminate in our main result in Chap. 5, the existence of competitive equilibrium of infinite time horizon economy with external increasing returns.

1.7. THE ORGANIZATION OF THE BOOK

The organization of this book is sketched by the table of contents of this book.

Chapter 2 is devoted to present the classical results on the exchange economies. In this chapter, we fix the notations and basic terminologies, and serve it as an introductory chapter for beginning graduate students.

Chapter 3 presents the theory of the large economy developed by Aumann, Hildenbrand, Mas-Colell and others. We will prove the existence and the core equivalence of the competitive equilibrium. We will also discuss the smoothing effects on the aggregated demand which are particularly interested in the model with many consumers. In Section 3.6, we will introduce the production into the economy for the first time in this book and discuss a problem on the existence of a production economy in which every consumer has a non-convex consumption sets.

In Chap. 4, we begin to study the production economies as a main theme. After discussing the classical competitive equilibria with decreasing returns to scale, we generalize the result to the case of (internal) increasing returns to scale and prove the existence of an equilibrium and the core following the classical paper by Scarf (1986). In Section 4.3, we will present the concept of the external increasing returns in most general form and prove the existence of the equilibrium on a finite-dimensional commodity space. This result will be used in the main result of the book of the next chapter. The Pareto optimality and the tax-subsidy policy in the presence of the increasing returns will be also discussed in Section 4.4.

Chapter 5 is devoted to study the equilibrium theory with infinite-dimensional commodity spaces which has been developed by Bewley, Mas-Colell, Jones, Zame, and many others. The main result which proves the existence of the competitive equilibrium in an infinite time horizon economy with external increasing returns will appear in Section 5.4.

We conclude the monograph by Chap. 6 in which the existence of Negishi type monopolistically competitive equilibrium with a large setup cost will be proved.

The author tried to take care of the balance between the mathematical generality and the economic idea. Therefore, the results in the text are not necessarily presented in completely general forms from the mathematical point of view. Rather, we hope to give educational proofs which give the readers mathematically essential points behind the economic ideas of those theorems. We found that in most cases, this requirement was achieved by the original proofs. Therefore, in many places of the book, we followed the proofs by the original authors of the results. For the mathematically general results in the literature, we refer them in the notes of each chapter.

Basically all mathematical techniques used in the text will be provided by mathematical appendices from A to H. Appendices A to D are the mathematical foundations for the book as a whole, while Appendix E is relevant to Sections 2.8 and 5.6. Appendices F and G are used in Chap. 3, and Appendix H will be needed for Chap. 5. References for the readers who are interested in the details including the proof of the mathematical results are found in Appendix I.

Definitions, theorems, and propositions are numbered by the section to which they belong and the order in which they appear. For instance, Theorem $a \cdot b$ means that it is the b-th theorem in the a-th section in some chapter. We do not indicate the chapter numbers in them, since we have no problems to refer them within the same chapter which you are now reading, and we do not want to take care of too many numbers. If we have to specify the chapter of theorems, we say such as "According to Theorem $a \cdot b$ of Chapter x ..." On the other hand, for the numbering of figures, we will indicate the chapter numbers rather than the section numbers, since we will have not too many figures in each chapter. For example, we mean by Fig $x \cdot a$ the a-th figure which appears in Chapter x.

1.8. NOTES

Section 1.1: For the modern discussions of the "limit theorem" of imperfectly competitive equilibria of Cournot, see Novshek (1980), Novshek and Sonnenschein (1983), and Mas-Colell (1983). See also the symposium issue of *Journal of Economic Theory*, 1980, volume 22 no. 2, A. Mas-Colell (Editor).

We are indebted for the descriptions of the work of Walras and Edgeworth for Arrow and Hahn (1971), Chap. 1.

For the controversies on the increasing returns, Chipman (1965) reports:

> "Marshall's only definition consists in the statement (*Principles*, p. 266 (p. 221 in the new edition)) that external economies are those economies of scale which are "dependent on the general development of the industry." The absence of any more elaborate formal definition in Marshall's writing is so conspicuous that it must be interpreted as deliberate; Robertson used the term "evasive" (cf. Newman (1960, p. 601)). In an earlier skeptical paper Robertson (1924, p. 26), after enumerating the usual examples (including the inevitable trade journal) sighed: "we have all at some time tried to memorize and to reproduce the formidable list." In the same year, Knight (1924, p. 597) set forth his famous objection to the concept of external economies in the words: "external economies in one business unit are internal economies in some other, within the industry." ... Knight's paper was largely a criticism of the concept of external economies — as was Robertson's 1924 paper — as used both by Pigou (1920) and by Graham (1923). Graham based his argument for protection on an analysis which took for granted the compatibility of perfect competition and increasing returns; this very assumption is what was challenged by Knight, and as long as Knight's objection stood, Graham's entire argument — whatever other defects it had, and there were several — was vitiated by having this as its premise. In his reply to Knight, Graham (1925) failed to come to grips with the main issue; and Knight (1925) in his rejoinder fairly placed the burden of proof on those who believed that competitive conditions could be reconciled with increasing returns. In saying with respect to external economies that "I have never succeeded in picturing then in my mind," Knight (1925, p. 323) was undoubtedly expressing a feeling that was widespread but suppressed, owing to the authority of Marshall and Pigou (Chipman (1965), pp. 740–41)."

Section 1.2: In a series of papers (1933–1934, 1934–1935), Wald studied the existence of competitive equilibria in alternative market models including a linear technology model and pure exchange model. Nuemann introduced a fixed-point theorem in a paper (1937) of balanced growth model. For the results of Wald and related works, see Arrow and Hahn (1971), Section 5 of Chap. 1.

It should be emphasized that the proof of the existence of Nash equilibrium by Nash himself (1950) had a close theoretical connection with Arrow and Debreu (1954). In fact, their method of proof was to invoke a version of Nash's theorem (Debreu, 1952) to an abstract game constructed from the given market model. See also Notes of Chap. 2.

Nishino (1971) is also an important contributor to the core limit theorem.

Section 1.3: For the textbook level of expositions of Walras' *tâtonnements*, see Arrow and Hahn (1971), Chaps. 11 and 12 and McKenzie (2002), Chap. 2. For the comparative statistics, see McKenzie (2002), Chap. 4. For the local uniqueness, see Section 2.8 of the present monograph and Notes for this section.

Section 1.4: We will discuss this topic in Chap. 3. For references, see Notes for this chapter.

Section 1.5: In a subsequent paragraph, Chipman also wrote:

> "To illustrate the case, an expansion in a certain industry may make possible a further division of labor, and give rise to new categories of technicians. The contribution of each individual firm to this process maybe so negligible that no single entrepreneur will take into account the effect of his own scale of operations on the development of new specialized skills. This element of cost therefore plays the same role as do market prices. It is curious indeed that Edgeworth, of all people, did not notice the analogy between this concept of external economies and his own limit theorem justifying the competitive price mechanism (cf. Edgeworth, 1881, pp. 240–243). All we need to assume is that a firm's size has a small effect (negligible from its point of view) on the organization of the industry (especially the labor market), and that the the firm consciously adjusts its organization to the changed condition of the industry (ibid, p. 740)."

Section 1.6: We will elaborate equilibrium theory with infinite-dimensional commodity spaces in Chap. 5. See Notes for this chapter.

CLASSICAL EXCHANGE ECONOMIES

Chapter

2

2.1. COMMODITIES AND MARKETS

In every economic phenomenon, we look at two major groups of "actors", namely "economic agents" and "commodities". The "stage" on which they act is called a market. The market is an abstract field in which the agents trade the commodities with each other.

If we continue this description, then it will contain a self-circulation and almost will not make any sense. In any case, the agents exchange or trade commodities in the market and whatever it is, the market is something which consists of the agents and commodities. This suggests that "the agents", "the commodities" and "the market" are the most fundamental concepts, which are to some extent, given a priori to our science. Therefore, any conceptually formal, independent, and self-contained definitions of these concepts are probably impossible and perhaps not useful, rather our understanding of these fundamental concepts is essentially based on our experience in the real economic life of every day. Moreover, even if we succeed to formalize an abstract economic theory which is suitably constructed from the mathematical point of view, the theory will implicitly assume a variety of social institutions outside the economic area, for instance, the law, politics, and history and other disciplines, which cannot be completely specified and usually not even mentioned at all. We hope that the readers bear in mind these remarks when they read the following explanations of the basic concepts.

We will start with the commodities. It is assumed that there exists a market open for each commodity. As it is well-known (see Debreu, 1959, for instance), a commodity is distinguished from each other by its physical character, the location, and the time at which it is available to traders. We can arrange the commodities in a suitable order as a form of finite series or vectors such as:

$$(x^1, x^2, \ldots, x^\ell)$$

when we assume that there are ℓ distinguishable commodities. In this case, the commodity vectors which are typically denoted by $\boldsymbol{x} = (x^1 \cdots x^\ell)$ are considered to be the elements of ℓ-dimensional Euclidean space $\mathbb{R}^\ell$. A commodity which is made available *to* an economic agent is called an input for him/her. What is made available *by* an economic agent is called an output for him/her. For the economic agents called the consumers (see Section 2.2), the input commodity is called the consumption commodity, and represented by positive quantity of the corresponding coordinate of the commodity space. For him/her, labor is an example of the output commodity and represented by the negative quantity. In Chap. 4, we introduce the economic agents called the firms. The output commodities of the firms are represented by the positive quantity, and the input negative quantity. The convenience of this notational convention will be clear in the subsequent chapters.

In Chap. 5, we will assume that there are infinitely many commodities, hence in this case, the commodity vectors are infinite sequences such as:

$$(x^0, x^1, \ldots, x^t, \ldots)$$

NB: Throughout this book, we define the set of natural numbers $\mathbb{N}$ as the set of nonnegative integers, or $\mathbb{N} = \{0, 1, 2, \ldots\}$. Then, it is notationally convenient to start the commodity index with zero when there are infinitely many commodities in the market.

The need for such infinite vectors come up with our motivation that commodities should be indexed sometimes by infinitely many indices. A typical case is that the commodity index means the time period at which the commodity is delivered to consumers. In such a theory, it seems natural not to specify the terminal period of trades (infinite time horizon model). We will use the commodity index t even if the number of commodities is finite.

We denote by L the set of all commodity vectors (commodity bundles) and call L, the commodity space. On the space L, the sum of $\boldsymbol{x}+\boldsymbol{y}$ and the scalar multiple $\alpha\boldsymbol{x}$, $\alpha \in \mathbb{R}$ are defined in an obvious way. Hence, L is either a finite or an infinite-dimensional vector space.

For each commodity t, we can observe a market price p^t at which the commodity t is traded among the economic agents (traders). The prices will also be made of a vector:

$$(p^1, p^2, \ldots, p^\ell)$$

when there exist ℓ commodities, and:

$$(p^0, p^1, \ldots, p^t, \ldots)$$

when there exist infinitely many commodities. The value of a commodity vector $\boldsymbol{x} = (x^1 \cdots x^\ell)$ evaluated at a price vector $\boldsymbol{p} = (p^1 \cdots p^\ell)$ is defined by:

$$\sum_{t=1}^{\ell} p^t x^t$$

and written by $\boldsymbol{px}$. When the number of commodities is infinite, we can define the value $\boldsymbol{px}$ similarly as:

$$\sum_{t=0}^{\infty} p^t x^t,$$

when this value is well defined.

Mathematically speaking, it is natural to define the price vector as a linear functional or a linear form (linear and real-valued function) on the vector space L of the commodity vectors. Let π be a price functional. The value of π at $\boldsymbol{x} \in L$ is denoted by $\pi(\boldsymbol{x})$. Then, for every $\boldsymbol{x}, \boldsymbol{y} \in L$ and every real number $\alpha, \beta \in \mathbb{R}$, it follows by definition of linearity that:

$$\pi(\alpha \boldsymbol{x} + \beta \boldsymbol{y}) = \alpha\pi(\boldsymbol{x}) + \beta\pi(\boldsymbol{y}).$$

In mathematical terminology, the set of all linear and continuous functionals on L is called the dual space of L and denoted by L^*. For the topologies and the theory of linear functionals on the vector spaces, we refer Appendices E and H for the readers who are not familiar with these mathematical concepts.

Let $\boldsymbol{e}_t = (0 \cdots 0, 1, 0, \ldots) \in L$ be the commodity vector, which contains 1 unit of the commodity t and 0 unit of the commodity for all $s \neq t$. Let $\pi \in L^*$ be a price functional and set $p^t = \pi(\boldsymbol{e}_t)$. Then, the price functional π has a vector representation which is denoted by $\boldsymbol{p} = (p^0, p^1, \ldots p^t, \ldots)$ as shown below. In this case, we call $\boldsymbol{p} = (p^t)$ the price vector.

Let $\boldsymbol{x} = (x^0, x^1, \ldots, x^t, \ldots) = \sum_{t=0}^{\infty} x^t \boldsymbol{e}_t$ be a commodity vector. For each T, we have:

$$\pi\left(\sum_{t=0}^{T} x^t \boldsymbol{e}_t\right) = \sum_{t=0}^{T} p^t x^t$$

by the linearity of π. Hence,

$$\pi(\boldsymbol{x}) = \pi\left(\lim_{T\to\infty} \sum_{t=0}^{T} x^t \boldsymbol{e}_t\right) = \lim_{T\to\infty} \sum_{t=0}^{T} p^t x^t = \sum_{t=0}^{\infty} p^t x^t = \boldsymbol{px}$$

as it should be under the suitable continuity assumption on π, which is not specified here.

To sum up:

The set of all commodities bundles form a finite or an infinite-dimensional vector space L. We call L the commodity space. A linear functional on L is called a price functional. The set of all price functionals is the dual space of L and is denoted by L^*.

2.2. THE PREFERENCE OF A CONSUMER

NB: In this book, we use the vector inequality notation that $\boldsymbol{x} = (x^t) \geq \mathbf{0} = (0,0,\dots)$ which means that $x^t \geq 0$ for all t, $\boldsymbol{x} > \mathbf{0}$ which means that $\boldsymbol{x} \geq \mathbf{0}$ and $\boldsymbol{x} \neq \mathbf{0}$. Moreover, we will sometimes use that $\boldsymbol{x} \gg \mathbf{0}$ which means that $x^t > 0$ for all t.

The consumers of an economy is a category of economic agents who purchase consumption goods and supply production factors such as labor. Let A be the set of all consumers in the economy. The set A may be either a finite or an infinite set. We usually denote an individual member of the set A by a.

A consumer $a \in A$ must decide his/her consumption plan vectors, or simply consumption vector $\boldsymbol{x}_a = (x_a^t) \in L$. It is usually assumed that if $x_a^t > 0$, then the commodity t is demanded by the consumer a, and if $x_a^t < 0$, then the commodity t is supplied by the consumer a (for example, the commodity t is labor measured by hours). The set of all consumption vectors possible for a consumer $a \in A$ is called the consumption set of a and denoted by X_a. It is a subset of the commodity space L. It seems that there are not many general conditions, which we can impose on the consumption set from the economic point of view. We will come back to this point in Chap. 3, but it seems relatively fair that the consumption set is bounded from below, since the ability of consumers to supply labor force is normally limited in time. That is to say, there exists a fixed vector (possibly depending on a) $\boldsymbol{b}_a \in L$ such that for all $\boldsymbol{x}_a \in X_a$, it follows that $\boldsymbol{b}_a \leq \boldsymbol{x}_a$. Furthermore, by a technical reason, we will assume that the consumption set is a closed subset of the commodity space L with respect to an appropriate topology, which will be specified in each case. This remark is important when the commodity spaces are infinite-dimensional as in Chap. 5. When the commodity space is finite-dimensional and $L = \mathbb{R}^\ell$, then we will use the usual topology on $\mathbb{R}^\ell$.

To sum up:

For every consumer $a \in A$, his/her consumption set is a closed subset X_a of the commodity space L which is bounded below.

We now introduce a preference of consumers. For a consumer $a \in A$, we assume that there exists a binary relation on X_a, denoted by $\prec_a$. Let $\boldsymbol{x},\boldsymbol{y} \in X_a$ be consumption vectors. Then, $(\boldsymbol{x},\boldsymbol{y}) \in \prec_a$ is read as "$\boldsymbol{y}$ is strictly preferred to $\boldsymbol{x}$ by the consumer a". We postulate that the preference relations satisfy the following conditions:

(IR) (Irrefrexivity) for every $\boldsymbol{x} \in X_a$, $(\boldsymbol{x},\boldsymbol{x}) \notin \prec_a$,
(TR) (Transitivity) for every $\boldsymbol{x},\boldsymbol{y}$, and $\boldsymbol{z} \in X_a$, $(\boldsymbol{x},\boldsymbol{y}) \in \prec_a$, and $(\boldsymbol{y},\boldsymbol{z}) \in \prec_a$ imply $(\boldsymbol{x},\boldsymbol{z}) \in \prec_a$, and
(CT) (Continuity) the set $\prec = \{(\boldsymbol{x},\boldsymbol{y}) \in X_a \times X_a | \boldsymbol{x} \prec_a \boldsymbol{y}\}$ is open in $X_a \times X_a$, as already mentioned, the topology is specified when the space L is infinite-dimensional, and otherwise, we mean the usual topology.

In the following, all preferences except the last part of Section 2.3 will be assumed to satisfy the irreflexivity (IR), transitivity (TR), and the continuity (CT).

$(\boldsymbol{x},\boldsymbol{y}) \in \prec_a$ is usually written as $\boldsymbol{x} \prec_a \boldsymbol{y}$. $(\boldsymbol{x},\boldsymbol{y}) \notin \prec_a$ is written as $(\boldsymbol{x},\boldsymbol{y}) \in \succsim_a$ or $\boldsymbol{x} \succsim_a \boldsymbol{y}$ and is read as "$\boldsymbol{x}$ is at least as desired as $\boldsymbol{y}$ by the consumer a". We define $\boldsymbol{x} \sim_a \boldsymbol{y}$ by $\boldsymbol{x} \succsim_a \boldsymbol{y}$ and $\boldsymbol{y} \succsim_a \boldsymbol{x}$, and read "$\boldsymbol{x}$ is indifferent with $\boldsymbol{y}$".

Sometimes, we assume the transitivity on $\succsim$, or equivalently:

(NTR) (Negative transitivity) $(\boldsymbol{x},\boldsymbol{y}) \notin \prec_a$ and $(\boldsymbol{y},\boldsymbol{z}) \notin \prec_a$ imply $(\boldsymbol{x},\boldsymbol{z}) \notin \prec_a$ for all $\boldsymbol{x},\boldsymbol{y}$, and $\boldsymbol{z} \in X_a$.

A topological condition which is sometimes required is:

(LNS) (Local nonsatiation) for every $\boldsymbol{x} \in X_a$ and every neighborhood U of $\boldsymbol{x}$, there, exists $\boldsymbol{y} \in U \cap X_a$ such that $\boldsymbol{x} \prec_a \boldsymbol{y}$,

with the usual remark on the topology when the space is infinite dimensional, and occasionally we will require an even stronger condition such as:

(MT) (Monotonicity) if $\boldsymbol{x},\boldsymbol{y} \in X_a$ and $\boldsymbol{x} < \boldsymbol{y}$, then $\boldsymbol{x} \prec_a \boldsymbol{y}$.

Finally, we introduce the convexity assumption, which says that the consumers prefer the mixed consumption vectors rather than each of the individual consumption bundle, or roughly speaking, they prefer mixed commodities rather than extreme ones.

(CV) (Convexity) for every $\boldsymbol{x} \in X_a$, the set $\{\boldsymbol{y} \in X_a | \boldsymbol{y} \succsim_a \boldsymbol{x}\}$ is convex.

It is a well-known history that before the consumers' preferences were represented by the binary relation, they were given by the real-valued function on the consumption sets called the utility function,

$$u_a : X_a \to \mathbb{R}, \quad a \in A,$$

which is related to the preference relation by that $u_a(\boldsymbol{x}) < u_a(\boldsymbol{y})$ if and only if $\boldsymbol{x} \prec_a \boldsymbol{y}$.

When this is the case, we say that the utility function u_a represents the preference relation $\prec_a$ (up to monotone transformations). Although, there existed some philosophical confusions with respect to the "meaning" of the values of the utility functions, we do not have to bother about these controversies among philosophers. Rather, the utility function is still a useful tool of economic analysis. Therefore, we will use it throughout this book. Note that when a preference relation is represented by a utility function, the irrefrexivity, the transitivity, and the negative transitivity conditions are automatically satisfied. The other assumptions in terms of the utility functions are:

(UCT) (Continuity) the utility function,

$$u_a: X_a \to \mathbb{R}$$

is continuous,

(LNS) (Local nonsatiation) for every $\boldsymbol{x} \in X_a$ and every neighborhood U of $\boldsymbol{x}$, there exists $\boldsymbol{y} \in U \cap X_a$ such that $u_a(\boldsymbol{x}) < u_a(\boldsymbol{y})$,

(UMT) (Monotonicity) if $\boldsymbol{x}, \boldsymbol{y} \in X_a$ and $\boldsymbol{x} < \boldsymbol{y}$, then $u_a(\boldsymbol{x}) < u_a(\boldsymbol{y})$,

(QCV) (Quasi-concavity) for every $\boldsymbol{x} \in X_a$, the set $\{\boldsymbol{y} \in X_a | u_a(\boldsymbol{y}) \geq u_a(\boldsymbol{x})\}$ is convex.

Of course in the assumptions (UCT) and (LNS), the usual caveat on the topology is understood.

To sum up:

The preference relation $\prec_a$ of the consumer a is an irreflexive and transitive binary relation on the consumption set X_a. When a real-valued function on X_a, $u_a: X_a \to \mathbb{R}$ satisfies $u_a(\boldsymbol{x}) < u_a(\boldsymbol{y})$ if and only if $\boldsymbol{x} \prec_a \boldsymbol{y}$, then we call u_a a utility function representing $\prec_a$.

2.3. THE DEMAND OF A CONSUMER

Till the end of this chapter, we assume that there exist m consumers in the economy so that $A = \{1 \cdots m\}$, the commodity space L is the ℓ-dimensional Euclidean space $\mathbb{R}^\ell$ and the consumption set X_a of each consumer $a \in A$ is a closed and convex subset of $\mathbb{R}^\ell$, which is bounded from below. Furthermore, in this section, we assume that each consumer $a = 1 \cdots m$ has an initial endowment vector, which satisfies the minimum income (MI) condition,

(MI) (Minimum income) for every price vector $\boldsymbol{p} \geq \boldsymbol{0}$ with $\boldsymbol{p} \neq \boldsymbol{0}$, $\boldsymbol{p}\omega_a > \inf \boldsymbol{p}X_a = \inf\{\boldsymbol{px} | \boldsymbol{x} \in X_A\}$, $a = 1 \cdots m$.

Given $\boldsymbol{p} \in \mathbb{R}^\ell_+$, the budget set of the consumer a is defined by:

$$\beta(a, \boldsymbol{p}) = \{\boldsymbol{x} \in X_a | \boldsymbol{px} \leq \boldsymbol{p}\omega_a\},$$

and the demand set of a is:

$$\phi(a,\boldsymbol{p}) = \{\boldsymbol{x} \in X_a | \boldsymbol{x} \in \beta(a,\boldsymbol{p}),\ \boldsymbol{x} \succsim_a \boldsymbol{z} \text{ for every } \boldsymbol{z} \in \beta(a,\boldsymbol{p})\}.$$

The demand set describes the consumer a's behavior, namely that he/she maximizes his/her utility subject to the budget constraint. By the definition of the set $\beta(a,\boldsymbol{p})$, it is clear that $\beta(\lambda \boldsymbol{p}) = \beta(a,\boldsymbol{p})$ for all $\lambda > 0$, or $\beta(a,\boldsymbol{p})$ is positively homogeneous of degree 0. Hence, so is the set $\phi(a,\boldsymbol{p})$. Therefore, without loss of generality, we can restrict the price vector $\boldsymbol{p}$ on the simplex,

$$\Delta = \left\{ \boldsymbol{p} = (p^t) \in \mathbb{R}^\ell \,\middle|\, \sum_{t=1}^{\ell} p^t = 1,\ p^t \geq 0,\ t = 1 \cdots \ell \right\}$$

and we are naturally led to think that β and ϕ are correspondences from Δ to X_a,

$$\beta(a,\cdot)\colon \Delta \to X_a, \quad \boldsymbol{p} \mapsto \beta(a,\boldsymbol{p}),$$

and

$$\phi(a,\cdot)\colon \Delta \to X_a, \quad \boldsymbol{p} \mapsto \phi(a,\boldsymbol{p}).$$

Then, our first task is to check that $\beta(a,\boldsymbol{p})$ and $\phi(a,\boldsymbol{p})$ are nonempty, if we insist that they are correspondences, and we are also interested in their continuity properties. Unfortunately, it is difficult to ensure the nonemptiness and the continuity of these correspondences unless we restrict their domains and/or ranges. We chose to restrict the ranges.

For sufficiently large $k > 0$, we can obtain $\omega_a \ll k\mathbf{1}$, where $\mathbf{1} = (1 \cdots 1)$. We define $K = \{\boldsymbol{x} \in \mathbb{R}^\ell \mid \boldsymbol{x} \leq k\mathbf{1} = (k \cdots k)\}$ and set:

$$\hat{X}_a = X_a \cap K = \{\boldsymbol{x} \in X_a \mid \boldsymbol{x} \leq k\mathbf{1}\}.$$

See Fig. 2.1.

It is obvious that the set $\hat{X}_a$ is convex. Since, $\hat{X}_a \subset B(0,M)$, where $M > \max\{\|\boldsymbol{b}_a\|, \sqrt{\ell}k\}$, the set $\hat{X}_a$ is bounded. Let $\{\boldsymbol{x}_n\} \subset \hat{X}_a$ be a converging sequence in $\hat{X}_a$ with $\boldsymbol{x}_n \to \boldsymbol{x}$. For each n, $\boldsymbol{b}_a \leq \boldsymbol{x}_n \leq k\mathbf{1}$, where $\boldsymbol{b}_a$ is the lower bound of X_a. Passing to the limit, we have $\boldsymbol{b}_a \leq \boldsymbol{x} \leq k\mathbf{1}$. Hence, $\hat{X}_a$ is closed. By Proposition C6, $\hat{X}_a$ is compact. We define the restricted budget correspondence and the restricted demand correspondence by:

$$\hat{\beta}(a,\cdot)\colon \Delta \to \hat{X}_a, \quad \boldsymbol{p} \mapsto \hat{\beta}(a,\boldsymbol{p}) = \{\boldsymbol{x} \in \hat{X}_a \,|\, \boldsymbol{p}\boldsymbol{x} \leq \boldsymbol{p}\omega_a\},$$

and

$$\begin{aligned}\hat{\phi}(a,\cdot)\colon \Delta \to \hat{X}_a, \quad &\boldsymbol{p} \mapsto \hat{\phi}(a,\boldsymbol{p}) \\ &= \{\boldsymbol{x} \in \hat{X}_a \,|\, \boldsymbol{x} \in \hat{\beta}(a,\boldsymbol{p}),\ \boldsymbol{x} \succsim_a \boldsymbol{z} \text{ for every } \boldsymbol{z} \in \hat{\beta}(a,\boldsymbol{p})\},\end{aligned}$$

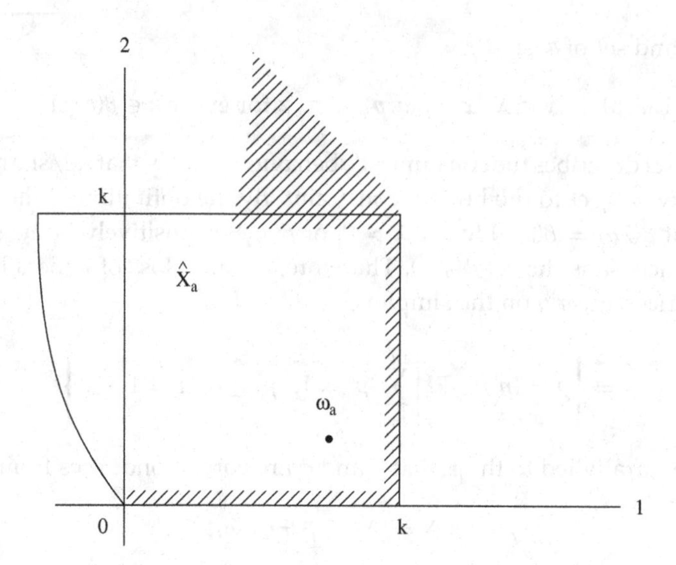

Figure 2.1. Truncated Consumption Set.

respectively. Since, there exists $z \in \hat{\beta}(a,p)$ for every $p \in \Delta$ by the assumption MI, the set $\hat{\beta}(a,p)$ is nonempty for all $p \in \Delta$. Next, the set $\hat{\beta}(a,p)$ is closed. Fix a vector $p \in \Delta$ and take a converging sequence $\{x_n\} \subset \hat{\beta}(a,p)$ with $x_n \to x$. Since $\hat{X}_a$ is closed, $x \in \hat{X}_a$. For every n, we have $px_n \leq p\omega_a$. Passing to the limit, one obtains $px \leq p\omega_a$, namely that $x \in \hat{\beta}(a,p)$. Hence, the set $\hat{\beta}(a,p)$ is a closed subset of $\hat{X}_a$ which is compact. Being a bounded and closed subset of $\mathbb{R}^\ell$, the set $\hat{\beta}(a,p)$ is compact for every $p \in \Delta$. We now prove that the demand relation is non-empty valued. For later use, we prove the proposition in a general form.

Proposition 3.1. *The set* $\hat{\phi}_a = \{x \in \hat{K}_a \mid x \in K, x \succsim z \text{ for every } z \in K\}$ *is nonempty for a compact set* K.

Proof. Suppose that $\hat{\phi}_a = \emptyset$. For each $x \in K$, the set $P(x) = \{z \in K | z \prec x\}$ is obviously open in K and we claim $K \subset \cup_{x \in K} P(x)$. For if not, then there exists $y \in K$ such that $y \notin P(x)$ for all x. This means that $y \succsim x$ for all $x \in K$, namely that $y \in \hat{\phi}_a$. So this is contradiction. Since K is compact, there exists a finite number of vectors $\{x_1 \cdots x_n\} \subset K$ such that $K \subset \cup_{i=1}^n P(x_i)$. By the transitivity of $\prec$, we have an $x* \in \{x_1 \cdots x_n\}$ such that $x* \succsim x_i$, $i = 1 \cdots n$. Hence, $x* \notin P(x_i)$, $i = 1 \cdots n$, contradicting $K \subset \cup_{i=1}^n P(x_i)$. □

Next, we show that the correspondence $\hat{\beta}(a,\cdot)\colon \Delta \to \hat{X}_a$ is continuous, or it is upper hemi-continuous and lower hemi-continuous. Since $\hat{X}_a$ is compact, $\hat{\beta}(a,\cdot)$ is *u.h.c* if its graph Graph$\hat{\beta}(a,\cdot) = \{(p,x) \in \hat{X}_a | x \in \hat{\beta}(a,p)\}$ is closed by Proposition D2. Let $(p_n, x_n)_{n \in \mathbb{N}}$ be a sequence in $\Delta \times \hat{X}_a$ such that $p_n x_n \leq p_n \omega_a$ for all n and

$(\boldsymbol{p}_n, \boldsymbol{x}_n) \to (\boldsymbol{p}, \boldsymbol{x})$. Passing to the limit, we have $\boldsymbol{px} \le \boldsymbol{p}\omega_a$. Since $\Delta \times \hat{X}_a$ is closed, we have $(\boldsymbol{p}, \boldsymbol{x}) \in \Delta \times \hat{X}_a$. This proves the upper hemi-continuity of $\hat{\beta}(a, \cdot)$.

Next, we show that $\hat{\beta}(a, \cdot)$ is lower hemi-continuous. Let $(\boldsymbol{p}_n)_{n\in\mathbb{N}}$ be a sequence in Δ converging to $\boldsymbol{p}$, and let $\boldsymbol{x} \in \hat{\beta}(a, \cdot)$. By Propositions D7 and D8, it is suffices to construct a sequence $(\boldsymbol{x}_n)_{n\in\mathbb{N}}$ converging to $\boldsymbol{x}$, and $\boldsymbol{x}_n \in \hat{\beta}(a, \boldsymbol{p}_n)$ for all n. We consider two cases. (Case 1): $\boldsymbol{px} < \boldsymbol{p}\omega_a$. Then, $\boldsymbol{p}_n\boldsymbol{x}_n < \boldsymbol{p}_n\omega_a$ for all n large enough, say for all $n \ge n_0$. Pick an arbitrary $\boldsymbol{x}_n$ from $\hat{\beta}(a, \boldsymbol{p}_n)$ for $n = 1 \cdots n_0$, and set $\boldsymbol{x}_n = \boldsymbol{x}$ for $n \ge n_0$. Then, we get a desired sequence. (Case 2): $\boldsymbol{px} = \boldsymbol{p}\omega_a$. By the assumption (MI) or $\boldsymbol{p}\omega_a > \inf \boldsymbol{p}X_a$, there exists a vector $\mathbf{z} \in X_a$ such that $\boldsymbol{p}\mathbf{z} < \boldsymbol{p}\omega_a$. Hence, $\boldsymbol{p}_n(\boldsymbol{x} - \mathbf{z}) > 0$ and $\boldsymbol{p}_n(\omega_a - \mathbf{z}) > 0$ for all n large enough. Set $t_n = \boldsymbol{p}_n(\omega_a - \mathbf{z})/\boldsymbol{p}_n(\boldsymbol{x} - \mathbf{z})$. Then, $t_n > 0$ for all n large enough and $t_n \to 1$. Let $\boldsymbol{y}_n = t_n\boldsymbol{x} + (1 - t_n)\mathbf{z},\ n = 1 \ldots$. Then $\boldsymbol{p}_n\boldsymbol{y}_n = \boldsymbol{p}_n\omega_a$ for all n and we define $\boldsymbol{x}_n = \boldsymbol{y}_n$ if $\boldsymbol{y}_n \in [\mathbf{z}, \boldsymbol{x}]$, and $\boldsymbol{x}_n = \boldsymbol{x}$ if $\boldsymbol{y}_n \notin [\mathbf{z}, \boldsymbol{x}]$. (Note that it could happen that $\boldsymbol{y}_n \notin \hat{X}_a$.) Since $\boldsymbol{p}_n\boldsymbol{x}_n \le \boldsymbol{p}_n\omega_a$ and $\boldsymbol{x}_n \to \boldsymbol{x}$, we get a desired sequence. Therefore, the proof of lower hemi-continuity is established.

Finally, we note that for every $\boldsymbol{p}$, the set $\hat{\beta}(a, \boldsymbol{p})$ is convex. Indeed, we denote the half space determined by the vector $\boldsymbol{p} \in \mathbb{R}^\ell$ and the scalar $w \in \mathbb{R}$ by

$$H(\boldsymbol{p}, w) = \{\boldsymbol{x} \in \mathbb{R}^\ell \mid \boldsymbol{px} \le w\},$$

the set $H(\boldsymbol{p}, w)$ is convex, hence so is the set $\hat{\beta}(a, \boldsymbol{p}) = H(\boldsymbol{p}, \boldsymbol{p}\omega_a) \cap \hat{X}_a$, as the intersection of convex sets.

Therefore, we have established:

Proposition 3.2. *Suppose that the consumer a satisfies the minimum income condition (MI). Then, his/her restricted budget correspondence:*

$$\hat{\beta}(a, \cdot) \colon \Delta \to \hat{X}_a, \quad \hat{\beta}(a, \boldsymbol{p}) = \{\boldsymbol{x} \in \hat{X}_a \mid \boldsymbol{px} \le \boldsymbol{p}\omega_a\}$$

is nonempty, compact and convex-valued, and continuous correspondence.

In Proposition 3.2, a remark on the role of the minimum income condition (MI) should be in order. In Fig. 2.2, the endowment vector is $\omega = (1, 0)$, and the condition MI is violated at $\boldsymbol{p} = (0, 1)$.

Consider the strictly positive price vector $\boldsymbol{p}_n$ converging to $\boldsymbol{p} = (0, 1)$. As long as $\boldsymbol{p}_n \ne \boldsymbol{p}$, the budget set $\hat{\beta}(a, \boldsymbol{p})$ is the shaded triangle area whose limit is the segment $[0, 1]$. However, actually in the limit, $\hat{\beta}(a, \boldsymbol{p}) = \mathbb{R}_+$, the segment $[1, +\infty]$ suddenly appears in the limit, and we cannot construct any sequence $(\boldsymbol{x}_n)_{n\in\mathbb{N}}$ in $\hat{\beta}(a, \boldsymbol{p}_n)$ converging to $\boldsymbol{x} = (x, 0)$ with $x > 1$. Hence, the lower hemi-continuity of $\hat{\beta}(a, \cdot)$ breaks down at $\boldsymbol{p} = (0, 1)$.

The upper hemi-continuity of the restricted demand correspondence is now immediate.

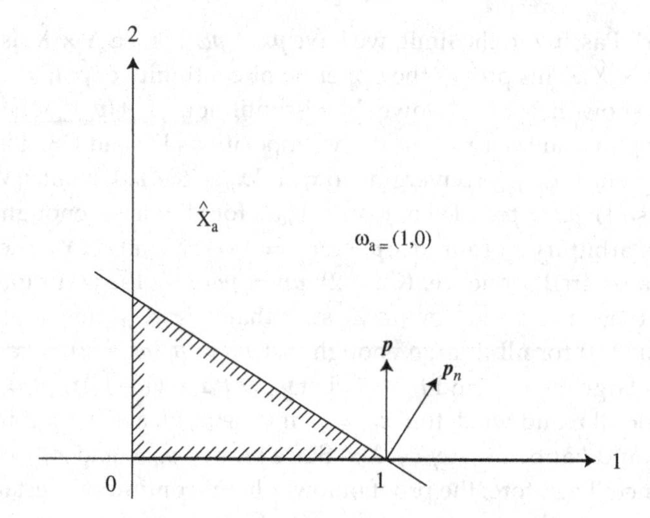

Figure 2.2. Lower Hemi-Continuity of the Budget Set.

Proposition 3.3. *Suppose that the consumer a satisfies the minimum income condition (MI). Then, the restricted demand correspondence:*

$$\hat{\phi}\colon \Delta \to \hat{X}_a,\ \boldsymbol{p} \mapsto \hat{\phi}(a,\boldsymbol{p}) = \{\boldsymbol{x} \in \hat{\beta}(a,\boldsymbol{p}) \mid \boldsymbol{x} \succsim_a \boldsymbol{y} \text{ whenever } \boldsymbol{y} \in \hat{\beta}(a,\boldsymbol{p})\}$$

is upper hemi-continuous.

Proof. Let $\boldsymbol{x}_n \in \hat{\phi}(a,\boldsymbol{p}_n)$, $n \in \mathbb{N}$, $\boldsymbol{p}_n \to \boldsymbol{p}$ and $\boldsymbol{x}_n \to \boldsymbol{x}$. Since the range of the correspondence $\hat{\phi}$ is compact, it is sufficient to show that $\boldsymbol{x} \in \hat{\phi}(a,\boldsymbol{p})$. By Proposition 3.2, we have $\boldsymbol{x} \in \hat{\beta}(a,\boldsymbol{p})$. Let $\boldsymbol{z} \in \hat{\beta}(a,\boldsymbol{p})$. By Proposition 3.2, we have a sequence $\boldsymbol{z}_n \in \hat{\beta}(a,\boldsymbol{p}_n)$ for all n and $\boldsymbol{z}_n \to \boldsymbol{z}$ as $n \to \infty$. Hence, $\boldsymbol{x}_n \succsim_a \boldsymbol{z}_n$ for all n, therefore, by the continuity (CT), $\boldsymbol{x} \succsim_a \boldsymbol{z}$ in the limit. This proves that $\boldsymbol{x} \in \hat{\phi}(a,\boldsymbol{p})$. □

The readers may not be satisfied with Propositions 3.1, 3.2 and 3.3, since they claim the nonemptiness and the (upper hemi-) continuity only for restricted (truncated) correspondence of $\hat{\beta}(a,\cdot)$ and $\hat{\phi}(a,\cdot)$, rather than for $\beta(a,\cdot)$ and $\phi(a,\cdot)$. However, the purpose of these results is to apply them to prove the existence of competitive equilibrium, which is to be demonstrated in Section 2.5, and for this purpose, our results for the truncated correspondences are adequate.

We conclude this section by including somewhat a surprising result obtained by Sonnenschein (1971). We have assumed that the preference relations are transitive which are considered to explain the fact that the consumers are rational enough to make their consumption decisions possible.

However, Sonnenschein proved that the transitivity is not necessary for the demand correspondence to be nonempty, when the preference is irreflexive and convex. Since the convexity of preference seems to represent the consumers' "taste" rather than the "rationality", namely that they prefer "mixed" commodities more than "extreme" commodities, his result is intuitively not obvious at all.

Let $\prec$ be a binary relation on a consumption set $X \subset \mathbb{R}^{\ell}$, which is irreflexive. We do not assume it to be transitive. We define the (strict) preference correspondence

$$P: X \to X, \quad \boldsymbol{x} \mapsto P(\boldsymbol{x}) = \{\boldsymbol{y} \in X \mid \boldsymbol{x} \prec \boldsymbol{y}\}.$$

The value $P(\boldsymbol{x})$ is called the preferred set of $\boldsymbol{x}$. We also define a correspondence:

$$L: X \to X, \quad \boldsymbol{x} \mapsto L(\boldsymbol{x}) = \{\boldsymbol{y} \in X \mid \boldsymbol{x} \in P(\boldsymbol{y})\}.$$

The value $L(\boldsymbol{x})$ is called the lower section of P at $\boldsymbol{x}$. In Fig. 2.3(a), the situation where $\boldsymbol{y} \in L(\boldsymbol{x})$ is illustrated and Fig. 2.3(b) illustrates the situation $\boldsymbol{y} \notin L(\boldsymbol{x})$.

Finally, we define the correspondence:

$$Q: X \to X, \quad \boldsymbol{x} \mapsto Q(\boldsymbol{x}) = \{\boldsymbol{y} \in X \mid \boldsymbol{x} \notin P(\boldsymbol{y})\}.$$

Note that the value $Q(\boldsymbol{x})$ is the complement of the lower section $L(\boldsymbol{x})$ in X. We now succeed to discard the transitivity of the preferences and the preference relation of consumers is represented by the (strict) preference correspondence $P(\cdot)$ till the end of this section.

We postulate the following assumptions on P:

($IRCV_{NT}$) (Irreflexivity and convexity) for all $\boldsymbol{x} \in X, \boldsymbol{x} \notin coP(\boldsymbol{x})$, where CoS denotes the convex hull of a set S,

(CT_{NT}) (Continuity) $L(\boldsymbol{x})$ is open relative to X for every $\boldsymbol{x} \in X$.

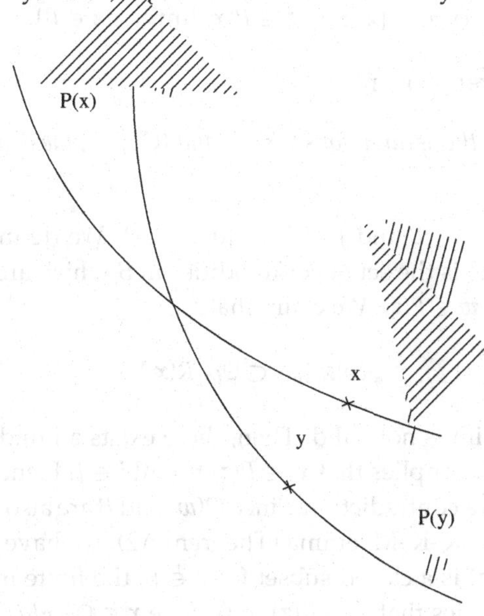

Figure 2.3(a). $y \in L(x)$.

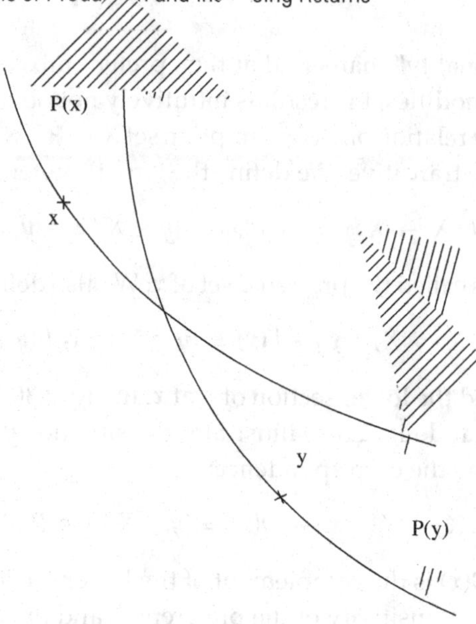

Figure 2.3(b). $y \notin L(x)$.

Since the set $Q(\boldsymbol{x})$ is the complement of the set $L(\boldsymbol{x})$, we have from (CT_{NT}) that $Q(\boldsymbol{x})$ is closed for every $\boldsymbol{x}$. Let B be a compact and convex subset of $\mathbb{R}^{\ell}$. Then, the demand set is defined by:

$$\phi_{NT} = \{\boldsymbol{x} \in B \mid \boldsymbol{z} \in P(\boldsymbol{x}) \text{ implies } \boldsymbol{z} \notin B\}.$$

Sonnenschein's theorem now reads:

Theorem 3.1. *Under the assumptions (IR_{NT}) and (CT_{NT}), the demand set is nonempty, $\phi_{NT} \neq \emptyset$.*

Proof. Take $\{\boldsymbol{x}_0 \cdots \boldsymbol{x}_r\} \subset B$ and $J \subset I = \{0, 1, \ldots r\}$. We define for $\boldsymbol{w} \in B$ the set $R(\boldsymbol{w}) = Q(\boldsymbol{w}) \cap B$. $R(\boldsymbol{w})$ is the set of commodities in B which are as good as $\boldsymbol{w}$. Note that $\boldsymbol{z} \notin R(\boldsymbol{w})$ implies $\boldsymbol{w} \in P(\boldsymbol{z})$. We claim that:

$$co\{\boldsymbol{x}_i\}_{i \in J} \subset \cup_{i \in J} R(\boldsymbol{x}_i).$$

Suppose that this claim is not valid. Then, there exists a bundle $\boldsymbol{z} \in co\{\boldsymbol{x}_i\}_{i \in J}$ such that $\boldsymbol{z} \notin \cup_{i \in J} R(\boldsymbol{x}_i)$. This implies that $\boldsymbol{x}_i \in P(\boldsymbol{z})$ for all $i \in J$. Hence, $co\{\boldsymbol{x}_i\}_{i \in J} \subset coP(\boldsymbol{z})$, and $\boldsymbol{z} \in coP(\boldsymbol{z})$. This is a contradiction. Since $Q(\boldsymbol{w})$ and B are also closed, the sets $R(\boldsymbol{x}_i)$ are also closed. By the K–K–M lemma (Theorem A2), we have $\cap_{i \in I} R(\boldsymbol{x}_i) \neq \emptyset$. Since B is compact and $Q(\boldsymbol{z})$ is a closed subset for $\boldsymbol{z} \in B$, the finite intersection property (see Appendix B) implies that $\cap_{\boldsymbol{z} \in B} R(\boldsymbol{z}) \neq \emptyset$. Take $\boldsymbol{x} \in \cap_{\boldsymbol{z} \in B} R(\boldsymbol{z})$. Then, $\boldsymbol{x} \in R(\boldsymbol{z})$ for all $\boldsymbol{z} \in B$, hence $\boldsymbol{z} \in P(\boldsymbol{x})$ implies that $\boldsymbol{z} \notin B$. This establishes that $\boldsymbol{x} \in \phi_{NT}$. □

2.4. THE DEMAND THEORY

In this section, we present the excellent theory of demand authored by McKenzie (1956–57) and (2002). He exploited the minimum income function (see below) neatly and has shown that the utility functions are not necessary to derive the main result (the Slutzky equation, Theorem 4.2) of the demand theory. He has also discarded the convexity assumption on the consumption set from the classical demand theory of Hicks (1939) and Samuelson (1947).

Let the consumption set X be a closed subset of $\mathbb{R}^\ell$, which is bounded from below. In this section, we omit the consumer's index a. The consumer's preference $\prec \subset X \times X$ is an irreflexive and transitive binary relation on X. As always, we impose the continuity of the preference. Moreover, we assume that the preference $\prec$ satisfies the negative transitivity and the local nonsatiation. Summing up, we assume in this section:

(NTR) (Negative transitivity) $(\boldsymbol{x},\boldsymbol{y}) \notin \prec$ and $(\boldsymbol{y},\boldsymbol{z}) \notin \prec$ imply $(\boldsymbol{x},\boldsymbol{z}) \notin \prec$ for all $\boldsymbol{x},\boldsymbol{y}$ and $\boldsymbol{z} \in X$,

(CT) (Continuity) the set $\prec = \{(\boldsymbol{x},\boldsymbol{y}) \in X \times X | \boldsymbol{x} \prec \boldsymbol{y}\}$ is open in $X \times X$, and

(LNS) (Local nonsatiation) for every $\boldsymbol{x} \in X$ and every neighborhood U of $\boldsymbol{x}$, there exists $\boldsymbol{y} \in U \cap X$ such that $\boldsymbol{x} \prec \boldsymbol{y}$.

For every $\boldsymbol{p} \in \mathbb{R}^\ell_+$ and $w \in \mathbb{R}$, the budget set and the demand set are defined by:

$$\beta(\boldsymbol{p}, w) = \{\boldsymbol{x} \in X \mid \boldsymbol{px} \leq w\},$$

and

$$\phi(\boldsymbol{p}, w) = \{\boldsymbol{x} \in X \mid \boldsymbol{x} \in \beta(\boldsymbol{p}, w), \text{ and } \boldsymbol{z} \in \beta(\boldsymbol{p}, w) \text{ implies } \boldsymbol{x} \succsim \boldsymbol{z}\},$$

respectively. Then, for $\boldsymbol{x} \in X$ and $\boldsymbol{p} \in \mathbb{R}^\ell_+$, we define the minimum income function $\mu_{\boldsymbol{x}}$ by:

$$\mu_{\boldsymbol{x}}: \mathbb{R}^\ell_+ \to \mathcal{R}, \quad \boldsymbol{p} \mapsto \mu_{\boldsymbol{x}}(\boldsymbol{p}) = \inf\{\boldsymbol{pz} \mid \boldsymbol{z} \succsim \boldsymbol{x}\}.$$

The basic properties of the minimum income function are:

Proposition 4.1. *The function $\mu_{\boldsymbol{x}}(\cdot)$ is positive homogeneous of degree 1 and concave.*

Proof. It is clear from the definition that $\mu_{\boldsymbol{x}}(t\boldsymbol{p}) = t\mu_{\boldsymbol{p}}(\boldsymbol{p})$ for all $t \geq 0$.

In order to prove the concavity, let $\boldsymbol{p} = t\boldsymbol{p}_0 + (1-t)\boldsymbol{p}_1, 0 \leq t \leq 1$. Given $\epsilon > 0$, we obtain $\boldsymbol{z} \in X$ with $\boldsymbol{z} \succsim \boldsymbol{x}$ and $\mu_{\boldsymbol{x}}(\boldsymbol{p}) > \boldsymbol{pz} - \epsilon$, namely that:

$$\mu_{\boldsymbol{x}}(\boldsymbol{p}) > t\boldsymbol{p}_0\boldsymbol{z} + (1-t)\boldsymbol{p}_1\boldsymbol{z} - \epsilon,$$

hence, by the definition of $\mu_{\boldsymbol{x}}$, we get:

$$\mu_{\boldsymbol{x}}(\boldsymbol{p}) > t\mu_{\boldsymbol{x}}(\boldsymbol{p}_0) + (1-t)\mu_{\boldsymbol{x}}(\boldsymbol{p}_1) - \epsilon.$$

Since this holds for any $\epsilon > 0$, it follows that:

$$\mu_{\boldsymbol{x}}(\boldsymbol{p}) \geq t\mu_{\boldsymbol{x}}(\boldsymbol{p}_0) + (1-t)\mu_{\boldsymbol{x}}(\boldsymbol{p}_1).$$

This proves the concavity of $\mu_{\boldsymbol{x}}(\cdot)$. □

By Theorem A4 of Appendix A, the concave function is continuous on the interior of its domain. Therefore, the function $\mu_{\boldsymbol{x}}(\cdot)$ is continuous for strictly positive price vector $\boldsymbol{p} \gg \boldsymbol{0}$.

Proposition 4.2. *For any $\boldsymbol{p} \gg \boldsymbol{0}$ and any $\boldsymbol{x} \in X$, $\mu_{\boldsymbol{x}}(\boldsymbol{p}) = \boldsymbol{pz}$ for some $\boldsymbol{z} \in X$ with $\boldsymbol{z} \succsim \boldsymbol{x}$.*

Proof. First, we show that for $\boldsymbol{p} \gg \boldsymbol{0}$, the budget set $\beta(\boldsymbol{p}, w) = \{\boldsymbol{x} \in X \mid \boldsymbol{px} \leq w\}$ is compact. It is closed as the intersection of closed sets X and the halfspace determined by $\boldsymbol{p}$ and w, $HS(\boldsymbol{p}, w) \equiv \{\boldsymbol{x} \in \mathbb{R}^\ell \mid \boldsymbol{px} \leq w\}$. Since X is bounded from below, so is $\beta(\boldsymbol{p}, w)$. Suppose that $\beta(\boldsymbol{p}, w)$ is not bounded above, and let $\{\boldsymbol{x}_n\}$ be a sequence in $\beta(\boldsymbol{p}, w)$ with $\|\boldsymbol{x}_n\| \to \infty$. Then, since $\boldsymbol{px}_n \leq w$ for all n, one obtains $\boldsymbol{px}_n/\|\boldsymbol{x}_n\| \leq w/\|\boldsymbol{x}_n\|$ for all n. Since $\{\boldsymbol{x}_n/\|\boldsymbol{x}_n\|\}$ is a sequence in the unit sphere which is compact, we can assume that $\boldsymbol{x}_n/\|\boldsymbol{x}_n\| \to \boldsymbol{y}$ and $\|\boldsymbol{y}\| = 1$. Since the sequence $\{\boldsymbol{x}_n\}$ is bounded from below, $\boldsymbol{y} \geq \boldsymbol{0}$. Since $\boldsymbol{p} \gg \boldsymbol{0}$, we have $\boldsymbol{px}_n/\|\boldsymbol{x}_n\| \to \boldsymbol{py} > 0$. But $w/\|\boldsymbol{x}_n\| \to 0$, which is a contradiction.

Let $R(\boldsymbol{x}) = \{\boldsymbol{z} \in X \mid \boldsymbol{z} \succsim \boldsymbol{x}\} \cap \beta(\boldsymbol{p}, \boldsymbol{px})$. Then $R(\boldsymbol{x}) \neq \emptyset$, since $\boldsymbol{x} \in R(\boldsymbol{x})$. Therefore, the function $\boldsymbol{pz}$ achieves its minimum value at some point in $R(\boldsymbol{x})$. □

Proposition 4.3. *Under the local nonsatiation (LNS), if $\boldsymbol{x} \in \phi(\boldsymbol{p}, w)$, then $w = \boldsymbol{px} = \mu_{\boldsymbol{x}}(\boldsymbol{p})$.*

Proof. By definition of $\phi(\boldsymbol{p}, w)$, we have $\boldsymbol{px} \leq w$. Suppose that $\boldsymbol{px} < w$. Then by (LNS), there exists $\boldsymbol{z} \in X$ which is close enough to $\boldsymbol{x}$ so that $\boldsymbol{pz} < w$ and $\boldsymbol{x} \prec \boldsymbol{z}$, which contradicts $\boldsymbol{x} \in \phi(\boldsymbol{p}, w)$. Hence, $\boldsymbol{px} = w$. If there is $\boldsymbol{z} \in X$ such that $\boldsymbol{z} \succsim \boldsymbol{x}$ and $\boldsymbol{pz} < w$, then by the same way as above, we obtain $\boldsymbol{w} \in X$ such that $\boldsymbol{z} \prec \boldsymbol{w}$ and $\boldsymbol{pw} < w$. Suppose that $\boldsymbol{x} \succsim \boldsymbol{w}$. Then, together with $\boldsymbol{z} \succsim \boldsymbol{x}$, it follows that $\boldsymbol{z} \succsim \boldsymbol{w}$ by the negative transitivity, contradicting $\boldsymbol{z} \prec \boldsymbol{w}$. Thus, we obtain $\boldsymbol{x} \prec \boldsymbol{w}$. Together with $\boldsymbol{pw} < w$, we have a contradiction to $\boldsymbol{x} \in \phi(\boldsymbol{p}, w)$. Therefore, $w = \mu_{\boldsymbol{x}}(\boldsymbol{p})$. □

Let the consumption bundle $\boldsymbol{x} \in X$ and the price vector $\boldsymbol{p} \gg \boldsymbol{0}$ be given. We define the compensated demand set $\phi_{\boldsymbol{x}}(\boldsymbol{p})$ by:

$$\phi_{\boldsymbol{x}}(\boldsymbol{p}) = \phi(\boldsymbol{p}, \mu_{\boldsymbol{x}}(\boldsymbol{p})).$$

In the following, we assume that $\phi(\boldsymbol{p}, w)$ and $\phi_{\boldsymbol{x}}(\boldsymbol{p})$ are single-element sets and they are functions over $\boldsymbol{p} \gg \boldsymbol{0}$ and $w > 0$.

Proposition 4.4. *Suppose that the local nonsatiation (LNS) holds and $\phi_{\boldsymbol{x}}(\boldsymbol{p})$ is differentiable at $\boldsymbol{p}(\gg \boldsymbol{0})$. Then, we have:*

$$\boldsymbol{p}\partial\phi_{\boldsymbol{x}}(\boldsymbol{p}) = 0,$$

where $\partial\phi_{\boldsymbol{x}}(\boldsymbol{p}) = (\partial_s \phi^t_{\boldsymbol{x}}(\boldsymbol{p}))^{\ell}_{s,t=1}$ is called the substitution matrix (see below).

Proof. Let $\boldsymbol{y} = \phi_{\boldsymbol{x}}(\boldsymbol{p})$ and take $\boldsymbol{q}$ close enough to $\boldsymbol{p} \gg \boldsymbol{0}$ so that $\boldsymbol{q} \gg \boldsymbol{0}$ and let $\boldsymbol{z} = \phi_{\boldsymbol{x}}(\boldsymbol{q})$. By Proposition 4.2, there exists $\boldsymbol{w}$ such that $\mu_{\boldsymbol{x}}(\boldsymbol{q}) = \boldsymbol{q}\boldsymbol{w}$ and $\boldsymbol{w} \succsim \boldsymbol{x}$. We claim that $\boldsymbol{z} \succsim \boldsymbol{x}$, for if not, $\boldsymbol{z} \prec \boldsymbol{x}$. Since $\boldsymbol{w} \succsim \boldsymbol{x}$, it follows from the negative transitivity (NTR) that $\boldsymbol{z} \prec \boldsymbol{w}$. But since $\boldsymbol{q}\boldsymbol{w} = \mu_{\boldsymbol{x}}(\boldsymbol{q})$ and $\boldsymbol{z} \prec \boldsymbol{w}$, one has $\boldsymbol{z} \neq \phi_{\boldsymbol{x}}(\boldsymbol{q})$, a contradiction. Thus, $\boldsymbol{z} \succsim \boldsymbol{x}$, $\boldsymbol{y} = \phi_{\boldsymbol{x}}(\boldsymbol{p}) = \phi(\boldsymbol{p}, \mu_{\boldsymbol{x}}(\boldsymbol{p}))$ implies that $\boldsymbol{p}\boldsymbol{y} \leq \mu_{\boldsymbol{x}}(\boldsymbol{p})$. On the other hand, $\boldsymbol{z} \succsim \boldsymbol{x}$ implies that $\mu_{\boldsymbol{x}}(\boldsymbol{p}) \leq \boldsymbol{p}\boldsymbol{z} = \boldsymbol{p}\phi_{\boldsymbol{x}}(\boldsymbol{q})$. By Proposition 4.3, we have $\boldsymbol{p}\boldsymbol{y} = \mu_{\boldsymbol{x}}(\boldsymbol{p})$. Hence, $\boldsymbol{p}\boldsymbol{y} = \boldsymbol{p}\phi_{\boldsymbol{x}}(\boldsymbol{p}) \leq \boldsymbol{p}\phi_{\boldsymbol{x}}(\boldsymbol{q})$ for all $\boldsymbol{q} \gg \boldsymbol{0}$ near $\boldsymbol{p}$. Differentiating $\boldsymbol{p}\phi_{\boldsymbol{x}}(\boldsymbol{q})$ with respect to $\boldsymbol{q}$ and evaluating at $\boldsymbol{p}$, we have the necessary condition for minimum,

$$\boldsymbol{p}\partial\phi_{\boldsymbol{x}}(\boldsymbol{p}) = 0.$$

This proves the proposition. □

The next proposition is the key of McKenzie's demand theory.

Proposition 4.5. *Under the assumption of local nonsatiation (LNS), we have:*

$$\frac{\partial}{\partial p^t}\mu_{\boldsymbol{x}}(\boldsymbol{p}) = \phi^t_{\boldsymbol{x}}(\boldsymbol{p}), \quad t = 1 \cdots \ell,$$

where we write $\phi_{\boldsymbol{x}}(\boldsymbol{p}) = (\phi^1_{\boldsymbol{x}}(\boldsymbol{p}) \cdots \phi^{\ell}_{\boldsymbol{x}}(\boldsymbol{p}))$.

Proof. Differentiating $\mu_{\boldsymbol{x}}$, we get:

$$\frac{\partial}{\partial p^t}\mu_{\boldsymbol{x}}(\boldsymbol{p}) = \frac{\partial}{\partial p^t}(\boldsymbol{p}\phi_{\boldsymbol{x}}(\boldsymbol{p})) = \phi^t_{\boldsymbol{x}}(\boldsymbol{p}) + \boldsymbol{p}\partial_t\phi_{\boldsymbol{x}}(\boldsymbol{p}) = \phi^t_{\boldsymbol{x}}(\boldsymbol{p})$$

by Proposition 4.4. □

As a corollary, we have under (LNS),

$$\frac{\partial^2}{\partial p^t \partial p^s}\mu_{\boldsymbol{x}}(\boldsymbol{p}) = \frac{\partial}{\partial p^s}\phi^t_{\boldsymbol{x}}(\boldsymbol{p}), \quad t, s = 1 \cdots \ell,$$

if the derivative exists. The main result of the demand theory is now immediate. We define the substitution matrix $S_{\boldsymbol{x}}(\boldsymbol{p})$ by:

$$S_{\boldsymbol{x}}(\boldsymbol{p}) = \partial\phi_{\boldsymbol{x}}(\boldsymbol{p}) = \left(\frac{\partial}{\partial p^s}\phi^t_{\boldsymbol{x}}(\boldsymbol{p})\right)^{\ell}_{s,t=1} = \left(\frac{\partial^2}{\partial p^t \partial p^s}\mu_{\boldsymbol{x}}(\boldsymbol{p})\right)^{\ell}_{s,t=1}.$$

Theorem 4.1. *Under the assumptions of (CT) and (LNS), for each $\boldsymbol{x} \in X$, $S_{\boldsymbol{x}}(\boldsymbol{p})$ exists for almost all $\boldsymbol{p} \gg \boldsymbol{0}$, and it satisfies:*

(i) *$S_{\boldsymbol{x}}(\boldsymbol{p})$ is symmetric and*
(ii) *$S_{\boldsymbol{x}}(\boldsymbol{p})$ is negative semi-definite.*

Proof. By Proposition 4.1, the function $\mu_{\boldsymbol{x}}(\boldsymbol{p})$ is concave. Therefore:

$$S_{\boldsymbol{x}}(\boldsymbol{p}) = \left(\frac{\partial^2}{\partial p^t \partial p^s}\mu_{\boldsymbol{x}}(\boldsymbol{p})\right)$$

is symmetric and negative semi-definite. □

Theorem 4.2 (Slutzky equation). *Under the assumptions of (CT) and (LNS), if the derivative exists for $\boldsymbol{p} \gg \boldsymbol{0}$,*

$$\frac{\partial}{\partial p^s}\phi^t_{\boldsymbol{x}}(\boldsymbol{p}) = \frac{\partial}{\partial p^s}\phi^t(\boldsymbol{p}, w) + \phi^s(\boldsymbol{p}, w)\frac{\partial}{\partial w}\phi^t(\boldsymbol{p}, w), \quad s, t = 1 \cdots \ell.$$

Proof. By Proposition 4.3, $\boldsymbol{px} = w = \mu_{\boldsymbol{x}}(\boldsymbol{p})$. Differentiating the compensated demand function $\phi^t_{\boldsymbol{x}}(\boldsymbol{p}) = \phi^t(\boldsymbol{p}, \mu_{\boldsymbol{x}}(\boldsymbol{p}))$, $t = 1 \cdots \ell$, we obtain:

$$\begin{aligned}\frac{\partial}{\partial p^s}\phi^t_{\boldsymbol{x}}(\boldsymbol{p}) &= \frac{\partial}{\partial p^s}\phi^t(\boldsymbol{p}, \mu_{\boldsymbol{x}}(\boldsymbol{p})) = \frac{\partial}{\partial p^s}\phi^t(\boldsymbol{p}, w) + \frac{\partial}{\partial w}\phi^t(\boldsymbol{p}, w)\frac{\partial \mu_{\boldsymbol{x}}(\boldsymbol{p})}{\partial p^s} \\ &= \frac{\partial}{\partial p^s}\phi^t(\boldsymbol{p}, w) + \phi^s(\boldsymbol{p}, w)\frac{\partial}{\partial w}\phi^t(\boldsymbol{p}, w), \quad s, t = 1 \cdots \ell.\end{aligned}$$

□

2.5. COMPETITIVE EQUILIBRIA OF A CLASSICAL EXCHANGE ECONOMY

In this section, we assume that the commodity space L is an ℓ-dimensional Euclidean space $\mathbb{R}^\ell$, and the set of consumers A is finite, $A = \{1 \cdots m\}$. Following McKenzie (1981, 2002), we call such an economy as the classical economy. Moreover, we restrict ourselves to an exchange economy in which there are no producers so that all trades of commodities are done among the consumers. This means that every consumer a owns his/her initial endowment vector $\omega_a \in L$, and the consumer undertakes his/her trades limited by the initial endowment. Hence, the total demand (and supply) is bounded by the total resource vector $\sum_{a\in A}\omega_a$. Recall that the consumption set X_a is a closed subset of $\mathbb{R}^\ell$, which is bounded from below. In this section, we also assume that X_a is convex. Let $\prec_a$ be a preference relation of the consumer a. An exchange economy is a $2m$-tuple of the preference relations and the initial endowments $(\prec_a, \omega_a)_{a=1}^m$ and denoted by $\mathcal{E}$.

An m-tuple of the consumption vectors $(\boldsymbol{x}_1 \ldots \boldsymbol{x}_m) \in \prod_{a\in A} X_a$ is called an allocation and an allocation is said to be feasible if:

$$\sum_{a\in A} \boldsymbol{x}_a \leq \sum_{a\in A} \omega_a.$$

A feasible allocation $(\boldsymbol{x}_1 \ldots \boldsymbol{x}_m) \in \prod_{a\in A} X_a$ is called Pareto optimal if and only if there exists no other feasible allocation $(\boldsymbol{y}_1 \ldots \boldsymbol{y}_m) \in \prod_{a\in A} X_a$ such that:

$$\boldsymbol{y}_a \succsim_a \boldsymbol{x}_a \quad \text{for all } a \in A,$$

and

$$\boldsymbol{x}_a \prec \boldsymbol{y}_a \text{ holds for at least one } a \in A.$$

We now state the definition of the fundamental equilibrium concept.

Definition 5.1. A pair $(\boldsymbol{p}, (\boldsymbol{x}_a))$ of a price vector $\boldsymbol{p} \in \mathbb{R}^\ell_+$ and an allocation $(\boldsymbol{x}_a)_{a\in A}$ is said to consist of a competitive equilibrium or a Walras equilibrium if and only if

(E-1) $\boldsymbol{p}\boldsymbol{x}_a \leq \boldsymbol{p}\omega_a$ and $\boldsymbol{x}_a \succsim_a \boldsymbol{y}$ for all $\boldsymbol{y} \in X_a$ such that $\boldsymbol{p}\boldsymbol{y} \leq \boldsymbol{p}\omega_a$, $a = 1 \cdots m$,
(E-2) $\sum_{a\in A} \boldsymbol{x}_a \leq \sum_{a\in A} \omega_a$.

The following easy Proposition 5.1 is known as the first fundamental theorem of welfare economics. In Chap. 4, we will see how this proposition should be modified when the external increasing returns are present.

Proposition 5.1. *Let $(\boldsymbol{p}, \boldsymbol{x}_1 \cdots \boldsymbol{x}_m) \in \mathbb{R}^\ell_+ \times \prod_{a\in A} X_a$ be a competitive equilibrium of an economy $(\prec_a, \omega_a)_{a=1}^m$ which satisfies the local nonsatiation (LNS) and the negative transitivity (NTR) of preferences for all $a \in A$. Then, the allocation $(\boldsymbol{x}_1 \cdots \boldsymbol{x}_m)$ is Pareto optimal.*

Proof. Suppose that $(\boldsymbol{x}_1 \cdots \boldsymbol{x}_m)$ is not Pareto optimal. Then, there exists a feasible allocation $(\boldsymbol{y}_1 \cdots \boldsymbol{y}_m) \in \prod_{a\in A} X_a$ such that:

$$\boldsymbol{y}_a \succsim_a \boldsymbol{x}_a \quad \text{for all } a \in A,$$

and

$$\boldsymbol{x}_a \prec \boldsymbol{y}_a \quad \text{holds for at least one } a \in A.$$

Then, it follows that $\boldsymbol{py}_a \geq \boldsymbol{p}\omega_a$ for all $a \in A$. For a such that $\boldsymbol{p}\omega_a \leq \inf \boldsymbol{p}X_a$, the inequality follows trivially. Suppose that $\boldsymbol{p}\omega_a > \inf \boldsymbol{p}X_a$ and $\boldsymbol{py}_a < \boldsymbol{p}\omega_a$ for some a. Then, by local nonsatiation (LNS), there exists a bundle $\boldsymbol{z} \in X_a$ which is close enough to $\boldsymbol{y}_a$ such that $\boldsymbol{y}_a \prec_a \boldsymbol{z}$ and $\boldsymbol{pz} < \boldsymbol{p}\omega_a$. Therefore, $\boldsymbol{x}_a \prec_a \boldsymbol{z}$ by the negative transitivity (NTR) and $\boldsymbol{pz} < \boldsymbol{p}\omega_a$, contradicting the condition (E-1) of Definition 5.1. Furthermore, for a such that $\boldsymbol{x}_a \prec_a \boldsymbol{y}_a$, we have $\boldsymbol{py}_a > \boldsymbol{p}\omega_a$. Summing these inequalities over a, one obtains $\boldsymbol{p}\sum_{a\in A} \boldsymbol{y}_a > \boldsymbol{p}\sum_{a\in A} \omega_a$. On the other hand, since the allocation $(\boldsymbol{y}_1 \cdots \boldsymbol{y}_m)$ is feasible, we have $\sum_{a\in A} \boldsymbol{y}_a \leq \sum_{a\in A} \omega_a$. Since $\boldsymbol{p} \geq \boldsymbol{0}$, we have $\boldsymbol{p}\sum_{a\in A} \boldsymbol{y}_a \leq \boldsymbol{p}\sum_{a\in A} \omega_a$, a contradiction. □

We now state our first result of the existence of the equilibrium. It is the most fundamental theorem of this book. Indeed, most parts of the book are, in a sense, variations on the theme of:

Theorem 5.1. *Suppose that an economy $\mathcal{E} = (\prec_a, \omega_a)_{a\in A}$ satisfies the continuity (CT), the convexity (CV), and the minimum income (MI) condition,*

$$\text{for every } \boldsymbol{p} \geq \boldsymbol{0} \quad \text{with} \quad \boldsymbol{p} \neq \boldsymbol{0}, \quad \boldsymbol{p}\omega_a > \inf \boldsymbol{p}X_a$$

for every $a \in A$. Then, there exists a competitive equilibrium $(\boldsymbol{p}^*, \boldsymbol{x}_1^* \cdots \boldsymbol{x}_m^*)$ for $\mathcal{E}$.

Proof. Take a positive number k such that $\| \sum_{a=1}^m \omega_a \| < k$, and set $K = \{\boldsymbol{x} \in \mathbb{R}^\ell | \boldsymbol{x} \leq k\boldsymbol{1}\}$, where $\boldsymbol{1} = (1 \cdots 1)$. As before, we define the restricted consumption sets,

$$\hat{X}_a(k) = X_a \cap K, \quad a = 1 \cdots m,$$

and we consider the budget correspondence on $\hat{X}_a(k)$,

$$\hat{\beta}(a, \cdot)\colon \Delta \to \hat{X}_a(k), \quad \hat{\beta}(a, \boldsymbol{p}) = \{\boldsymbol{x} \in \hat{X}_a(k) \mid \boldsymbol{px} \leq \boldsymbol{p}\omega_a\}, \quad a = 1 \cdots m,$$

and the demand correspondence:

$$\hat{\phi}(a, \cdot)\colon \Delta \to \hat{X}_a(k), \quad \hat{\phi}(a, \boldsymbol{p}) = \{\boldsymbol{x} \in \hat{\beta}(a, \boldsymbol{p}) \mid \boldsymbol{x} \succsim_a \boldsymbol{y} \text{ for all } \boldsymbol{y} \in \hat{\beta}(a, \boldsymbol{p})\},$$
$$a = 1 \cdots m.$$

Then, we define the aggregate excess demand correspondence,

$$\hat{\zeta}\colon \Delta \to \mathbb{R}^\ell, \quad \hat{\zeta}(\boldsymbol{p}) = \sum_{a=1}^m \hat{\phi}(a, \boldsymbol{p}) - \sum_{a=1}^m \omega_a.$$

Set $\hat{X}(k) = \sum_{a=1}^{m} \hat{X}_a(k) - \sum_{a=1}^{m} \omega_a$. Since $\hat{X}_a(k) - \{\omega_a\}$ is compact and convex subset of $\mathbb{R}^\ell$ for each a, the set $\hat{X}(k)$ is also compact and convex, and $\hat{\zeta}(p) \subset \hat{X}(k)$ for every $p \in \Delta$. By Proposition 3.3 and the assumption (CV), the correspondence $\hat{\zeta}$ is upper hemi-continuous and convex-valued and satisfies $p\hat{\zeta}(p) \leq 0$ by the budget condition of each a. Applying Gale–Nikaido–Debreu lemma (Theorem D3) of Appendix D, we obtain $p(k) \in \Delta$ such that:

$$\hat{\zeta}(p(k)) = \sum_{a=1}^{m} \hat{\phi}(a, p(k)) - \sum_{a=1}^{m} \omega_a \leq 0.$$

Take $x_a(k) \in \hat{\phi}(p_k)$, $a = 1 \cdots m$. Then we have:

(E-1_k) $p(k)x_a(k) \leq p(k)\omega_a$ and $x_a(k) \succsim_a y$ for every $y \in \hat{X}_a(k)$ such that:

$$p(k)y \leq p(k)\omega_a,$$

and

(E-2_k) $\sum_{a=1}^{m} x_a(k) \leq \sum_{a=1}^{m} \omega_a$.

Since $p(k) \in \Delta$ and $0 \leq x_a(k) \leq \sum_a \omega_a$ for all k, we can assume that $p(k) \to p^* \in \Delta$, and $x_a(k) \to x_a^*$.

We claim that $(p^*, x_1^* \cdots x_m^*)$ is a desired competitive equilibrium. Passing to the limit in the above market condition (E-2_k), we obtain:

(E-2) $\sum_{a=1}^{m} x_a^* \leq \sum_{a=1}^{m} \omega_a$.

Suppose there exists $y \in X_a$ such that $p^*y \leq p^*\omega_a$ and $x_a^* \prec_a y$. For k sufficiently large, we have $y \in \hat{X}_a(k)$. By the continuity (CT) of the preference relation, we can take $z \in \hat{X}_a(k)$, which is close enough to y such that $x_a^* \prec_a z$ and $p^*z < p^*\omega_a$. Hence, for k large enough, we have $x_a(k) \prec_a z$ and $p(k)z < p(k)\omega_a$, contradicting the condition (E-1_k). Therefore, the condition (E-1) is met. □

We now investigate welfare properties of the competitive equilibria. In Proposition 5.1, we saw that the competitive equilibrium is Pareto optimal. A partial converse of this result, which is known as the second fundamental theorem of welfare economics has been obtained. In order to present this result, we state a new equilibrium concept.

Definition 5.2. Suppose a price vector $p \in \mathbb{R}_+^\ell$ is given. An allocation $(x_a)_{a\in A}$ is said to be an equilibrium relative to the price vector p if and only if

(R-1) $py \leq px_a$ implies that $x_a \succsim_a y$ for all $a \in A$ and
(E-2) $\sum_{a\in A} x_a \leq \sum_{a\in A} \omega_a$.

It is obvious from the definitions that if the $m+1$-tuple $(p^*, (x_a^*))$ is a competitive equilibrium of the economy $\mathcal{E} = (\prec_a, \omega_a)_{a=1}^m$, then the allocation (x_a^*) is the equilibrium relative to $\boldsymbol{p}^*$ of the economy $\mathcal{E}$. On the other hand, let $(\boldsymbol{x}_a^*)$ be an equilibrium relative to the price vector $\boldsymbol{p}^*$ in the economy $\mathcal{E}$. Then, $(\boldsymbol{p}^*, (\boldsymbol{x}_a^*))$ is a competitive equilibrium of the economy $\mathcal{E}' = (\prec_a, \omega_a')$, where $\omega_a' = \boldsymbol{x}_a^*,\ a = 1 \cdots m$. For this reason, some authors call this equilibrium relative to $\boldsymbol{p}$ as the competitive equilibrium with re-distribution in the economy $\mathcal{E}$.

Somewhat a weaker equilibrium concept is the following.

Definition 5.3. Suppose a price vector $\boldsymbol{p} \in \mathbb{R}_+^\ell$ is given. An allocation $(\boldsymbol{x}_a)_{a\in A}$ is said to be a quasi-equilibrium relative to the price vector $\boldsymbol{p}$ if and only if

(Q-1) $\boldsymbol{x}_a \prec_a \boldsymbol{y}$ implies that $\boldsymbol{p}\boldsymbol{x}_a \leq \boldsymbol{p}\boldsymbol{y}$ for all $a \in A$ and
(E-2) $\sum_{a\in A} \boldsymbol{x}_a \leq \sum_{a\in A} \omega_a$.

Clearly, if an allocation $(\boldsymbol{x}_a)$ is an equilibrium relative to $\boldsymbol{p}$, then it is the quasi-equilibrium. But the converse does not hold as the following counter example authored by Arrow shows. In Fig. 2.4, the consumption bundle $\boldsymbol{x}$ and $\boldsymbol{z}$ have the same value evaluated by the price vector $\boldsymbol{p}$, or $\boldsymbol{p}\boldsymbol{x} = \boldsymbol{p}\boldsymbol{z}$, but $\boldsymbol{z}$ is strictly preferred to $\boldsymbol{x}$, namely $\boldsymbol{x} \prec \boldsymbol{z}$. Therefore, the condition (Q-1) holds but not the condition (R-1). This kind of phenomenon appears when the allocation bundle $\boldsymbol{x}$ satisfies $\boldsymbol{p}\boldsymbol{x} = \inf\{\boldsymbol{p}\boldsymbol{z}|\ \boldsymbol{z} \in X\}$. The second fundamental theorem of welfare economics reads as follows.

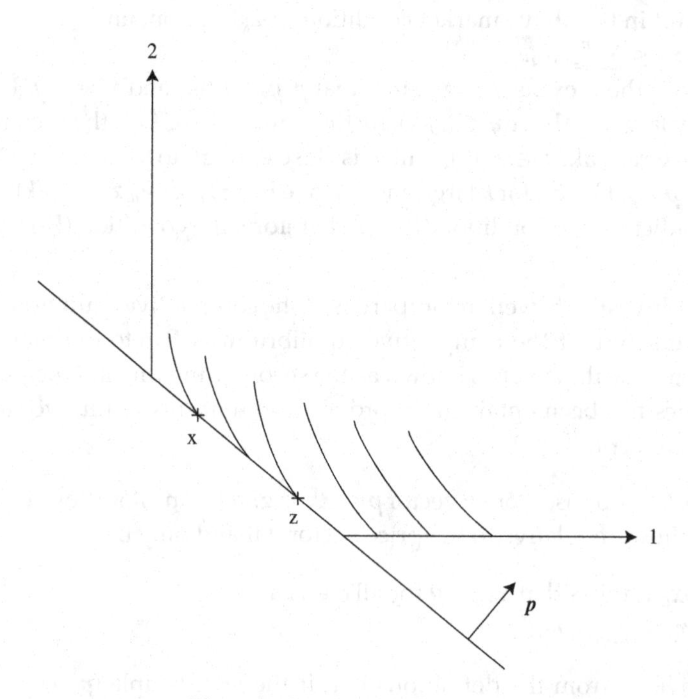

Figure 2.4. Arrow's Example.

Theorem 5.2. *For every consumer $a \in A$, suppose that the preference relation satisfies the negative transitivity (NTR), the continuity (CT), the convexity (CV), and the monotonicity (MT), then, every Pareto-optimal allocation is a quasi-equilibrium relative to some price vector $\boldsymbol{p} \in \mathbb{R}^{\ell}_{+}$.*

Proof. Let $(\boldsymbol{x}_a)_{a=1}^m$ be a Pareto-optimal allocation. Then, since it is a feasible allocation, the condition (E-2) is met. For each consumer a, define the strictly preferred set $P_a(\boldsymbol{x})$ by:

$$P_a(\boldsymbol{x}) = \{\mathbf{z} \in X_a \mid \boldsymbol{x} \prec_a \mathbf{z}\}.$$

The set $P_a(\boldsymbol{x})$ is nonempty for all $\boldsymbol{x} \in X_a$ by the assumption (MT). We claim that the set $P_a(\boldsymbol{x})$ is convex for all $\boldsymbol{x} \in X_a$. Let $\mathbf{z}_1, \mathbf{z}_2 \in P_a(\boldsymbol{x})$ and take $0 \leq t \leq 1$. Without loss of generality, we may assume that $\mathbf{z}_1 \succsim_a \mathbf{z}_2$. By the assumption (CV), the set $\{\mathbf{z} \in X_a \mid \mathbf{z} \succsim_a \boldsymbol{x}\}$ is convex for every $\boldsymbol{x} \in X_a$. Hence, $t\mathbf{z}_1 + (1-t)\mathbf{z}_2 \succsim_a \mathbf{z}_2$ and $\boldsymbol{x} \prec_a \mathbf{z}_2$, therefore $\boldsymbol{x} \prec_a t\mathbf{z}_1 + (1-t)\mathbf{z}_2$. For if not, we have $\boldsymbol{x} \succsim_a t\mathbf{z}_1 + (1-t)\mathbf{z}_2$. Then, it follows that $\boldsymbol{x} \succsim_a \mathbf{z}_2$ by the negative transitivity. This contradicts $\mathbf{z}_2 \in P_a(\boldsymbol{x})$. Therefore, $t\mathbf{z}_1 + (1-t)\mathbf{z}_2 \in P_a(\boldsymbol{x})$ and $P_a(\boldsymbol{x})$ is convex.

Let $Q = \sum_{a=1}^m P_a(\boldsymbol{x}_a)$. Then, Q is convex as the sum of the convex sets. Since the allocation $(\boldsymbol{x}_a)_{a=1}^m$ is Pareto optimal, $\sum_{a=1}^m \omega_a \notin Q$. Hence, we can apply the separation theorem (Theorem A1) of Appendix A, and we obtain a vector $\boldsymbol{p} \in \mathbb{R}^{\ell}$ with $\boldsymbol{p} \neq \mathbf{0}$ and:

$$\boldsymbol{p} \sum_{a=1}^m \omega_a \leq \boldsymbol{p} \sum_{a=1}^m \mathbf{z}_a$$

for every $\mathbf{z}_a \in P_a(\boldsymbol{x}_a)$, $a = 1 \cdots m$. By the monotonicity (MT), it follows that $\boldsymbol{p} \geq \mathbf{0}$. Fix an a. For $b \neq a$, we can take $\mathbf{z}_b$ arbitrary close to $\boldsymbol{x}_b$. Hence,

$$\boldsymbol{p} \sum_{a=1}^m \omega_a \leq \boldsymbol{p} \sum_{b \neq a}^m \boldsymbol{x}_b + \boldsymbol{p}\mathbf{z}_a \leq \boldsymbol{p} \sum_{a=1}^m \omega_a - \boldsymbol{p}\boldsymbol{x}_a + \boldsymbol{p}\mathbf{z}_a,$$

and from this, we obtain:

$$\boldsymbol{p}\boldsymbol{x}_a \leq \boldsymbol{p}\mathbf{z}_a \quad \text{for every } \mathbf{z}_a \in P_a(\boldsymbol{x}_a), \quad a = 1 \cdots m.$$

Therefore, the condition (Q-1) is met, and the theorem is proved. □

If the situation that $\boldsymbol{p}\boldsymbol{x}_a = \inf \boldsymbol{p} X_a$ is excluded, then we have an equilibrium rather than a quasi-equilibrium.

Corollary. *Under the assumptions of Theorem 5.2, if the Pareto-optimal allocation* $(\boldsymbol{x}_a)_{a=1}^m$ *satisfies* $\boldsymbol{p}\boldsymbol{x}_a > \inf \boldsymbol{p}X_a$*, then it is an equilibrium relative to* $\boldsymbol{p}$.

Proof. Suppose not. Then, there exists a consumption bundle $\boldsymbol{z}$ such that $\boldsymbol{p}\boldsymbol{z} \leq \boldsymbol{p}\boldsymbol{x}_a$ and $\boldsymbol{x}_a \prec_a \boldsymbol{z}$. We can take $\boldsymbol{y} \in X_a$ such that $\boldsymbol{x}_a \prec_a \boldsymbol{y}$ and $\boldsymbol{p}\boldsymbol{y} < \boldsymbol{p}\boldsymbol{x}_a$. Since the set $Q_a(\boldsymbol{x}_a) = \{\boldsymbol{z} \in X_a | \ \boldsymbol{x}_a \prec_a \boldsymbol{z}\}$ is convex, it follows that:

$$\boldsymbol{x}_a \prec_a t\boldsymbol{z} + (1-t)\boldsymbol{y} \quad \text{for every } 0 \leq t \leq 1.$$

But $\boldsymbol{p}(t\boldsymbol{z} + (1-t)\boldsymbol{y}) < \boldsymbol{p}\boldsymbol{x}_a$, contradicting that $(\boldsymbol{x}_a)$ is a quasi-equilibrium. □

2.6. A LIMIT THEOREM OF THE CORE OF A CLASSICAL EXCHANGE ECONOMY

In this section, we assume that each consumer $a(= 1 \cdots m)$ has the consumption set $X_a = \mathbb{R}_+^\ell$, the non-negative orthant of $\mathbb{R}^\ell$, and the preference of the consumer is represented by the utility function:

$$u_a\colon X_a \to \mathbb{R}, \quad \boldsymbol{x} \mapsto u_a(\boldsymbol{x}).$$

The convexity assumption on the preference is strengthened.

(SCV) (Strong convexity) Let $\boldsymbol{x}, \boldsymbol{y} \in X_a$ be commodity bundles such that $\boldsymbol{x} \neq \boldsymbol{y}$ and $u_a(\boldsymbol{x}) \geq u_a(\boldsymbol{y})$. Then, $u_a(t\boldsymbol{x} + (1-t)\boldsymbol{y}) > u_a(\boldsymbol{y})$ for every $0 < t < 1$.
We also assume that for each $a = 1 \cdots m$,
(CT) (Continuity) the function $u_a(\,\cdot\,)$ is continuous on X_a,
(LNS) (Local nonsatiation) for every $\boldsymbol{x} \in X_a$ and every neighborhood U of $\boldsymbol{x}$, there exists $\boldsymbol{y} \in U \cap X_a$ such that $u_a(\boldsymbol{y}) > u_a(\boldsymbol{x})$, and finally,
(PE) (Positive endowment) $\omega_a \gg \boldsymbol{0}$.

Let $A = \{1 \cdots m\}$ be the set of consumers. A non-empty subset C of the set A is called a coalition. Recall that an allocation $(\boldsymbol{x}_1 \cdots \boldsymbol{x}_m) \in \prod_{a=1}^m X_a$ is said to be feasible if and only if

$$\sum_{a=1}^m \boldsymbol{x}_a \leq \sum_{a=1}^m \omega_a,$$

where ω_a is the initial endowment vector of the consumer a.

Let $C \subset A$ be a coalition. A feasible allocation is said to be blocked by the coalition C if and only if there exists an allocation $(\boldsymbol{y}_1 \cdots \boldsymbol{y}_m) \in \prod_{a=1}^{m} X_a$ such that:

$$u_a(\boldsymbol{y}_a) \geq u_a(\boldsymbol{x}_a) \quad \text{for all } a \in C,$$

$$u_a(\boldsymbol{y}_a) > u_a(\boldsymbol{x}_a) \quad \text{for some } a \in C,$$

and

$$\sum_{a \in C} \boldsymbol{x}_a \leq \sum_{a \in C} \omega_a.$$

Definition 6.1. A feasible allocation is called a core allocation if no coalition can block it. The set of all core allocations is called the core.

By definition, it is clear that the core allocations are Pareto optimal, since they cannot be blocked by the grand coalition $C = A$. Indeed, the first welfare theorem of welfare economics, Proposition 5.3 is extended to the next proposition.

Proposition 6.1. *Under the local non-satiation (LNS), every competitive equilibrium allocation is in the core.*

Proof. Let $(\boldsymbol{x}_1 \cdots \boldsymbol{x}_m) \in \prod_{a=1}^{m} X_a$ be a competitive equilibrium allocation, namely that there exists a price vector $\boldsymbol{p} \in \Delta$ such that $(m+1)$-tuple $(\boldsymbol{p}, (\boldsymbol{x}_a)) \in \Delta \times \prod_{a=1}^{m} X_a$ is a competitive equilibrium. Hence, $u_a(\boldsymbol{y}) > u_a(\boldsymbol{x}_a)$ implies that $\boldsymbol{py} > \boldsymbol{p}\omega_a$. We claim that $u_a(\boldsymbol{y}) \geq u_a(\boldsymbol{x}_a)$ implies that $\boldsymbol{py} \geq \boldsymbol{p}\omega_a$. For otherwise, $\boldsymbol{py} < \boldsymbol{p}\omega_a$ and by the assumption (LNS), we can find that $\boldsymbol{z} \in X_a$ close enough to $\boldsymbol{y}$ so that $\boldsymbol{pz} < \boldsymbol{p}\omega_a$ and $u_a(\boldsymbol{z}) > u_a(\boldsymbol{y}) \geq u_a(\boldsymbol{x}_a)$, contradicting the condition (E-1) of Definition 5.1. for the consumer a. Suppose that $C \subset A$ was a blocking coalition, so that one has an allocation $(\boldsymbol{y}_a)$ such that $\sum_{a \in C} \boldsymbol{y}_a \leq \sum_{a \in C} \omega_a$, and $u_a(\boldsymbol{y}_a) \geq u_a(\boldsymbol{x}_a)$ for all $a \in C$ and $u_a(\boldsymbol{y}_a) > u_a(\boldsymbol{x}_a)$ for at least one $a \in C$. Therefore, we have $\boldsymbol{py}_a \geq \boldsymbol{p}\omega_a$ for all $a \in C$ and $\boldsymbol{py}_a > \boldsymbol{p}\omega_a$ for at least one $a \in C$. Hence, $\sum_{a \in C} \boldsymbol{p}(\boldsymbol{y}_a - \omega_a) > \boldsymbol{0}$, contradicting $\sum_{a \in C} (\boldsymbol{y}_a - \omega_a) \leq \boldsymbol{0}$. □

Note that the only continuity (CT) and the local non-satiation (LNS) are used in the proof of Proposition 6.1, as in the proof of Proposition 5.3. The negative transitivity (NTR) holds, since the preference is represented by a utility function.

The intuitive relationship among Pareto-optimal allocations, core allocations, and competitive equilibrium allocations can be depicted by Edgeworth diagram (Fig. 2.5).

For the rest of this section, we will study the theoretical relationship between the competitive equilibria and the core. As we saw, in the competitive equilibrium, consumers watch the signal of the market price only and they do not care about other consumers' consumption decisions. That is to say, they act independently of other consumers' actions of buying and selling, or in the game-theoretic terms, they behave noncooperatively with each other.

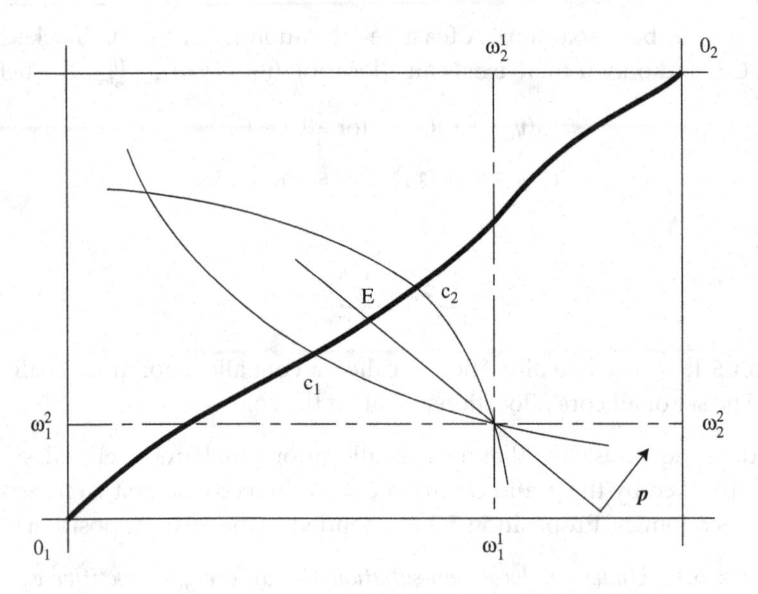

Figure 2.5. Edgeworth Box.

On the other hand, in the concept of the core (Definition 6.1), we consider bargaining among individual consumers. No bargains reach the final one until no coalition conclude that they prefer a new bargain to their existing bargains. In the game-theoretic terms, the concept of the core is cooperative one.

Edgeworth (1881) showed that in the case of two commodities and two types of consumers, the core allocation will approach the competitive equilibrium allocation as the number of consumers becomes larger and larger in such a way that identical consumers of each type increase proportionally, or each type of consumers is "replicated".

Debreu and Scarf (1963) then generalized this result to the case of arbitrary (finite) numbers of commodities and consumers, and we will present their result in Theorem 6.1.

We consider m types of consumers, indexed by $a = 1 \cdots m$, with r consumers of each type indexed by $q = 1 \cdots r$. Therefore, each individual consumer is indexed by a pair of integers (a, q), and there exist mr consumers in the economy.

The r-replicated economy of the economy $\mathcal{E} = (u_a, \omega_a)_{a=1}^m$ is an mr-tuple $(u_{a,q}, \omega_{a,q})$, $1 \leq a \leq m$, $1 \leq q \leq r$ such that $u_{a,1} = u_{a,2} = \cdots = u_{a,r} \equiv u_a$ and $\omega_{a,1} = \omega_{a,2} = \cdots = \omega_{a,r} \equiv \omega_a$. The r-replicated economy of $\mathcal{E}$ is denoted by $\mathcal{E}(r)$.

An allocation is a list of mr bundles, $(\boldsymbol{x}_{a,q}), 1 \le a \le m,\ \ 1 \le q \le r$, and it is said to be feasible if and only if

$$\sum_{a=1}^{m}\sum_{q=1}^{r} \boldsymbol{x}_{a,q} \le r \sum_{a=1}^{m} \omega_a.$$

Note that r consumers of each type, $(a, q),\ \ q = 1 \cdots r$, have the same initial endowment ω_a.

The next proposition states that in the core allocation, the same type of consumers take the same consumption bundle.

Proposition 6.2. *Let* $(\boldsymbol{x}_{a,q}),\ \ 1 \le a \le m,\ \ 1 \le q \le r$ *be a core allocation of the economy* $\mathcal{E}(r)$. *Then* $\boldsymbol{x}_{a,1} = \cdots \boldsymbol{x}_{a,r}$ *for each* $a = 1 \cdots m$.

Proof. For any type of fixed a, let q_a-th individual take the least-desired bundle among r consumers, namely:

$$u_a(\boldsymbol{x}_{a,q_a}) \le u_a(\boldsymbol{x}_{a,q}) \quad \text{for all } q = 1 \cdots r.$$

and suppose for some b, $\boldsymbol{x}_{b,s} \ne \boldsymbol{x}_{b,t}$, $1 \le s, t \le r$. We denote $\boldsymbol{x}_{a,q_a}$ by $\boldsymbol{x}_a$. Then, by the strong convexity assumption (SCV),

$$u_a\left(\frac{1}{r}\sum_{q=1}^{r} \boldsymbol{x}_{a,q}\right) \ge u_a(\boldsymbol{x}_a) \quad \text{for all } a,$$

and

$$u_b\left(\frac{1}{r}\sum_{q=1}^{r} \boldsymbol{x}_{b,q}\right) > u_b(\boldsymbol{x}_b).$$

However, we have:

$$\sum_{a=1}^{m}\left(\frac{1}{r}\sum_{q=1}^{r} \boldsymbol{x}_{a,q} - \omega_a\right) = 0,$$

hence, the coalition which consists of one consumer of each type a who receives $\boldsymbol{x}_a = \boldsymbol{x}_{a,q_a}$ would be a blocking coalition. □

By Proposition 6.2, we can describe core allocations of the economy $\mathcal{E}(r), r \ge 1$, by m-tuples of consumption bundles $(\boldsymbol{x}_1 \cdots \boldsymbol{x}_m)$ such that $\sum_{a=1}^{m} \boldsymbol{x}_a \le \sum_{a=1}^{m} \omega_a$. Also by Proposition 6.1, every competitive equilibrium allocation of the economy $\mathcal{E}(r)$ is a core allocation of $\mathcal{E}(r)$. Therefore, competitive equilibrium allocations are also described by the m-tuples of feasible consumption bundles.

Let $\mathcal{W}(r)$ and $\mathcal{C}(r)$ be the set of equilibrium allocations and the core allocations of the economy $\mathcal{E}(r)$, respectively. We have shown that $\mathcal{W}(r) \subset \mathcal{C}(r)$ for each r. Furthermore, we have:

Proposition 6.3. $\mathcal{C}(r+1) \subset \mathcal{C}(r)$ *for all* $r \geq 1$.

Proof. Let $A(r) = \{(1,1) \cdots (1,r), (2,1) \cdots (m,r)\}$ be the set of the consumers of the economy $\mathcal{E}(r)$. Obviously, we have $A = A(1) \subset A(2) \subset \cdots \subset A(r) \subset A(r+1) \subset \cdots$.

If an allocation of the economy $\mathcal{E}(r)$ is blocked by a coalition $C \subset A(r)$, then C will block that allocation in the economy $\mathcal{E}(r+1)$. Therefore, if an allocation does not belong to $\mathcal{C}(r)$, it does not belong to $\mathcal{C}(r+1)$. Hence, the proposition follows. □

If $(\boldsymbol{p}, (\boldsymbol{x}_a))_{a=1}^m$ is a competitive equilibrium of the economy $\mathcal{E} = (u_a, \omega_a)_{a=1}^m$, then it is also a competitive equilibrium for all $\mathcal{E}(r), r \geq 1$. For under the equilibrium price $\boldsymbol{p}$ in $\mathcal{E}$, the consumer (a,s) and the consumer (a,t), $1 \leq s,t \leq r$, have the same utility function u_a, which is strictly convex and the same initial endowment ω_a. Thus, they demand the same consumption bundle $\boldsymbol{x}_a (= \boldsymbol{x}_{a,s} = \boldsymbol{x}_{a,t})$. In other words, we have $\mathcal{W} = \mathcal{W}(1) = \cdots \mathcal{W}(r) = \cdots$. On account of Proposition 6.3, it follows that $\mathcal{W} \subset \cdots \mathcal{C}(r+1) \subset \mathcal{C}(r) \subset \cdots \subset \mathcal{C}$, or $\mathcal{W} \subset \cap_{r=1}^{\infty} \mathcal{C}(r)$, where $\mathcal{W}$ and $\mathcal{C}$ are the set of competitive equilibrium allocations and the core allocations of the economy $\mathcal{E}$, respectively. The limit theorem of Debreu and Scarf asserts the converse.

Theorem 6.1. *Let* $\mathcal{W}$ *be the set of competitive equilibrium allocations of an economy* $\mathcal{E} = (u_a, \omega_a)_{a=1}^m$, *and* $\mathcal{C}(r)$ *be the set of core allocations of r-replicated economy* $\mathcal{E}(r)$. *Then, we have:*

$$\mathcal{W} = \cap_{r=1}^{\infty} \mathcal{C}(r)$$

Proof. $\mathcal{W} \subset \cap_{r=1}^{\infty} \mathcal{C}(r)$ has been already shown. We prove the converse. Let $(\boldsymbol{x}_1 \cdots \boldsymbol{x}_m) \in \cap_{r=1}^{\infty} \mathcal{C}(r)$, and define the set Γ_a by:

$$\Gamma_a = \{\mathbf{z} \in \mathbb{R}^{\ell} \mid u_a(\mathbf{z} + \omega_a) > u_a(\boldsymbol{x}_a)\}, \quad a = 1 \cdots m,$$

and the set Γ by:

$$\Gamma = co\left(\cup_{a=1}^m \Gamma_a\right),$$

where coA denotes the convex hull of a set A. Note that the set Γ_a is convex by the convexity assumption on the utility function u_a, so that Γ can be written as:

$$\Gamma = \left\{ \sum_{a=1}^m \alpha_a \mathbf{z}_a \middle| \; \alpha_a \geq 0, \sum_{a=1}^m \alpha_a = 1, u_a(\mathbf{z}_a + \omega_a) > u_a(\boldsymbol{x}_a) \right\}.$$

We claim that $\{\mathbf{0}\} \notin \Gamma$. Suppose that this is not. Then, we have bundles $\mathbf{z}_1 \cdots \mathbf{z}_m$ such that $\sum_{a=1}^{m} \alpha_a \mathbf{z}_a = \mathbf{0}$ with $\alpha_a \geq 0$, $\sum_{a=1}^{m} \alpha_a = 1$, $u_a(\mathbf{z}_a + \omega_a) > u_a(\boldsymbol{x}_a)$. Let $k \in \mathbb{N}$ be an integer and let α_a^k be the smallest integer such that $\alpha_a^k \geq k\alpha_a$. We denote $I = \{a|\ \alpha_a > 0\}$. For each $a \in I$, let $\mathbf{z}_a^k = (k\alpha_a/\alpha_a^k)\mathbf{z}_a$. Then, it follows that $\omega_a \leq \mathbf{z}_a^k + \omega_a \leq \mathbf{z}_a + \omega_a$, and $\mathbf{z}_a^k \to \mathbf{z}_a$ as $k \to \infty$. By the continuity of u_a, $u_a(\mathbf{z}_a^k + \omega_a) > u_a(\boldsymbol{x}_a)$ for k sufficiently large, and:

$$\sum_{a\in I} \alpha_a^k \mathbf{z}_a^k = k \sum_{a\in I} \alpha_a \mathbf{z}_a = \mathbf{0}.$$

Consider the coalition which consists of α_a^k members of the type $a \in I$ to each one of whom receives $\omega_a + \mathbf{z}_a^k$. This coalition would block the core allocation $(\boldsymbol{x}_1 \cdots \boldsymbol{x}_m)$ of the economy $\mathcal{E}(\max\{\alpha_a^k|a \in I\})$. A contradiction. Hence, $\{\mathbf{0}\} \notin \Gamma$. Then, by the separation hyperplane theorem A1 of Appendix A, one obtains the non-zero vector $\boldsymbol{p}$ such that $\boldsymbol{p}\mathbf{z} \geq 0$ for all $\mathbf{z} \in \Gamma$.

If $u_a(\boldsymbol{y}) > u_a(\boldsymbol{x}_a)$, then $\boldsymbol{y} - \omega_a \in \Gamma_a \subset \Gamma$, so that $\boldsymbol{p}\boldsymbol{y} \geq \boldsymbol{p}\omega_a$, $a = 1 \cdots m$. By the local nonsatiation assumption (LNS), we have $\boldsymbol{y}$ which is strictly preferred to $\boldsymbol{x}_a$ and arbitrarily close to $\boldsymbol{x}_a$. Thus, it follows that $\boldsymbol{p}\boldsymbol{x}_a \geq \boldsymbol{p}\omega_a$, $a = 1 \cdots m$. But:

$$\sum_{a=1}^{m} (\boldsymbol{x}_a - \omega_a) \leq \mathbf{0},$$

so that $\boldsymbol{p}\boldsymbol{x}_a = \boldsymbol{p}\omega_a$, $a = 1 \cdots m$. Therefore, we have demonstrated that the vector $\boldsymbol{p}$ and the allocation $(\boldsymbol{x}_a)$ consist of the quasi-equilibrium, namely the conditions:

(QE-1) $u_a(\boldsymbol{x}_a) < u_a(\boldsymbol{y})$ implies that $\boldsymbol{p}\omega_a \leq \boldsymbol{p}\boldsymbol{y}$ for all $a = 1 \cdots m$,

(E-2) $\sum_{a\in A} \boldsymbol{x}_a \leq \sum_{a\in A} \omega_a$.

As in the proof of the corollary of Theorem 5.2, we can show that they are indeed competitive equilibria,

(E-1) $u_a(\boldsymbol{x}_a) < u_a(\boldsymbol{y})$ implies $\boldsymbol{p}\omega_a < \boldsymbol{p}\boldsymbol{y}$ for all $a = 1 \cdots m$,

by the assumption (PE) $\omega_a \gg \mathbf{0}$ for all a. □

2.7. NASH EQUILIBRIA AND THE CORE OF GAMES

In this section, we will explain some applications of fixed-point theory to game theory. A (strategic form of) non-cooperative N-person game Γ is formally described as follows.

Let $A = \{1 \cdots N\}$ be the set of players. The player $a \in A$ has a choice set X_a which is a non-empty subset of $\mathbb{R}^\ell$, a constraint correspondence $\mathcal{A}_a$: $\prod_{b=1}^{N} X_b \to X_a$, and a preference correspondence P_a: $\prod_{b=1}^{N} X_b \to X_a$.

The $3N$-tuple $\Gamma = (X_a, \mathcal{A}_a, P_a)_{a=1}^N$ is called an N-person (non-cooperative) abstract game. An interpretation of the correspondence $P_a(\boldsymbol{x})$ is the set of actions, which is preferred by a to the action, $\boldsymbol{x}_a$ when the profile of actions $\boldsymbol{x} = (\boldsymbol{x}_1 \cdots \boldsymbol{x}_a \cdots \boldsymbol{x}_N)$ is given.

Definition 7.1. An N-tuple of strategies $\boldsymbol{x}^* = (\boldsymbol{x}_1^* \cdots \boldsymbol{x}_N^*) \in \prod_{a=1}^N X_a$ is called a Nash equilibrium of the abstract game $\Gamma = (X_a, \mathcal{A}_a, P_a)_{a=1}^N$ if and only if for each $a = 1 \cdots N$,

$$\boldsymbol{x}_a^* \in \mathcal{A}_a(\boldsymbol{x}^*) \quad \text{and} \quad P_a(\boldsymbol{x}^*) \cap \mathcal{A}_a(\boldsymbol{x}^*) = \emptyset.$$

Suppose that an exchange economy $\mathcal{E} = (X_a, u_a, \omega_a)_{a=1}^{N-1}$ with $N-1$ consumers is given. We can interpret $\mathcal{E}$ together with a fictitious "market player" which is explained below as an N-person game Γ, and the Nash equilibrium of Γ is reduced to the competitive equilibrium of $\mathcal{E}$ as follows.

The first $N-1$ players have the choice sets X_a, which are the consumption sets of the consumer $a = 1 \cdots N-1$, and the last player N called the market player has the choice set Δ, the unit simplex. The player $a(=1 \cdots N-1)$ has the constraint correspondence:

$$\mathcal{A}_a \colon \prod_{b=1}^{N-1} X_b \times \Delta \to X_a, \quad \mathcal{A}_a((\boldsymbol{x}_b), \boldsymbol{p}) = \{\boldsymbol{y} \in X_a \mid \boldsymbol{p}\boldsymbol{y} \leq \boldsymbol{p}\omega_a\},$$

which is the budget correspondence of the consumer a, and the market player's constraint correspondence is given by $\mathcal{A}_N((\boldsymbol{x}_b), \boldsymbol{p}) = \Delta$.

The first $N-1$ players' preference is given by the preference correspondence (see Section 2.3),

$$P_a \colon \prod_{b=1}^{N-1} X_b \times \Delta \to X_a, \quad ((\boldsymbol{x}_b), \boldsymbol{p}) \mapsto P_a(\boldsymbol{x}_a),$$

and the market player's preference correspondence is:

$$P_N \colon \prod_{b=1}^{N-1} X_b \times \Delta \to X_a,$$

$$P_N((\boldsymbol{x}_b), \boldsymbol{p}) = \left\{ \boldsymbol{q} \in \Delta \,\middle|\, \boldsymbol{q} \sum_{b=1}^{N-1} (\boldsymbol{x}_b - \omega_b) > \boldsymbol{p} \sum_{b=1}^{N-1} (\boldsymbol{x}_b - \omega_b) \right\}.$$

It is easy to check that a Nash equilibrium of the game Γ is a competitive equilibrium of the economy $\mathcal{E}$.

Shafer and Sonnenschein (1975) proved the existence of Nash equilibrium based on the works of Nash (1950) and Debreu (1952).

Theorem 7.1. *Let* $\Gamma = (X_a, \mathcal{A}_a, P_a)_{a=1}^N$ *be an N-person game satisfying for each* $a = 1 \cdots N$,

(i) X_a *is a non-empty compact and convex subset of* $\mathbb{R}^\ell$;
(ii) *the correspondence* $\mathcal{A}_a$: $\prod_{b=1}^N X_b \to X_a$ *is nonempty and convex;*
(iii) *for each* $\boldsymbol{x} = (\boldsymbol{x}_b) \in \prod_{b=1}^N X_b, \boldsymbol{x}_a \in \mathcal{A}_a(\boldsymbol{x})$ *is nonempty and convex;*
(iv) *the graph of* P_a, $\text{Graph}P_a = \{((\boldsymbol{x}_b), \boldsymbol{y}) \in \prod_{b=1}^N X_b \times X_a \mid \boldsymbol{y} \in P_a((\boldsymbol{x}_b))\}$ *is open in* $\prod_{b=1}^N X_b \times X_a$; *and*
(v) *for each* $\boldsymbol{x} = (\boldsymbol{x}_b) \in \prod_{b=1}^N X_b, \boldsymbol{x}_a \notin coP_a(\boldsymbol{x})$, *where coA is the convex hull of the set A.*

Then Γ *has a Nash equilibrium.*

Proof. Denoting $X = \prod_{b=1}^N X_b$, we define a real-valued function u_a on $X \times X_a$ by:

$$u_a(\boldsymbol{y}, \boldsymbol{x}_a) = \inf\{d((\boldsymbol{y}, \boldsymbol{x}_a), \boldsymbol{z}) \mid \boldsymbol{z} \notin \text{Graph}P_a\},$$

where $d(\boldsymbol{z}_1, \boldsymbol{z}_2) = \|\boldsymbol{z}_1 - \boldsymbol{z}_2\|$ for $\boldsymbol{z}_1, \boldsymbol{z}_2 \in X \times X_a$. Then, u_a is continuous and $u_a(\boldsymbol{y}, \boldsymbol{x}_a) > 0$, if and only if $\boldsymbol{x}_a \in P_a(\boldsymbol{y})$. Indeed, take a converging subsequence $(\boldsymbol{y}(n), \boldsymbol{x}_a(n)) \to (\boldsymbol{y}, \boldsymbol{x}_a)$. Since $X \times X_a$ is compact by the assumption *(i)* and the complement of the graph of P_a is closed by the assumption *(iv)*, the infimum of the distance between $(\boldsymbol{y}(n), \boldsymbol{x}(n))$ and a compact set $X \times X_a \backslash \text{Graph}P_a$ is achieved by $\boldsymbol{z}(n)$. We can assume that $\boldsymbol{z}(n) \to \boldsymbol{z}$. It is easy to verify that $u_a(\boldsymbol{y}(n), \boldsymbol{x}_a(n)) = d((\boldsymbol{y}(n), \boldsymbol{x}_a(n)), \boldsymbol{z}(n)) \to d((\boldsymbol{y}, \boldsymbol{x}_a), \boldsymbol{z}) = u_a(\boldsymbol{y}, \boldsymbol{x}_a)$. This proves the continuity of u_a. The second claim that $u_a(\boldsymbol{y}, \boldsymbol{x}_a) > 0$ if and only if $\boldsymbol{x}_a \in P_a(\boldsymbol{y})$ follows from the definition of u_a.

For each a, we define the correspondence F_a: $X \to X_a$ by:

$$F_a(\boldsymbol{y}) = \{\boldsymbol{x}_a \in \mathcal{A}_a(\boldsymbol{y}) \mid u_a(\boldsymbol{y}, \boldsymbol{x}_a) \geq u_a(\boldsymbol{y}, \boldsymbol{z}) \text{ for every } \boldsymbol{z} \in \mathcal{A}_a(\boldsymbol{y})\}.$$

Since u_a is continuous and $\mathcal{A}_a$ is continuous and non-empty compact-valued correspondence, it follows from Theorem D2 of Appendix D, that F_a is upper hemi-continuous and compact-valued correspondence. Then, we define the correspondence $G: X \to X, \boldsymbol{y} = (\boldsymbol{y}_a) \mapsto G(\boldsymbol{y})$ by:

$$G(\boldsymbol{y}) = \prod_{a=1}^N coF_a(\boldsymbol{y}_a).$$

Then, by Propositions D5 and D6, G is upper hemi-continuous, compact, and convex-valued correspondence. By Kakutani's fixed-point theorem (Theorem D1 of Appendix D), there exists a fixed point $\boldsymbol{x}^* \in G(\boldsymbol{x}^*)$, or $\boldsymbol{x}_a^* \in coF_a(\boldsymbol{x}_a^*)$ for every a. We claim that $\boldsymbol{x}^*$ is a Nash equilibrium of Γ.

Since $F_a(\boldsymbol{x}^*) \subset \mathcal{A}_a(\boldsymbol{x}^*)$ and $\mathcal{A}_a(\boldsymbol{x}^*)$ is convex by the assumption (ii), $coF_a(\boldsymbol{x}^*) \subset \mathcal{A}_a(\boldsymbol{x}^*)$. Hence, $\boldsymbol{x}_a^* \in \mathcal{A}_a(\boldsymbol{x}^*)$. If there is a point $\boldsymbol{z}_a \in P_a(\boldsymbol{x}^*) \cap \mathcal{A}_a(\boldsymbol{x}^*)$, then $u_a(\boldsymbol{x}^*, \boldsymbol{z}_a) > 0$. Thus, $u_a(\boldsymbol{x}^*, \boldsymbol{y}_a) > 0$ for all $\boldsymbol{y}_a \in F_a(\boldsymbol{x}^*)$. Therefore, $\boldsymbol{z}_a \in P_a(\boldsymbol{x}^*) \cap \mathcal{A}_a(\boldsymbol{x}^*)$ implies that $F_a(\boldsymbol{x}^*) \subset P_a(\boldsymbol{x}^*)$, which yields that $\boldsymbol{x}_a^* \in coF_a(\boldsymbol{x}_a^*) \subset coP_a(\boldsymbol{x}^*)$, a contradiction. □

Two remarkable points of Shafer–Sonnenschein theorem are (1) the transitivity of preference is not assumed and (2) the correspondences $\mathcal{A}_a$ and P_a admit externalities, namely that they include other players' action in their variables. We will apply the Shafer–Sonnenschein theorem in Chap. 4.

Next, we will discuss the core of a game and its relations with that of an exchange economy. In Section 2.6, we saw that the competitive equilibria are in the core and since we proved the existence of equilibrium in Section 2.5, we have also obtained the nonemptiness of the core of the exchange economies. However, the competitive equilibrium and the core are mutually independent concepts, so that the proof of the existence of the core is important and interesting in its own right.

We will present the classical existence result of Scarf (1967). First, we will give definitions of a (non-side payment) game and its core, and explain how the exchange economy can be interpreted as the game. We will see that the core of the economy is reduced to that of the game. The proof of the existence of the core will follow.

Let $A = \{1 \cdots N\}$ be the set of players and $\mathcal{A} = \{C \subset A \mid C \neq \emptyset\}$ be the family of non-empty subsets of A, which we call as coalitions.

For each coalition $C \in \mathcal{A}$, we define the subspace $\mathbb{R}^C$ of $\mathbb{R}^N$ by:

$$\mathbb{R}^C = \{\boldsymbol{x} = (x^a) \in \mathbb{R}^N \mid x^a = 0 \text{ for } a \notin C\}.$$

A (non-side payment) game in characteristic function form is a correspondence $\tilde{V}: \mathcal{A} \to \mathbb{R}^N$ such that $\tilde{V}(C) \subset \mathbb{R}^C$ for every $C \in \mathcal{A}$. The intended interpretation is that each element $(v^a) \in V(C)$ represents the utility allocation of the members of C which brought to them by their cooperation. In the following, however, we embed $\tilde{V}(C)$ into $\mathbb{R}^N$ as a cylinder to make our argument smooth, namely that for each $C \in \mathcal{A}$, define the subset $V(C)$ of $\mathbb{R}^N$ by:

$$V(C) = \{\boldsymbol{v} = (v^a) \in \mathbb{R}^N \mid \text{proj}_C(\boldsymbol{v}) \in \tilde{V}(C)\},$$

where $\text{proj}_C(\boldsymbol{v}) = (w^a)$ is the projection of the vector $\boldsymbol{v}$ to $\mathbb{R}^C$, that is $w^a = v^a$ for $a \in C$, and $w^a = 0$ otherwise. From now on, we will work with V rather than $\tilde{V}$ and call it a game.

A utility allocation $(v^a)_{a=1}^N \in \mathbb{R}^N$ is called feasible if $(v^a) \in V(A)$. A coalition $C \in \mathcal{A}$ is said to block or improve upon a utility allocation (v^a) if there exists utility allocation $(w^a) \in V(C)$ such that $v^a < w^a$ for all $a \in C$. The definition of the core of the game V, denoted by $\mathcal{C}(V)$ is now obtained immediately.

Definition 7.2. The core of a game V, $\mathcal{C}(V)$ is the set of feasible utility allocations which are not improved upon by any coalition.

We now explain the relationship between the game V and the exchange economy $\mathcal{E} = (u^a, \omega_a)_{a \in A}$.

For each coalition $C \in \mathcal{A}$, define:

$$V(C) = \left\{ v = (v^a) \in \mathbb{R}^N \,\middle|\, \text{there exists an allocation } (y_a) \in \prod_{a=1}^{N} X_a \text{ such that:} \right.$$

$$\left. v^a \leq u_a(y_a) \text{ for all } a \in C \text{ and } \sum_{a \in C} y_a = \sum_{a \in C} \omega_a \right\}.$$

The economic meaning of the set $V(C)$ is now clear; it is the set of utility allocations, which can be obtained by the re-distribution of the endowments among the members of C, and it is obvious that under the assumptions of the continuity (CT) and the monotonicity (MT), the core of the exchange economy $\mathcal{E}$ coincides with that of the reduced game V.

A family $\mathcal{B}$ of subsets of $\mathcal{A}$ is balanced if there exist non-negative "balanced weights" λ_C for $C \in \mathcal{B}$ such that $\sum_{C \in \mathcal{B}_a} \lambda_C = 1$ for all $a \in A$, where $\mathcal{B}_a = \ldots \{C \in \mathcal{B} | a \in C\}$ is the family of subsets of A containing a.

A (non-side payment) game V is called balanced if for every balanced subfamily $\mathcal{B}$ of $\mathcal{A}$, it follows that $\cap_{C \in \mathcal{B}} V(C) \subset V(A)$. We will show that essentially, the balancedness condition is sufficient for nonemptiness of the core of the game V. Before we state the main theorem of the existence of the core, we show that the game derived from the exchange economy is balanced.

Proposition 7.1. *Let $\mathcal{E} = (u_a, \omega_a)$ be an exchange economy and suppose that for every $a \in A$, the consumption set X_a is a convex subset of $\mathbb{R}^\ell$, $\omega_a \in X_a$ and u_a is quasi-concave, or the set $\{\mathbf{z} \in X_a | \ u_a(\mathbf{z}) \geq u_a(\mathbf{x})\}$ is convex for all $\mathbf{x} \in X_a$. Then, the associated game V is balanced.*

Proof. Let $\mathcal{B} \subset \mathcal{A}$ be a balanced subfamily of $\mathcal{A}$ with the weights $\{\lambda_C | \ C \in \mathcal{B}\}$, and take any $v = (v^a) \in \cap_{C \in \mathcal{B}} V(C)$. Then, for each $C \in \mathcal{B}$, there exists an allocation $(y_a) \in \prod_{a \in A} X_a$ such that $\sum_{a \in C} y_a = \sum_{a \in C} \omega_a$ and $v^a \leq u_a(y_a)$ for all $a \in C$. Define $x_a = \sum_{C \in \mathcal{B}_a} \lambda_C y_a$ for each $a \in C$. Since X_a is convex and $u_a(\cdot)$ is quasi-concave, we have $x_a \in X_a$ and $v^a \leq u_a(x_a)$. Moreover, it follows that:

$$\sum_{a \in A} x_a = \sum_{a \in A} \sum_{C \in \mathcal{B}_a} \lambda_C y_a = \sum_{C \in \mathcal{B}_a} \sum_{a \in A} \lambda_C y_a$$

$$= \sum_{C \in \mathcal{B}_a} \sum_{a \in A} \lambda_C \omega_a = \sum_{a \in A} \sum_{C \in \mathcal{B}_a} \lambda_C \omega_a = \sum_{a \in A} \omega_a.$$

This proves $v = (v^a) \in V(A)$. □

Define the real number $b_a = \sup\{v^a \in \mathbb{R} \mid (v^a) \in V(\{a\})\}$ for each $a \in A$.

Scarf's core existence theorem now reads:

Theorem 7.2. *Let $V: \mathcal{A} \to \mathbb{R}^N$ be a (non-sidepayment) game. The core of V is nonempty if:*

(i) $V(C) - \mathbb{R}^N_+ = V(C)$ *for every* $C \in \mathcal{A}$;
(ii) *there exists* $M \in \mathbb{R}$ *such that for each* $C \in \mathcal{A}$;
if $\boldsymbol{u} = (u^a) \in V(C) \cap \{(b^a) + \mathbb{R}^N_+\}$, *then* $u^a < M$ *for all* $a \in C$;
(iii) $V(C)$ *is closed subset of* $\mathbb{R}^N$ *for every* $C \in \mathcal{A}$; *and*
(iv) *the game* V *is balanced.*

Proof. Without loss of generality, we may assume that $b^a = 0$ for all $a \in A$. Given the positive number M of the assumption (ii), we define $\Delta^C = co\{-MN\boldsymbol{e}^t | t \in C\}$ for every $C \in \mathcal{A}$, where $\boldsymbol{e}^t = (0 \cdots 0, 1, 0 \cdots 0) \in \mathbb{R}^N$ and 1 is in the t-th coordinate. We define the function $\tau: \Delta^A \to \mathbb{R}$ by $\tau(\boldsymbol{u}) = \sup\{t \in \mathbb{R} \mid \boldsymbol{u} + t\mathbf{1} \in \cup_{C \in \mathcal{A}} V(C)\}$, where $\mathbf{1} = (1 \cdots 1) \in \mathbb{R}^N$. By Theorem D2 of Appendix D (i), the function τ is continuous. Hence, the function $f: \Delta^A \to \mathbb{R}^N$ defined by $f(\boldsymbol{u}) = \boldsymbol{u} + \tau(\boldsymbol{u})\mathbf{1}$ is also continuous. Note that $f(\boldsymbol{u}) \geq 0$. Let $G_C = \{\boldsymbol{u} \in \Delta^A | f(\boldsymbol{u}) \in V(C)\}$. Since the set G_C is the inverse image under f of the closed set $V(C)$, G_C is closed. We claim if $\Delta^C \cap G_B \neq \emptyset$ for $B, C \in \mathcal{A}$, then $B \subset C$. This claim trivially holds when $C = A$. Thus, assume that $\sharp C < N$, and take any $\boldsymbol{u} \in \Delta^C \cap G_B$.

On one hand, $\sum_{a \in C} u^a = -MN$, so that there exists $a \in C$ such that $u^a \leq -MN/\sharp C < -M$. Since $f(\boldsymbol{u}) = \boldsymbol{u} + \tau(\boldsymbol{u})\mathbf{1} \geq 0$, $u^a + \tau(\boldsymbol{u}) \geq 0$. Hence, $\tau(\boldsymbol{u}) > M$. On the other hand, $f(\boldsymbol{u}) \in V(B)$, so that for every $a \in B$, $u^a + \tau(\boldsymbol{u}) < M$. Thus, $u^a < 0$ for every $a \in B$. Since $\boldsymbol{u} = (u^a) \in \Delta^C$ implies that $u^a = 0$ for every $a \notin C$, one obtains that $B \subset C$, and the claim is established. Then, the conditions of the Knaster–Kuratowski–Mazurkiewicz–Shapley (K–K–M–S) lemma (Theorem A3 of Appendix A) are satisfied; hence there exists a balanced family $\mathcal{B}$ and a point $\boldsymbol{u}^* \in \Delta^A$ such that $\boldsymbol{u}^* \in \cap_{C \in \mathcal{B}} G_C$. Thus, $f(\boldsymbol{u}^*) \in \cap_{C \in \mathcal{B}} V(C) \subset V(A)$. By the definition of τ, $f(\boldsymbol{u}^*)$ is on the boundary of $\cap_{C \in \mathcal{A}} V(C)$, and according to (i), it cannot be improved upon by any coalition. Therefore, $f(\boldsymbol{u}^*) \in \mathcal{C}(V)$. □

From the Scarf's theorem, the noteworthiness of the core of the exchange economy $\mathcal{E} = (u_a, \omega_a)_{a \in A}$ is obtained immediately.

Theorem 7.3. *Let $\mathcal{E} = (u_a, \omega_a)_{a \in A}$ be a pure exchange economy such that for every $a \in A$, the consumption set X_a is a closed and convex subset of $\mathbb{R}^\ell$, which is bounded from below, $\omega_a \in X_a$ and the utility function u_a is quasi-concave. Then, the core of the economy $\mathcal{E}$, $\mathcal{C}(\mathcal{E})$ is nonempty.*

Proof. We shall show that the conditions from (i) to (iv) of Theorem 7.2 are satisfied. The condition (i) follows from the definition of $V(C)$. To show the conditions

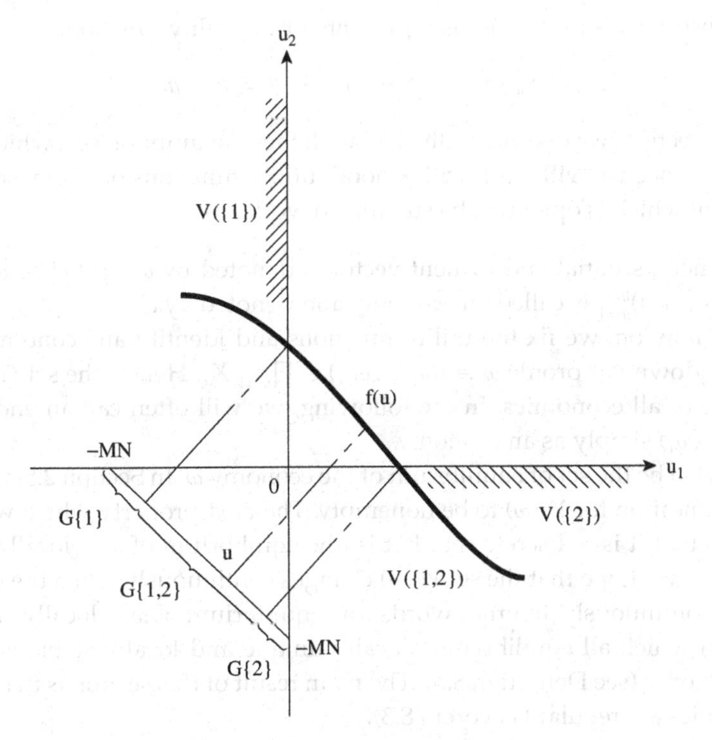

Figure 2.6. The sets $V(C)$, and G_C.

(ii) and (iii), define the set $A(C)$ of attainable allocations for the coalition C by:

$$A(C) = \left\{ (\boldsymbol{x})_{a \in C} \in \prod_{a \in C} X_a \middle| \sum_{a \in C} \boldsymbol{x}_a = \sum_{a \in C} \omega_a \right\}.$$

Since X_a is bounded from below and closed, $A(C)$ is bounded and closed, hence it is compact. Then, $V(C) = \{\boldsymbol{v} = (v^a) \in \boldsymbol{R}^N \mid v^a \leq u_a(\boldsymbol{x}_a)$ for some $(\boldsymbol{x}_a)_{a \in C} \in A(C)\}$. The conditions (ii) and (iii) follow from the compactness of $A(C)$. Finally, the condition (iv) is a consequence of Proposition 7.1. □

2.8. THE LOCAL UNIQUENESS OF EQUILIBRIA

Consider an exchange economy with m consumers. The commodity space of our economy is ℓ-dimensional Euclidean space $\mathbb{R}^\ell$.

The consumer a's consumption set X_a is defined by:

$$X_a = \{\boldsymbol{x} = (x)^t \in \mathbb{R}^\ell \mid x^t > 0 \ t = 1 \cdots \ell\} \equiv \mathbb{R}^\ell_{++}, \quad a = 1 \cdots m.$$

and the consumer a's preference is represented by a utility function:

$$u_a: X_a \to \mathbb{R}, \quad \boldsymbol{x} \mapsto u_a(\boldsymbol{x}), \; a = 1 \cdots m.$$

NB: In this section, we exceptionally deal with the consumption sets which are not closed, since we will work with smooth utility functions on them, so that the domain which is open will be suitable to work with.

The consumer a's initial endowment vector is denoted by $\omega_a = (\omega_a^1 \cdots \omega_a^\ell) \in X_a$. The list $((u_a), (\omega_a))_{a=1}^m$ is called an economy and denoted by $\mathcal{E}$.

From now on, we fix the utility functions and identify an economy $\mathcal{E}$ with its initial endowment profile $\omega \equiv (\omega_1 \dots \omega_m) \in \prod_{a=1}^m X_a$. Hence, the set $\Omega \equiv \prod_a X_a$ is the space of all economies. In the following, we will often call an endowment profile $\omega = (\omega_a)$ simply as an economy.

Let $\mathcal{W}(\omega)$ be the set of equilibrium of the economy ω. In Section 2.5, we gave a sufficient condition for $\mathcal{W}(\omega)$ to be nonempty. The next property which we expect for $\mathcal{W}(\omega)$ is that it is a discrete set, that is, the equilibrium of ω is locally unique. Furthermore, we hope that the set $\mathcal{W}(\omega)$ changes continuously when the economy ω changes continuously. In other words, the equilibrium of ω is locally stable. An economy in which all equilibria are locally unique and locally stable is called a regular economy (see Definition 8.3). The main result of this section is that "almost all" economies are regular (Theorem 8.3).

We assume that the utility function satisfies the following conditions.

(U-1) u_a is twice continuously differentiable, namely that of class C^2 on X_a.

Let $Du_a(\boldsymbol{x}) = (\partial_1 u_a(\boldsymbol{x}) \cdots \partial_\ell u_a(\boldsymbol{x}))$ be the derivative (tangent map) of u_a at $\boldsymbol{x} \in X_a$, and we have:

$$D^2 u_a(\boldsymbol{x}) = \begin{pmatrix} \partial_1 \partial_1 u_a(\boldsymbol{x}) & \cdots & \partial_1 \partial_\ell u_a(\boldsymbol{x}) \\ \cdots & \cdots & \cdots \\ \partial_\ell \partial_1 u_a(\boldsymbol{x}) & \cdots & \partial_\ell \partial_\ell u_a(\boldsymbol{x}) \end{pmatrix}$$

be the second derivative, where $\partial_t u_a(\boldsymbol{x}) = \lim_{h \to 0} (1/h)(u_a(x^1 \cdots x^t + h \cdots x^\ell) - u_a(x^1 \cdots x^t \cdots x^\ell)$, and so on. For every $\boldsymbol{x} \in X_a$, the Hessian $D^2 u_a(\boldsymbol{x})$ is considered to be a linear map from $\mathbb{R}^\ell$ to itself. We assume for every a,

(U-2) u_a is strictly differentiably monotone, i.e., $Du_a(\boldsymbol{x}) \gg 0$ for every $\boldsymbol{x} \in X_a$.

(U-3) u_a is strictly differentiably concave, i.e., $D^2 u_a(\boldsymbol{x})$ is a nondegenerate, negative definite bilinear form on $\mathbb{R}^\ell$, namely that:

$\boldsymbol{y} D^2 u_a(\boldsymbol{x}) \boldsymbol{y} = \sum_{s,t=1}^\ell \partial_s \partial_t u_a(\boldsymbol{x}) y^s y^t \leq 0$[1] for every $\boldsymbol{y} \in \mathbb{R}^\ell$ and the equality holds only when $\boldsymbol{y} = (y^t) = \boldsymbol{0}$.

[1]Note that in this section, we do not distinguish the row vectors and the column vectors.

(U-4) for every sequence, $\boldsymbol{x}_n = (x_n^t) \in X_a$ such that $x_n^t \to 0$ for some t, it follows that $u_a(\boldsymbol{x}_n) \to -\infty$.

Note that the assumption (U-3) implies that the linear map $D^2 u_a(\boldsymbol{x})$ is a linear isomorphism of $\mathbb{R}^\ell$ to itself. The assumption (U-4) will be sometimes called Inada condition.

A price vector in the economies is an ℓ-vector $\boldsymbol{p} = (p^t) \gg 0$. An m-tuple of consumption vectors $(\boldsymbol{x}_a) \in \prod_{a=1}^m X_a$ is called an allocation. As usual, the allocation is said to be feasible if $\sum_{a=1}^m \boldsymbol{x}_a = \sum_{a=1}^m \omega_a$. The concept of the Pareto optimality (Section 2.5) can be re-formulated as follows.

Definition 8.1. An allocation $(\boldsymbol{x}_a)$ is said to be Pareto optimal if it is a solution of the following social welfare maximization problem:

$$\text{Given } \lambda = (\lambda_2 \cdots \lambda_m) \in \mathbb{R}_{++}^{m-1},$$

$$\text{maximize } u_1(\boldsymbol{x}_1) + \sum_{a=2}^m \lambda_a u_a(\boldsymbol{x}_a) \text{ subject to } \sum_a \boldsymbol{x}_a = \sum_a \omega_a.$$

It is easy to verify that the solution of this maximization problem exists. As it will be shown in the proof of Proposition 8.1, the solution $\boldsymbol{x}_a$ associated with λ is a smooth function of $\lambda = (\lambda_2 \cdots \lambda_m)$. Therefore, we will denote them by $\boldsymbol{x}_a(\lambda)$.

On account of the second fundamental theorem of welfare economics (Theorem 5.2 and its corollary), a Pareto-optimal allocation which satisfies the budget constraints of all consumers is a competitive equilibrium. The following Negishi equation (Balasko, 1997a) then express this property of the competitive equilibrium.

$$v_a(\lambda, \omega) = Du_a(\boldsymbol{x}_a(\lambda))(\boldsymbol{x}_a(\lambda) - \omega_a) = 0, \quad a = 2 \cdots m.$$

Note that the first consumer's budget equation follows from the feasibility condition: $\sum_{a=1}^m (\boldsymbol{x}_a - \omega_a) = 0$ and the first-order conditions of the maximization problem, namely that $Du_1(\boldsymbol{x}_1) = \lambda_a Du_a(\boldsymbol{x}_a)$, $a = 2 \cdots m$. See the proof of Proposition 1. The next definition is of Balasko (1997a).

Definition 8.2. A pair $(\lambda, \omega) \in \mathbb{R}_{++}^{m-1} \times \Omega$ is a λ-equilibrium if it is a solution of the Negishi equation systems,

$$v_a(\lambda, \omega) = Du_a(\boldsymbol{x}_a(\lambda))(\boldsymbol{x}_a(\lambda) - \omega_a) = 0, \quad a = 2 \cdots m.$$

Example 8.1. There exist two consumers. The utility functions are of Cobb–Douglas or log-linear form,

$$u_a(\boldsymbol{x}) = \sum_{t=1}^{\ell} \alpha_a^t \log x^t, \quad \alpha_a^t > 0, \ a = 1, 2.$$

Then we have,

$$Du_a(\boldsymbol{x}) = \left(\frac{\alpha_a^1}{x_a^1} \cdots \frac{\alpha_a^\ell}{x_a^\ell}\right), \quad a = 1,2$$

and

$$D^2u_a(\boldsymbol{x}) = -\begin{pmatrix} \frac{\alpha_a^1}{(x^1)^2} & 0 & 0 & \dots & 0 \\ 0 & \frac{\alpha_a^2}{(x^2)^2} & 0 & \dots & 0 \\ \dots & \dots & \dots & & \\ 0 & 0 & 0 & \dots & \frac{\alpha_a^\ell}{(x_\ell)^2} \end{pmatrix}, \quad a = 1,2.$$

Let $\mathcal{G} \subset \mathbb{R}^{m-1}_{++} \times \Omega$ denotes the set of λ-equilibria. The projection map $\pi: \mathcal{G} \to \Omega$ is the restriction to $\mathcal{G}$ of the natural projection $(\lambda, \omega) \to \omega$.

Proposition 8.1. *The Negishi function defined by:*

$$v: \mathbb{R}^{m-1}_{++} \times \Omega \to \mathbb{R}^{m-1}, \quad (\lambda, \omega) \mapsto v(\lambda, \omega) = (v_2(\lambda, \omega) \cdots v_m(\lambda, \omega))$$

is of class C^1.

Proof. First, we shall show that the map $\lambda \to \boldsymbol{x}_a(\lambda)$, $a = 2 \cdots m$ is smooth. Note that the function is determined implicitly by the first-order condition of the social welfare maximization problem with a resource constraint:

$$\sum_{a=1}^{m} x_a - \sum_{a=1}^{m} \omega_a = 0,$$

$$Du_1(\boldsymbol{x}_1) - \lambda_b Du_b(\boldsymbol{x}_b) = 0, \quad b = 2 \cdots m.$$

It is sufficient to check that the Jacobian matrix of the above equation system:

$$\begin{pmatrix} I & I & I & \dots & I \\ -D^2u_1 & \lambda_2 D^2u_2 & 0 & \dots & 0 \\ -D^2u_1 & 0 & \lambda_3 D^2u_3 & \dots & 0 \\ \dots & \dots & \dots & & \\ \dots & \dots & \dots & \dots & \dots \\ -D^2u_1 & 0 & 0 & \dots & \lambda_m D^2u_m \end{pmatrix}$$

is invertible (Theorem E2 of Appendix E). This matrix is equivalent to:

$$\begin{pmatrix} I & 0 & 0 & \cdots & 0 \\ -D^2u_1 & D^2u_1 + \lambda_2 D^2u_2 & D^2u_1 & \cdots & D^2u_1 \\ -D^2u_1 & D^2u_1 & D^2u_1 + \lambda_3 D^2u_3 & \cdots & D^2u_1 \\ \dots & \dots & \dots & & \\ \dots & \dots & \dots & \dots & \dots \\ -D^2u_1 & D^2u_1 & D^2u_1 & \cdots & D^2u_1 + \lambda_m D^2u_m \end{pmatrix}.$$

Therefore, it suffices to show that the matrix:

$$M = \begin{pmatrix} D^2u_1 + \lambda_2 D^2u_2 & D^2u_1 & \cdots & D^2u_1 \\ D^2u_1 & D^2u_1 + \lambda_3 D^2u_3 & \cdots & D^2u_1 \\ \cdots\cdots & \cdots\cdots & & \\ \cdots\cdots & \cdots\cdots & \cdots & \cdots \\ D^2u_1 & D^2u_1 & \cdots & D^2u_1 + \lambda_m D^2u_m \end{pmatrix}$$

is invertible. Let $\boldsymbol{v} = (\boldsymbol{v}_2 \cdots \boldsymbol{v}_m) \in (\mathbb{R}^\ell)^{m-1}$ and consider the equation $M\boldsymbol{v} = 0$. Then, we have:

$$(D^2u_1 + \lambda_b D^2u_b)\boldsymbol{v}_b + D^2u_1 \left(\sum_{c \neq b} \boldsymbol{v}_c \right) = 0, \quad b = 2 \cdots m.$$

Making the inner product with the vector $\boldsymbol{v}_b$, then yields:

$$\boldsymbol{v}_b(D^2u_1 + \lambda_b D^2u_b)\boldsymbol{v}_b + \boldsymbol{v}_b D^2u_1 \left(\sum_{c \neq b} \boldsymbol{v}_c \right) = 0, \quad b = 2 \cdots m,$$

which simplifies into:

$$\lambda_b \boldsymbol{v}_b D^2u_b \boldsymbol{v}_b + \boldsymbol{v}_b D^2u_1 \left(\sum_{c=2}^{m} \boldsymbol{v}_c \right) = 0, \quad b = 2 \cdots m.$$

We add up these equations for b from 2 to m,

$$\sum_{b=2}^{m} \lambda_b \boldsymbol{v}_b D^2u_b \boldsymbol{v}_b + \left(\sum_{b=2}^{m} \boldsymbol{v}_b \right) D^2u_1 \left(\sum_{b=2}^{m} \boldsymbol{v}_b \right) = 0.$$

Since D^2u_b are negative definite, each term in the sum of the left-hand side of the above equation is ≤ 0, hence each term must be zero,

$$\left(\sum_{b=2}^{m} \boldsymbol{v}_b \right) D^2u_1 \left(\sum_{b=2}^{m} \boldsymbol{v}_b \right) = 0,$$

$$\lambda_b \boldsymbol{v}_b D^2u_b \boldsymbol{v}_b = 0, \quad b = 2 \cdots m.$$

Therefore, we have $\boldsymbol{v}_2 = \cdots = \boldsymbol{v}_m = 0$ as desired. □

Therefore, by construction, the Negishi map:

$$\nu_b(\lambda,\omega) = Du_b(\boldsymbol{x}_b(\lambda))(\boldsymbol{x}_b(\lambda) - \omega_b), \quad b = 2\cdots m,$$

is of class C^1 with respect to $(\lambda,\omega) = ((\lambda_b),(\omega_a)), a = 1\cdots m, b = 2\cdots m$.

Example 8.2. The utility functions for the consumers $a = 1,2$ are given as in Example 8.1. Then, the first-order conditions for the social optimization are:

$$x_1^t + x_2^t = \omega_1^t + \omega_2^t \equiv \omega^t,$$

$$\frac{\alpha_1^t}{x_1^t} - \lambda\frac{\alpha_2^t}{x_2^t} = 0, \quad t = 1\cdots\ell.$$

From this, we obtain:

$$x_1^t = \frac{\alpha_1^t\omega^t}{\alpha_1^t + \lambda\alpha_2^t}, \quad \text{and} \quad x_2^t = \frac{\lambda\alpha_2^t\omega^t}{\alpha_1^t + \lambda\alpha_2^t}.$$

Certainly, $\boldsymbol{x}_1 = (x_1^t)$ and $\boldsymbol{x}_2 = (x_2^t)$ are smooth functions of λ.

Proposition 8.2. *The set of equilibria $\mathcal{G}$ is a smooth submanifold of $\mathbb{R}^{m-1}\times\Omega$ of co-dimension $m-1$.*

Proof. We shall prove this proposition by applying the regular value theorem (Theorem E5, Appendix E), and it amounts to prove that $\mathbf{0} \in \mathbb{R}^{m-1}$ is a regular value of the map ν, which is smooth by Proposition 8.1. Let $D\nu(\lambda,\omega)\colon \mathbb{R}^{m-1} \times T_\omega\Omega \to \mathbb{R}^{m-1}$ denote the tangent map (derivative) of the map ν at $(\lambda,\omega) \in \mathcal{G}$, where $T_\omega\Omega$ is the tangent space of Ω at ω. We denote the tangent map $D\nu(\lambda,\omega)$ as $(D\nu_b(\lambda,\omega))_{b=2}^m$, which is defined by the $m-1$ coordinate mappings of the derivatives of ν_b with respect to $(\lambda_2\cdots\lambda_m,\omega_1\cdots\omega_m)$. Note that the partial derivative of ν_b with respect to ω_c $(c \neq b)$ is 0. The derivative of ν_b is then written as:

$$D\nu_b(\lambda,\omega)(\dot{\lambda},(\dot{\omega}_a)) = \sum_{c=2}^{m}\frac{\partial\nu_b(\lambda,\omega)}{\partial\lambda_c}\dot{\lambda}_c + \sum_{t=1}^{\ell}\frac{\partial\nu_b(\lambda,\omega)}{\partial\omega_b^t}\dot{\omega}_b^t$$

where $\dot{\lambda} = (\dot{\lambda}_c) \in \mathbb{R}^{m-1}$ and $\dot{\omega}_b = (\dot{\omega}_b^t) \in \mathbb{R}^\ell$ are tangent vectors. In order to show that the $D\nu(\lambda,\omega)$ is onto, it suffices to prove that a restriction of this map to some subspace of $\mathbb{R}^{m-1}\times T_\omega\Omega$ is onto. Pick an arbitrary period t. The Jacobian matrix of $D\nu$ with respect to $(\omega_2^t\cdots\omega_m^t)$ is:

$$\begin{pmatrix} \partial_t u_2(\boldsymbol{x}_2(\lambda)) & 0 & \cdots & 0 \\ 0 & \partial_t u_3(\boldsymbol{x}_3(\lambda)) & \cdots & 0 \\ \cdots & \cdots & \cdots & \cdots \\ 0 & 0 & \cdots & \partial_t u_m(\boldsymbol{x}_m(\lambda)) \end{pmatrix},$$

which is obviously of rank $m-1$ by the assumption (U-2). □

By Proposition 8.2, it follows that the projection map $\pi: \mathcal{G} \to \Omega, (\lambda, \omega) \mapsto \omega$ is smooth, since it is a composition of smooth maps, the canonical embedding $\mathcal{G} \to \mathbb{R}^{m-1}_{++} \times \Omega$ and the canonical projection map $\mathbb{R}^{m-1}_{++} \times \Omega \to \Omega$. We call $\mathcal{G}$ the equilibrium manifold.

We now come up with the definition of our main theme of this section.

Definition 8.3. A λ-equilibrium $(\lambda, \omega) \in \mathcal{G}$ is called regular if it is a regular point of the projection map $\pi: \mathcal{G} \to \Omega$. The regular value $\omega \in \Omega$ is called a regular economy. An economy which is not regular is called critical.

An important property of the map π is the following.

Proposition 8.3. *The projection map $\pi: \mathcal{G} \to \Omega$ is proper, that is, its inverse $\pi^{-1}(K)$ of every compact set $K \subset \Omega$ is compact.*

Proof. Let K be a compact subset of Ω and K_a be the image of K by the restriction of the natural projection map $(\omega_a) \mapsto \omega_a$. The set K_a is therefore a compact subset of $X_a = \mathbb{R}^{\ell}_{++}$. The utility function $u_a: X_a \to \mathbb{R}$ being smooth, hence continuous, so that the image $u_a(K_a)$ is compact, hence bounded by some $b_a \in \mathbb{R}$, $a = 1 \cdots m$. Define a subset L of $\mathbb{R}^{\ell m}$ by:

$$\to L = \left\{ (\boldsymbol{x}_a) \in \mathbb{R}^{\ell m}_{++} \,\middle|\, \sum_{a=1}^{m} \boldsymbol{x}_a = \sum_{a=1}^{m} \omega_a, \ \ u_a(\boldsymbol{x}_a) \geq b_a, a = 1 \cdots m \right\}$$

We claim that the set L is compact. First, for each $a = 1 \cdots m$, define the set $L_a \subset \mathbb{R}^{\ell}_{++}$ by $L_a = \{\boldsymbol{x}_a \in \mathbb{R}^{\ell}_{++} \mid \boldsymbol{x}_a \leq \sum_a \omega_a, u_a(\boldsymbol{x}_a) \geq b_a\}$. The set L_a is bounded since zero is a lower bound of L_a and $\sum_a \omega_a$ is an upper bound. Next the set L_a is closed. For, let $(\boldsymbol{x}_n)$ $n = 1, 2 \ldots$ be a sequence in L_a with $\boldsymbol{x}_n \to \boldsymbol{x}_*$. Clearly, $\boldsymbol{x}_* \leq \sum_{a=1}^{m} \omega_a$ and by the continuity of u_a, we have $u_a(\boldsymbol{x}_*) \geq b_a$. It remains to show that $\boldsymbol{x}_* \in \mathbb{R}^{\ell}_{++}$. Obviously, $\boldsymbol{x}_* \geq 0$. Since $u_a(\boldsymbol{x}_*) \geq b_a$, it is impossible that $x^t_* = 0$ for any t by assumption (U-2). Therefore, L_a is compact, hence the product $\prod_{a=1}^{m} L_a$ is also compact by the Tychonoff theorem (Theorem B1 of Appendix B). As a closed subset of $\prod_{a=1}^{m} L_a$, the set L is compact.

The individual rationality means that if $\boldsymbol{x} \in \Omega$ is an equilibrium allocation associated with $\omega \in \Omega$, then the inequality $u_a(\boldsymbol{x}_a) \geq u_a(\omega_a)$ is satisfied for all $a = 1 \cdots m$. This implies that all equilibrium allocations $(\boldsymbol{x}_a)$ associated with endowments $\omega \in K$, the inequality $u_a(\boldsymbol{x}_a) \geq u_a(\omega_a) \geq b_a$ is satisfied for all $a = 1 \cdots m$. In other words, if $(\omega_a) \in K$, then $(\boldsymbol{x}_a) \in L$. We now claim that the set of Pareto-optimal allocations associated with endowments $(\omega_a) \in K$ is a closed subset of a compact set L, hence the set is compact.

Let $(\boldsymbol{x}^n_a), n = 1, 2, \ldots$ be a sequence of the solutions of social welfare maximization problem associated with the endowments (ω^n_a) in K such that $\boldsymbol{x}^n_a \to \boldsymbol{x}^*_a$. Then for each n, there exists a social welfare weight $(\lambda^n_a)^m_{a=2}$ which satisfies the first-order conditions:

$$\sum_{a=1}^{m} \boldsymbol{x}_a^n - \sum_{a=1}^{m} \omega_a^n = 0,$$

$$Du_1(\boldsymbol{x}_1^n) - \lambda_a^n Du_a(\boldsymbol{x}_a^n) = 0, \quad a = 2 \cdots m, \; n = 1, 2, \ldots.$$

Since K is compact, we may assume $\omega_a^n \to \omega_a^*, a = 1 \cdots m$. For each n, we have $\lambda_a^n = \partial_1 u_1(\boldsymbol{x}_1^n)/\partial_1 u_a(\boldsymbol{x}_a^n)$. Since the partial derivative are continuous and $\boldsymbol{x}_a^n \to \boldsymbol{x}_a^*$, it follows that $\lambda_a^n \to \lambda_a^* = \partial_1 u_1(\boldsymbol{x}_1^*)/\partial_1 u_a(\boldsymbol{x}_a^*)$ by the assumption (U-2). Clearly one has:

$$\sum_{a=1}^{m} \boldsymbol{x}_a^* - \sum_{a=1}^{m} \omega_a^* = 0,$$

$$Du_1(\boldsymbol{x}_1^*) - \lambda_a^* Du_a(\boldsymbol{x}_a^*) = 0, \quad a = 2 \ldots m.$$

This shows that the set of Pareto-optimal allocations associated with endowments in K is a closed subset of L, as desired. The set of associated welfare weights is then necessarily a compact subset H of $\mathbb{R}_{++}^{m-1}$. Since $\pi^{-1}(K) = (H \times K) \cap \mathcal{G}$ and $\mathcal{G}$ is closed in $\mathbb{R}^{m-1} \times \Omega$, it follows that $\pi^{-1}(K)$ is compact. □

The next proposition which characterizes the regular and/or singular equilibria will be useful for the subsequent analysis.

Proposition 8.4. *The λ-equilibrium $(\lambda, \omega) \in \mathcal{G}$ is critical if and only if:*

$$\det \frac{Dv(\lambda, \omega)}{D\lambda} = 0,$$

where $\det A$ *means the determinant of a matrix A.*

Proof. The tangent space of $\mathcal{G}$ at (λ, ω) and the derivative map of the projection π are written as:

$$T_{(\lambda,\omega)}\mathcal{G} = \left\{ (\dot{\lambda}, \dot{\omega}) \in \mathbb{R}^{m-1} \times \mathbb{R}^{\ell m} \middle| \; \frac{Dv(\lambda, \omega)}{D\lambda}\dot{\lambda} + \frac{Dv(\lambda, \omega)}{D\omega}\dot{\omega} = 0 \right\},$$

and

$$D\pi \colon T_{(\lambda,\omega)}\mathcal{G} \to \mathbb{R}^{\ell m}, (\dot{\lambda}, \dot{\omega}) \mapsto \dot{\omega},$$

where $\dot{\lambda} \in \mathbb{R}^{m-1}$ and $\dot{\omega} \in \mathbb{R}^{\ell m}$ are the tangent vectors.

Since dimension $T_{(\lambda,\omega)}\mathcal{G} = \ell m$ by Theorem E5 of Appendix E, the map $D\pi$ fails to be onto if and only if the linear subspace of $T_{(\lambda,\omega)}\mathcal{G}$ of the form $\{(\dot{\lambda}, \dot{\omega}) \in T_{(\lambda,\omega)}\mathcal{G} | \; \dot{\omega} = \mathbf{0}\}$ contains the linear space other than $\{0\}$. The necessary and sufficient condition for this is that the linear equation $\frac{Dv(\lambda,\omega)}{D\lambda}\dot{\lambda} = 0$ has a non-zero solution $\dot{\lambda} \neq 0$, namely that $\det \frac{Dv(\lambda,\omega)}{D\lambda} = 0$. □

The fundamental properties of the regular economy are that they have finitely many (regular) equilibria, which are locally stable and unique.

Theorem 8.1. *If ω is a regular economy, then the set of λ-equilibria associated with ω is finite.*

Proof. By Proposition 8.4, a λ-equilibrium $(\lambda_0, \omega_0) \in \mathcal{G}$ is regular if and only if $\det \frac{D\nu(\lambda,\omega)}{D\lambda} \neq 0$. Then, by the implicit function theorem (Theorem E2 of Appendix E), we can take a neighborhood $\mathcal{U}_0$ of ω_0 such that λ is a C^1 function of ω and $\nu(\lambda(\omega), \omega) = 0$ for all $\omega \in \mathcal{U}_0$. Let $\mathcal{G}_0 = \{(\lambda, \omega) \in \mathcal{G} \mid \omega \in \mathcal{U}_0\}$ and define the map $\rho \colon \mathcal{U}_0 \to \mathcal{G}_0$ by $\rho(\omega) = (\lambda(\omega), \omega)$. We then have $\pi \circ \rho =$ identity on $\mathcal{U}_0$, which means that π restricted to $\mathcal{G}_0$ is a diffeomorphism between $\mathcal{U}_0$ and $\mathcal{G}_0$ with the inverse ρ.

Let ω be a regular value of π. Since π is a proper map by Proposition 8.3, the set $\pi^{-1}(\omega)$ is compact. In order to show that the set $\pi^{-1}(\omega)$ is finite, it suffices to show that $\pi^{-1}(\omega)$ is discrete. Take open sets $\mathcal{U}_0$ and $\mathcal{G}_0$ of the preceding discussion. We claim that $\pi^{-1}(\omega) \cap \mathcal{G}_0$ is one point set $\{(\lambda(\omega), \omega)\}$. If not, $\{(\lambda(\omega), \omega)\}$ contains at least two distinct points both of which are mapped to ω by the projection π. This contradicts the fact that π restricted to $\mathcal{G}_0$ is bijective. □

Let $\mathcal{R}$ be the set of regular economies. We will show in Theorem 8.3 that $\mathcal{R}$ is open and dense subset of $\Omega = \mathbb{R}^{\ell m}_{++}$. The next theorem shows that the regular equilibrium is locally unique and moves continuously when the economy changes continuously (locally stable).

Theorem 8.2. *For every $\omega \in \mathcal{R}$, there exists an open neighborhood V of ω in $\mathcal{R}$ such that the pre-image $\pi^{-1}(V)$ is the disjoint union of a family of finite number of open subsets $\mathcal{U}_i$ of $\pi^{-1}(\mathcal{R})$, $i = 1 \cdots k$ and the restriction $\pi_i \colon \mathcal{U}_i \to V$ of π to each $\mathcal{U}_i$ is a homeomorphism.*

Proof. By Theorem 8.2, $\pi^{-1}(\omega)$ is a finite set $\{(\lambda_1, \omega) \cdots (\lambda_k, \omega)\}$. As in the proof of Theorem 1, we can take open disjoint neighborhoods $\mathcal{U}'_1 \cdots \mathcal{U}'_k$ of regular equilibria $(\lambda_1, \omega) \cdots (\lambda_k, \omega)$ such that the restriction of π to $\mathcal{U}'_i$ is a diffeomorphism with $V_i = \pi(\mathcal{U}'_i)$, $i = 1 \cdots k$.

We claim that the image of a closed set by the proper map π is closed. For let C be a closed set and take a converging sequence $\{\boldsymbol{y}_n\}$ in $\pi(C)$, $\boldsymbol{y}_n \to \boldsymbol{y}_*$. The set $Y = \{\boldsymbol{y}_n\} \cup \{\boldsymbol{y}_*\}$ being compact, $\pi^{-1}(Y)$ is a compact subset of C by the properness of π. Take $\boldsymbol{x}_n \in \pi^{-1}(\boldsymbol{y}_n)$ for each n. Then, we can assume that $\boldsymbol{x}_n \to \boldsymbol{x}_* \in \pi^{-1}(Y) \subset C$. Since the projection is continuous, we have $\boldsymbol{y}_* = \pi(\boldsymbol{x}_*) \in \pi(C)$, as desired.

Since the set $\mathcal{G} \backslash (\mathcal{U}'_1 \cup \cdots \cup \mathcal{U}'_k)$ is closed in $\mathcal{G}$, its image by the map π is closed. Define the set V by:

$$V = (V_1 \cap \cdots \cap V_k) \backslash \pi(\mathcal{G} \backslash (\mathcal{U}'_1 \cup \cdots \cup \mathcal{U}'_k)).$$

Clearly, the set V is open in Ω. We show that $\omega \in V$. Since $\omega \in \bigcap_{i=1}^{k} V_i$, it suffices to show that $\omega \notin \pi(\mathcal{G}\backslash(U_1' \cup \cdots \cup U_k'))$. This follows from the fact that $\pi^{-1}(\omega) \subset U_1' \cup \cdots \cup U_k'$.

Define $U_i = U_i' \cap \pi^{-1}(V)$. The restriction $\pi_i = \pi|_{U_i}$ is then a homeomorphism between U_i and $\pi(U_i) = V$. It remains to prove that $\pi^{-1}(V) = \bigcup_{i=1}^{k} U_i$. $\bigcup_{i=1}^{k} U_i \subset \pi^{-1}(V)$ is clear. Suppose that $\pi^{-1}(V) \subset \bigcup_{i=1}^{k} U_i$ does not hold. Then, there exists $\boldsymbol{x}' \in \pi^{-1}(V)$ such that $\boldsymbol{x}' \notin U_i$ for all i. Then, $\boldsymbol{x}'$ must belong to $\mathcal{G}\backslash(U_1' \cup \cdots \cup U_k')$, which implies that $\omega' = \pi(x') \in \pi(\mathcal{G}\backslash(U_1' \cup \cdots \cup U_k'))$. Therefore, $\omega' \notin V$, contradicting the choice of $\boldsymbol{x}' \in \pi^{-1}(V)$. □

Finally, we prove that the size of the regular economies is "big" in the space of all economics, in other words, the size of singular economies is "small" in the set Ω.

Theorem 8.3. *The set of singular economies is a closed subset of* Ω *which has zero measure.*

Proof. It follows from Sard's theorem (Theorem E6 of Appendix E) that the set of critical economies, as the set of critical values of a smooth map $\pi: \mathcal{G} \to \Omega$, is of measure zero, since dimension $\mathcal{G}$ = dimension Ω. By Proposition 8.4, the set of critical equilibria is a closed subset of $\mathcal{G}$, since the function det $\frac{Dv(\lambda,\omega)}{D\lambda}$ is continuous. The set of critical economies is therefore closed, since it is the image of a proper map π. □

2.9. NOTES

Classical monographs of the general equilibrium theory are Debreu (1959), Nikaido (1968), Arrow and Hahn (1971), and McKenzie (2002). These monographs are written by the founders of the theory and any serious graduate students should have a copy of them. We now give some remarks for each of the sections.

Section 2.1: The basic reference is Debreu (1959), Chap. 2. See also Chap. 7 of this book for the market structure with uncertainty, which is out of the scope of the present monograph. In this monograph, we introduce at first the commodity space including the infinite-dimensional cases and do not restrict the set of economic agents to be finite in contrast to the classical monographs given above. We hope that the readers do not take this as pedantic; we did so just because we need them in later chapters.

Section 2.2: We refer Debreu (1959, Chap. 4) and Hildenbrand (1974, Sections 1.1 and 1.2) as basic references for this section. In particular, following Hildenbrand, we postulate the strict preference relations $\prec$ on which the negative transitivity is not assumed rather than (weak) preference relations $\precsim$ as the basic concept. This is because the space of strict preference relations has simpler topological structure, which will be clear in Section 3.2.

For various kinds of the convexity of preferences and their relationships, see Debreu (1959), Chap. 4.

Section 2.3: Basic references of this section are Debreu (1959) Chap. 4, Arrow and Hahn (1971) Chap. 4, Hildenbrand (1974) Section 1.2, and McKenzie (2002) Chap. 1. For the idea of proof of Theorem 3.1 in which K–K–M lemma was applied, we follow McKenzie (2002), Section 1.2. A surprising result which was originally proved by Sonnenschein (1972) and (1973) and developed by Mantel (1974) and Debreu (1974) should be mentioned. In Section 2.3, we saw that an individual excess demand correspondence is upper hemi-continuous, hence continuous if it is a function (Proposition 3.3), and satisfies the budget equation (under the local nonsatiation); $\boldsymbol{p}(\phi(a,\boldsymbol{p}) - \omega_a) = 0$. Therefore the aggregate excess demand $\sum_a (\phi(a,\boldsymbol{p}) - \omega_a)$ is also continuous and fulfills the Walras law; $\boldsymbol{p} \sum_a (\phi(a,\boldsymbol{p}) - \omega_a) = 0$. Sonnenschein showed that essentially they are the only properties which all aggregate excess demand functions should necessarily have, namely that for each continuous function (market excess demand function) which satisfies the Walras law, one can construct a finite exchange economy in which the sum of the individual excess demand functions coincides on the unit sphere (up to $\epsilon > 0$ qualification, see below) with the given excess demand function. Debreu (1974) gave a beautiful geometric proof for the next result;

Theorem 9.1. *Let* $S_\epsilon = \{\boldsymbol{p} = (p^t) \in \mathbb{R}^\ell \mid \|\boldsymbol{p}\| = 1, p^t \geq \epsilon \text{ for all } t\}$ *and* $\Phi : S_\epsilon \to \mathbb{R}^\ell$ *be a continuous function satisfying* $\boldsymbol{p}\Phi(\boldsymbol{p}) = 0$. *Then for every* $\epsilon > 0$, *there exist* ℓ *consumers* $(\succsim_a, \omega_a)_{a=1}^{\ell}$ *such that the sum of their individual excess demand functions is equal to* $\Phi(\boldsymbol{p})$ *on* S_ϵ, $\sum_{a=1}^{\ell} (\phi(a,\boldsymbol{p}) - \omega_a) = \Phi(\boldsymbol{p})$ *for* $\boldsymbol{p} \in S_\epsilon$.

An educational and readable exposition of this result can be found in Mas-Colell, Whinston and Green (1995) Chap. 17.

Section 2.4: For all of the expositions, we follow McKenzie (1956–1957), and (2002).

Section 2.5: A basic reference is Debreu (1959), Chap. 5. See also Arrow and Hahn (1971), Chap. 5, McKenzie (2002), Chap. 6, and Nikaido (1968) Chap. 5. A general procedure to prove the existence of equilibria is to construct some devices (a mapping or a game) to which the Kakutani's fixed-point theorem will be applied. At least three alternative constructions have been known:

(i) An abstract game constructed from market model and invoke an existence theorem of Nash equilibrium. This method was initiated by Arrow and Debreu (1954) who used a result of Debreu (1952), which generalized the original result of Nash (1950).
(ii) Mapping constructed from the market excess demand correspondences to which Kakutani's fixed-point theorem is applied. This was started by

McKenzie (1954) and Nikaido (1956) and Debreu (1959). We have followed this procedure in this section.

(iii) A fixed-point mapping which exploits the first welfare theorem. This strategy of proof was initiated by Negishi (1960) and Arrow and Hahn (1971).

We will utilize the procedure (i) In Chap. 4 and (iii) In Chap. 5. Mas-Colell (1974) proved an astonishing result that the competitive equilibrium exists without completeness or transitivity of the preferences. Without the completeness or the transitivity, the demand correspondences are not generally convex-valued, and his proof is very sophisticated. Gale and Mas-Colell (1975, see also corrections 1979) presented a more readable proof.

The concept of the (quasi)-equilibrium relative to the price $\boldsymbol{p}$ is in Debreu (1959), Chap. 6. McKenzie (2002) called it as the equilibrium with re-distribution. For the proof of the second welfare theorem (Theorem 5.3), we followed Debreu (1959), Chap. 6. See also Arrow and Hahn (1971), Section 4 of Chap. 4.

Section 2.6: The proof of Theorem 6.1 has followed the original paper of Debreu and Scarf (1963). See also Arrow and Hahn (1971), Chap. 8 and McKenzie (2002), Section 5.2.

Section 2.7: For the proof of Theorem 7.1, we followed Shafer and Sonnenschein (1975). We followed Ichiishi (1983) for the presentation of the model and the proof of Theorem 7.2.

Section 2.8: A clear and readable text book for this topic is Dierker (1974). Advanced readers will be interested in Mas-Colell (1985). The exposition of the present section follows Balasko (1997a, b). The advantage of Balasko's approach which parametrizes an economy by social welfare weights (Negishi's approach) will be apparent in the infinite-dimensional cases. See Balasko (1997c). The most far-reaching result on this subject by now is that of Shannon and Zame (2002).

Chapter 3

ECONOMIES WITH A CONTINUUM OF TRADERS

3.1. MARKETS WITH A MEASURE SPACE OF CONSUMERS

In this chapter, we introduce equilibrium models with a measure space of economic agents. In Chap. 2, we discussed the competitive equilibria of exchange economies in which each consumer behaves as a price taker, namely that consumers have no power to affect market prices, and they take market prices as given when they make decisions on their consumption plans.

From the logical point of view, however, this hypothesis of price-taking behavior of each consumer is not consistent with an assumption that there exist only finitely many consumers in the market. Even if the number of consumers is very large, each of them has some size or weight in the market, and the changes of their behavior will have some effect on the determination of equilibrium market price systems as long as the population is finite.

A consistent model with perfectly competitive equilibria is considered to be achieved by an equilibrium model with a continuum of traders. In such a model, each individual should be negligible or has zero weight compared to the whole market. In order to give a precise meaning to "weight" or "mass" of traders, we assume that the set of traders is a measure space $(A, \mathcal{A}, \mu)$ (see Appendix G for the definition of the measure space). Then, integrating over A the individual demand $\phi(a, \boldsymbol{p})$ which is obtained from the price-taking and utility maximizing behavior, we get the total or aggregated demand $\Phi(\boldsymbol{p}) = \int_A \phi(a, \boldsymbol{p}) d\mu$, and the idea of competitive equilibrium is obtained by postulating the total demand equal to the total supply $\int_A \omega(a) d\mu$, where $\omega(a)$ is the initial endowment of the consumer a. Sometimes, we call the economy with the measure space of consumers as a *large economy*.

In Chap. 2, we discussed the limit theorem of the core which says that the core of the sequence of the replica economies converges with the competitive equilibrium of the original economy as long as the number of the consumers tends to infinity. In the economy with a continuum of consumers, an equivalence theorem of the core and the competitive equilibria holds good in such a way, that in the large economy, the sets of the core allocations and the competitive equilibrium allocations are equal. This theorem will be discussed in Section 3.4.

Moreover in the large economy, we can observe remarkable effects which come from the aggregation of the individual behaviors. One of these effects is the convexing effect of total demand correspondence, which will be discussed in Section 3.3. This says that in markets with an atomless measure space of consumers (the atom is a subset of consumers, which is indivisible and has a measure greater than zero, see Appendix G), the total demand correspondence is convex valued even if the individual demand correspondences are not. Mathematically, this is a straightforward consequence of Liapunov's theorem (Theorem G1 of Appendix G). Thanks to this result, we can prove the existence of equilibrium of exchange economies with atomless measure space of consumers without the convexity of preferences (Theorem 3.1).

There exists one more effect of the aggregation which is relevant to the existence of equilibria. As we will see in Section 3.5, when the consumption set of a consumer is not convex, the demand correspondence of the consumer exhibits discontinuous behaviors with respect to the continuous change of price system. If the set of the consumers with these discontinuous behaviors has a positive measure, the total demand correspondence will also behave discontinuously. However, we will observe that if the distribution of the consumers' characteristics (for example, the distribution of the initial endowment vectors) is "dispersed" in an appropriate sense, the total demand correspondence will be upper hemi-continuous, even if the individual demand correspondences are not.

In Section 3.6, we will introduce the production activity into the model for the first time in this book, and observe that "the aggregation effect" for the exchange economy in Section 3.5 could disappear in the presence of the production sector. The reason is as follows. In the exchange economies, the distribution of the income levels is solely determined by the distribution of the initial endowments. Hence, the dispersed endowment distribution brings the dispersed income distribution from which the "regularizing effect" of the total demand correspondence comes out. However, in the production economies, the income level of a consumer is, by definition, the market value of his/her initial endowment plus the profit share of the firms which he/she owns. Therefore, the dispersed endowment could be vanished by the counter act of the distribution of the profit shares. We will see this kind of phenomenon by a simple example. However, we can prove that every production economies with possibly empty equilibrium can be approximated by economies with equilibria. In terms of topology on the space of economies, which were explored in Section 2.8 in the context of the local uniqueness of the equilibria, the set of economies with equilibria is dense in the space of all economies with dispersed endowments. Unfortunately, we will see by example that the set is not an open subset of the whole space, so that the set of economies without equilibria could also be "large".

Finally, in Section 3.7, we will discuss the effect of aggregation to the law of demand, namely that the demand curve will be downward sloping under a suitable condition on the income distribution, even if the individual demand curves are not.

To sum up:
In this chapter, the set of economic agents is a measure space $(A, \mathcal{A}, \mu)$. For each measurable subset $C \in \mathcal{A}$, $\mu(C)$ is a mass or weight of the consumers belonging to the set C.

3.2. SPACES OF PREFERENCES

Recall that the consumption set X is a non-empty closed subset of $\mathbb{R}^\ell$. Let $\mathcal{X}$ be the set of all (admissible) consumption sets. In this chapter, we assume that there exists a real number $b \in \mathbb{R}$ and a compact set $K \subset \mathbb{R}^\ell$ such that for all $X \in \mathcal{X}$,

$$b \leq X \quad \text{and} \quad X \cap K \neq \emptyset.$$

Let $\prec$ be the (strict) preference relation on $X \in \mathcal{X}$, which is irreflexive (IR), transitive (TR), and continuous (CT). A pair $(X, \prec)$ consisting of the consumption set and the preference relation is called the preference-consumption set pair. We sometimes abuse the term and notation loosely, in such a way that we call the pair $(X, \prec)$ simply as the (strict) preference (on X) and denote it by $\prec$ when the consumption set is understood.

Definition 2.1. $\mathcal{P}$ is the set of all preference-consumption set pairs $(X, \prec)$, such that $\prec$ is a IR and TR binary relation on $X \in \mathcal{X}$ and $\prec$ is open relative to $X \times X$, or satisfies the condition CT.

The first task of this section is to endow some nice topology with $\mathcal{P}$. Recall that to every preference relation $(X, \prec) \in \mathcal{P}$, we associate the relation $(X, \precsim)$ or simply $\precsim$ defined by $(X, \precsim) = \{(\boldsymbol{x}, \boldsymbol{y}) \in X \times X | (\boldsymbol{x}, \boldsymbol{y}) \notin \prec\} = X \times X \setminus \prec$. Since the relations $\prec$ and $\precsim$ are the complements of each other, they are in one-to-one correspondences and $\precsim$ is a closed subset of $\mathbb{R}^{2\ell}$, if X is closed in $\mathbb{R}^\ell$ and $\prec$ is open in $X \times X$.

Consequently, the set $\mathcal{P}$ of preference relations can be considered as a subset of $\mathcal{F}(\mathbb{R}^{2\ell})$; the set of all closed subsets of $\mathbb{R}^{2\ell}$ endowed with the topology of closed convergence τ_c (see Appendix F).

We then have:

Theorem 2.1. *The set $\mathcal{P}$ is compact and metrizable, and a sequence $(X_n, \prec_n)$ converges to $(X, \prec)$ in the topology τ_c, if and only if $Li(\precsim_n) = \precsim = Ls(\precsim_n)$, where $\precsim_n$ and $\precsim$ are the* *associated relations to* $(X_n, \prec_n)$ *and* $(X, \prec)$, *respectively. Moreover, the topology* τ_c *is the weakest Hausdorrf topology on* $\mathcal{P}$ *such that the set:*

$$\{(X, \prec, \boldsymbol{x}, \boldsymbol{y}) \in \mathcal{P} \times \mathbb{R}^\ell \times \mathbb{R}^\ell \,|\, \boldsymbol{x}, \boldsymbol{y} \in X \text{ and } \boldsymbol{x} \precsim \boldsymbol{y}\}$$

is closed.

Proof. By Theorem F2 of Appendix F, the set $\mathcal{F}(\mathbb{R}^{2\ell})$ endowed with the topology τ_c is compact and metrizable. In order to show that $(\mathcal{P}, \tau_c)$ is compact, it suffices from Proposition B4 of Appendix B that $\mathcal{P}$ is a closed subset of $(\mathcal{F}(\mathbb{R}^{2\ell}), \tau_c)$.

Let $\{(X_n, \prec_n)\}_{n \in \mathbb{N}}$ be a sequence in $\mathcal{P}$ and $\{\precsim_n\}_{n \in \mathbb{N}}$ be a sequence of the corresponding complement relations $\precsim_n = \{(\boldsymbol{x}, \boldsymbol{y}) \in X_n \times X_n \mid (\boldsymbol{x}, \boldsymbol{y}) \in X_n \times X_n \setminus \prec_n\}$ such that $\precsim_n \to \precsim$ in the topology τ_c, where $\precsim = \{(\boldsymbol{x}, \boldsymbol{y}) \in X \times X \mid (\boldsymbol{x}, \boldsymbol{y}) \in X \times X \setminus \prec\}$. We have to show that $(X, \prec)$ belongs to $\mathcal{P}$. First, we show that $\prec$ is irreflexive. Let $\boldsymbol{x} \in X$. Since $\prec_n$ is irreflexive for each n, $(\boldsymbol{x}, \boldsymbol{x}) \in \precsim_n$ for all n. Hence, $(\boldsymbol{x}, \boldsymbol{x}) \in \precsim = Li(\precsim_n)$, so that $(\boldsymbol{x}, \boldsymbol{x}) \notin \prec$. Therefore, $\prec$ is irreflexive. Next, we show that $\prec$ is transitive. Let $\boldsymbol{x} \prec \boldsymbol{y}$ and $\boldsymbol{y} \prec \boldsymbol{z}$. Suppose that $(\boldsymbol{x}, \boldsymbol{z}) \notin \prec$, or $(\boldsymbol{x}, \boldsymbol{z}) \in \precsim$. Since $Li(\precsim_n) = \precsim$, there exists a sequence $(\boldsymbol{x}_n, \boldsymbol{z}_n) \in \precsim_n$ with $(\boldsymbol{x}_n, \boldsymbol{z}_n) \to (\boldsymbol{x}, \boldsymbol{z})$. Now for n large enough, we have $(\boldsymbol{x}_n, \boldsymbol{y}_n) \notin \precsim_n$ and $(\boldsymbol{y}_n, \boldsymbol{z}_n) \notin \precsim_n$, where $\{\boldsymbol{y}_n\}$ is a sequence in X converging to $\boldsymbol{y}$. For if not, it would follow that $(\boldsymbol{x}, \boldsymbol{y}) \in Ls(\precsim_n) = \precsim$ or $(\boldsymbol{y}, \boldsymbol{z}) \in \precsim$, contradicting that $\boldsymbol{x} \prec \boldsymbol{y}$ and $\boldsymbol{y} \prec \boldsymbol{z}$. Hence, by the transitivity of $\prec$, we obtain $(\boldsymbol{x}_n, \boldsymbol{z}_n) \notin \precsim_n$, a contradiction. By Theorem F2 of Appendix F, we have that $\prec_n \to \prec$ in the topology τ_c if and only if $Li(\precsim_n) = \precsim = Ls(\precsim_n)$.

Finally, since $(\mathcal{P}, \tau_c)$ is a compact space, the identity map on $\mathcal{P}$, $id: (\mathcal{P}, \tau_c) \to (\mathcal{P}, \sigma)$ is continuous if and only if the topology σ is weaker than τ_c. Hence, by Proposition B7 of Appendix B, every Hausdorff topology σ on $\mathcal{P}$ which is weaker than τ_c coincides with τ_c. Therefore, it remains to show that the set $\{(X, \prec, \boldsymbol{x}, \boldsymbol{y}) \in \mathcal{P} \times \mathbb{R}^\ell \times \mathbb{R}^\ell \mid \boldsymbol{x}, \boldsymbol{y} \in X$ and $\boldsymbol{x} \precsim \boldsymbol{y}\}$ is closed in $\mathcal{P} \times \mathbb{R}^\ell \times \mathbb{R}^\ell$. Let $(X_n, \prec_n, \boldsymbol{x}_n, \boldsymbol{y}_n) \to (X, \prec, \boldsymbol{x}, \boldsymbol{y})$, where $\boldsymbol{x}_n, \boldsymbol{y}_n \in X_n$ and $\boldsymbol{x}_n \precsim_n \boldsymbol{y}_n$. Hence, $(\boldsymbol{x}_n, \boldsymbol{y}_n) \in \precsim_n$, which implies that $(\boldsymbol{x}, \boldsymbol{y}) \in Li(\precsim)$, or $\boldsymbol{x}, \boldsymbol{y} \in X$ and $\boldsymbol{x} \precsim \boldsymbol{y}$. □

We, sometimes, denote $(X_n, \prec_n) \to (X, \prec)$ by $\prec_n \to \prec$ or even simply by $\precsim_n \to \precsim$ when they would not be misunderstood.

Recall that for a topological space, the Borel σ-field on X, denoted by $\mathcal{B}(X)$ is the σ-field generated by open sets in X. A set $B \in \mathcal{B}(X)$ is called a Borel set or Borellian. Since the space $\mathcal{P}$ is compact (and metrizable), it is Borellian.

Sometimes, it is convenient to work with the relation $\precsim$ instead of $\prec$, and we usually postulate the following conditions on $\precsim$.

(RF) (Reflexivity) for all $\boldsymbol{x} \in X$, $\boldsymbol{x} \precsim \boldsymbol{x}$, and
(TR) (Transitivity) for all $\boldsymbol{x}, \boldsymbol{y}, \boldsymbol{z} \in X$, $\boldsymbol{z} \precsim \boldsymbol{y}$ and $\boldsymbol{y} \precsim \boldsymbol{x}$ imply that $\boldsymbol{z} \precsim \boldsymbol{x}$.

Note that the reflexiveness on $\precsim$ is equivalent to the irreflexivity on $\prec$, and the transitivity on $\precsim$ is equivalent to the negative transitivity on $\prec$,

(NTR) (Negative transitivity) for all $x,y,z \in X$, $(z,y) \notin \prec$ and $(y,x) \notin \prec$ imply that $(z,x) \notin \prec$.

Let $\mathcal{P}*$ be the subset of $\mathcal{P}$ of all negative transitive preference relations.

We recall the other conditions which will be occasionally used in the subsequent analysis.

(LNS) (Local nonsatiation) for every $x \in X$ and every neighborhood U of x, there exists $y \in U \cap X$ such that $x \prec y$.

Let $\mathcal{P}_{lns}$ be the subset of $\mathcal{P}$ of all locally non-satiated preferences.

(MT) (Monotonicity) if $x,y \in X$ and $x < y$, then $x \prec y$.

We denote by $\mathcal{P}_{mo}$, the subset of $\mathcal{P}$ of all monotonic preference relations, and finally:

(CV) (Convexity) for every $x \in X$ the set $\{y \in X | y \succsim x\}$ is convex, and let $\mathcal{P}_{co}$ be the subset of $\mathcal{P}$ of all convex preference relations. We are interested in the topological and measurable properties of $\mathcal{P}*$, $\mathcal{P}_{lns}$, $\mathcal{P}_{mo}$, and $\mathcal{P}_{co}$. A set Q is said to be G_δ, if it is an intersection of countably many open sets. It is called F_σ, if it is a union of countably many closed sets. Obviously, the G_δ sets and the F_σ sets are Borel-measurable sets.

We shall prove that:

Theorem 2.2. *The sets $\mathcal{P}*, \mathcal{P}_{lns}, \mathcal{P}_{mo}$, and $\mathcal{P}_{co}$ are G_δ sets.*

Proof. First, we show that the set $\mathcal{P}*$ is G_δ. For $S, T \subset X$, we denote by $S \prec T$ that for all $x \in S$ and all $y \in T, x \prec y$.

For positive integers m and n, we define the set $\mathcal{P}*^{m,n} \subset \mathcal{P}$ by:

$$\mathcal{P}*^{m,n} = \{(X,\prec) \in \mathcal{P} | \text{ there exist } x,y,z \in X \text{ such that } \|x\| \le m,\ \|y\| \le m,$$
$$\|z\| \le m,\ z \succsim y,\ y \succsim x,\ B(z,1/n) \cap X \prec B(x,1/n) \cap X\}.$$

Then, clearly we have $\mathcal{P} \backslash \mathcal{P}* = \cup_{m=1}^{\infty} \cup_{n=1}^{\infty} \mathcal{P}*^{m,n}$. Therefore, it suffices to show that $\mathcal{P}*^{m,n}$ is closed for all m, n. Let $\{(X_k, \prec_k)\}_{k\in\mathbb{N}}$ be a sequence in $\mathcal{P}*^{m,n}$ such that $(X_k, \prec_k) \to (X, \prec)$ in the topology τ_c. We want to show that $\prec \in \mathcal{P}*^{m,n}$. For each k, there exists a sequence $(x_k, y_k, z_k)_{k\in\mathbb{N}}$ such that $x_k, y_k, z_k \in X_k$ for all k, $\|x_k\| \le m$, $\|y_k\| \le m$, $\|z_k\| \le m$, $z_k \succsim_k y_k$, $y_k \succsim_k x_k$, and $B(z_k, 1/n) \cap X_k \prec_k B(x_k, 1/n) \cap X_k$. Since the sequence $(x_k, y_k, z_k)_{k\in\mathbb{N}}$ is bounded, we can assume that $x_k \to x \in X$, $y_k \to y \in X$, and $z_k \to z \in X$. Obviously, we have $\|x\| \le m$, $\|y\| \le m$, $\|z\| \le m$, and $z \succsim y$, $y \succsim x$. Suppose that $B(z,1/n) \cap X \prec B(x,1/n) \cap X$ does not hold.

Then, there exist $x' \in B(z, 1/n) \cap X$ and $z' \in B(x, 1/n) \cap X$ such that $z' \precsim x'$. Since $Li(\precsim_k) = \precsim$, there exists a sequence $(z'_k, x'_k) \in \precsim_k$ with $(z'_k, x'_k) \to (z', x') \in \precsim$. For k sufficiently large, $\|x'_k - x_k\| \leq 1/n$, $\|z'_k - z_k\| \leq 1/n$ and $z'_k \precsim x'_k$. This contradicts the fact that $B(z_k, 1/n) \cap X \prec B(x_k, 1/n) \cap X$. This proves that $\mathcal{P}*^{m,n}$ is closed.

Similarly, for each positive integer m, n, we define $\mathcal{P}^{m,n}_{lns} \subset \mathcal{P}$ by:

$$\mathcal{P}^{m,n}_{lns} = \{(X, \prec) \in \mathcal{P} | \text{ there exists } x \in X \text{ such that } \|x\| \leq m \text{ and if } y \in B(x, 1/n) \cap X, \text{ then } x \precsim y.\}.$$

Then, it is obvious that $\mathcal{P} \backslash \mathcal{P}_{lns} = \cup_{m=1}^{\infty} \cup_{n=1}^{\infty} \mathcal{P}^{m,n}_{lns}$. In order to prove that $\mathcal{P}_{lns}$ is G_δ, it suffices to show that $\mathcal{P}^{m,n}_{lns}$ is closed for all m, n. Let $\{(X_k, \prec_k)\}_{k \in \mathbb{N}}$ be a sequence in $\mathcal{P}^{m,n}_{lns}$ such that $(X_k, \prec_k) \to (X, \prec)$ in the topology τ_c. Then, for each k, there exists $x_k \in X_k$ such that $\|x_k\| \leq m$ and if $y \in B(x_k, 1/n) \cap X$, then $x_k \precsim y$. We can assume that $x_k \to x \in X$. Then, $\|x\| \leq m$. Suppose that for some $y \in B(x, 1/n) \cap X$, $x \prec y$. Since $x_k \to x$, $\|x_k - y\| < 1/n$ for k large enough. For such a k, we have $y \in B(x_k, 1/n) \cap X$ and $x_k \prec_k y$, a contradiction. This proves that $\prec \in \mathcal{P}^{m,n}_{lns}$.

In order to prove that $\mathcal{P}_{mo}$ is G_δ, for positive integers m, n, we define:

$$\mathcal{P}^{m,n}_{mo} = \{(X, \prec) \in \mathcal{P} \mid \text{ there exists } x, y \in X \text{ such that } \|x\| \leq m, \|y\| \leq m, y \precsim x, x \geq y \text{ and } \|x - y\| \geq 1/n\}.$$

Since $\mathcal{P} \backslash \mathcal{P}_{mo} = \cup_{m=1}^{\infty} \cup_{n=1}^{\infty} \mathcal{P}^{m,n}_{mo}$, $\mathcal{P}_{mo}$ is G_δ if each $\mathcal{P}^{m,n}_{mo}$ is closed.

Let $\{(X_k, \prec_k)\}_{k \in \mathbb{N}}$ be a sequence in $\mathcal{P}^{m,n}_{mo}$ such that $(X_k, \prec_k) \to (X, \prec)$ in the topology τ_c. Then, for each k, there exist $x_k, y_k \in X_k$ such that $\|x_k\| \leq m$, $\|y\| \leq m$, $y_k \precsim_k x_k$, $x_k \geq y_k$, and $\|y_k - x_k\| \geq 1/n$. We can assume that $x_k \to x \in X$ and $y_k \to y \in X$. Then passing to the limit, we have $\|x\|, \|y\| \leq m$, $y \precsim x$, $y \geq x$, and $\|x - y\| \geq 1/n$. This proves that $\prec \in \mathcal{P}^{m,n}_{mo}$, hence $\mathcal{P}^{m,n}_{mo}$ is a closed set.

Finally, in order to show that $\mathcal{P}_{co}$ is a G_δ set, we define for each positive integer m, n, r, s, the set $\mathcal{P}^{mnrs}_{co} \subset \mathcal{P}$ by:

$$\begin{aligned}\mathcal{P}^{mnrs}_{co} = \{(X, \prec) \in \mathcal{P} \mid & \text{ there exists } x, y, z \in X \text{ and } \alpha \in \mathbb{R} \text{ such that:} \\ & \|x\| \leq m, \|y\| \leq m, \|z\| \leq m, \|x - y\|, \|x - z\| \geq 1/n, \\ & \text{and } (1/r) \leq \alpha \leq 1 - (1/r), x \precsim \alpha y + (1 - \alpha) z, \\ & B(x, 1/s) \cap X \prec B(y, 1/s) \cap X \text{ and} \\ & B(x, 1/s) \cap X \prec B(z, 1/s) \cap X\}.\end{aligned}$$

By definition, $\mathcal{P} \backslash \mathcal{P}_{co} = \cup_{m=1}^{\infty} \cup_{n=1}^{\infty} \cup_{r=1}^{\infty} \cup_{s=1}^{\infty} \mathcal{P}^{mnrs}_{co}$. It remains to show that each $\mathcal{P}^{mnrs}_{co}$ is a closed set. Let $(X_k, \prec_k) \in \mathcal{P}_{mnrs}$ and $(X_k, \prec_k) \to (X, \prec)$. Then, for each k, there exist $x_k, y_k, z_k \in X_k$ and $\alpha_k \in \mathbb{R}$ such that $\|x_k\| \leq m$, $\|y_k\| \leq m$, $\|z_k\| \leq m$, $\|x_k - y_k\|$, $\|x_k - z_k\| \geq 1/n$, and $(1/r) \leq \alpha_k \leq 1 - (1/r)$, $x_k \precsim_k \alpha_k y_k + (1 - \alpha_k) z_k$, $B(x_k, 1/s) \cap X_k \prec B(y_k, 1/s) \cap X_k$, and $B(x_k, 1/s) \cap X_k \prec B(z_k, 1/s) \cap X_k$.

We can assume that $(x_k, y_k, z_k, \alpha_k) \to (x, y, z, \alpha)$. Then, we have $x, y, z \in X$ and $\|x\| \leq m$, $\|y\| \leq m$, $\|z\| \leq m$, $\|x - y\|, \|x - z\| \geq 1/n$, and $(1/r) \leq \alpha \leq 1 - (1/r)$, $x \precsim \alpha y + (1 - \alpha)z$. Suppose that $B(x, 1/s) \cap X \prec B(y, 1/s) \cap X$ does not hold good. Then, there exist $x' \in B(x, 1/s) \cap X$ and $y' \in B(y, 1/s) \cap X$ such that $x' \succsim y'$. Since $Li(\succsim_k) = \succsim$ and $(x', y') \in \succsim$, we have a sequence $(x'_k, y'_k) \in \succsim_k$ with $(x'_k, y'_k) \to (x', y')$. Since $\|x_k - x'_k\| \leq \|x_k - x\| + \|x - x'_k\|$, one has $\|x_k - x'_k\| \leq 1/s$ for k large enough. Similarly, we have $\|y_k - y'_k\| \leq 1/s$ for k sufficiently large. This contradicts the fact that $B(x_k, 1/s) \cap X_k \prec B(y_k, 1/s) \cap X_k$. By similar arguments, we can show that $B(x, 1/s) \cap X \prec B(z, 1/s) \cap X$. Hence, $\prec \in \mathcal{P}_{co}^{mnrs}$ and the proof is complete. □

Summing up, by the above theorem, it follows that the sets $\mathcal{P}*$, $\mathcal{P}_{lns}$, $\mathcal{P}_{mo}$, and $\mathcal{P}_{co}$ are Borel-measurable sets.

3.3. EXISTENCE OF COMPETITIVE EQUILIBRIA

Let $(A, \mathcal{A}, \mu)$ be an atomless measure space of consumers. We assume that the measure μ is a probability measure, that is, $\mu(A) = 1$. The economic implication will be explained below.

Definition 3.1. An (exchange) economy is a Borel-measurable mapping:

$$\mathcal{E}: (A, \mathcal{A}, \mu) \to (\mathcal{P} \times \mathbb{R}^\ell, \mathcal{B}(\mathcal{P} \times \mathbb{R}^\ell)), \quad a \mapsto ((X_a, \prec_a), \omega_a)$$

such that $\int_A \omega(a) d\mu < +\infty$.

A measurable mapping f from A to $\mathbb{R}^\ell$ is called an allocation if $f(a) \in X_a$, a.e. The allocation $f: A \to \mathbb{R}^\ell$ is said to be feasible if and only if it is integrable and satisfies the resource-constraint condition:

$$\int_A f(a) d\mu \leq \int_A \omega(a) d\mu.$$

The allocation $f(a)$ is sometimes denoted by $x(a)$. We now give a remark on the economic interpretation of the integration of an allocation over the set of consumers A. First, consider the case in which the set A is finite, or $\#A < +\infty$.[1] Then, we have a natural probability measure on A, namely, the counting measure μ defined by:

$$\mu(B) = \sharp B / \sharp A \quad \text{for every } B \subset A.$$

[1] $\#A$ means the number of elements of the set A.

Therefore the integration of an allocation $f: A \to \mathbb{R}^\ell$ with respect to the counting measure μ is:

$$\int_A f(a)d\mu = \left(\frac{1}{\#A}\right)\sum_{a\in A} f(a)$$

which is the mean or average of the allocation f, in other words, the per capita total allocation.

By the analogy of the finite case, we interpret the integration of the allocation $f: A \to \mathbb{R}^\ell$ over the atomless probability measure space of the traders $(A, \mathcal{A}, \mu)$:

$$\int_A f(a)d\mu$$

as the mean (per capita total) allocation. Hence, the value of the integration is represented by the per capita term, rather than the simple aggregated term. For the same reason, we call the integral $\int_A \omega(a)d\mu \in \mathbb{R}^\ell$, the mean endowment vector, or simply the mean endowment.

A feasible allocation $f: A \to \mathbb{R}^\ell$ is said to be improved upon or blocked, if there exists another feasible allocation $g: A \to \mathbb{R}^\ell$ such that:

$$g(a) \succsim_a f(a) \quad \text{a.e.,}$$

and

$$\mu(\{a \in A \mid f(a) \prec_a g(a)\}) > 0.$$

In other words, the allocation f is improved upon by a feasible allocation g if it is at least as desired as f by almost all consumers, and it is strictly preferred to f by a group of consumers with positive measure.

A feasible allocation f is said to be Pareto optimal if it is not improved upon by any other feasible allocations.

We now state the definition of the competitive equilibrium of an exchange economy with the measure space of consumers.

Definition 3.2. A pair consisting of a price vector $\boldsymbol{p} \in \mathbb{R}^\ell_+$ and a feasible allocation $f(a) = \boldsymbol{x}(a) = (x^t(a))$, $(\boldsymbol{p}, \boldsymbol{x}(a))$ is called a competitive equilibrium of an (exchange) economy $\mathcal{E}: (A, \mathcal{A}, \mu) \to (\mathcal{P} \times \mathbb{R}^\ell, \mathcal{B}(\mathcal{P} \times \mathbb{R}^\ell))$, $a \mapsto ((X_a, \prec_a), \omega_a)$ if the following conditions hold good.

(E-1) $\boldsymbol{px}(a) \leq \boldsymbol{p}\omega(a)$ and for every $\boldsymbol{y} \in X_a$, if $\boldsymbol{py} \leq \boldsymbol{p}\omega(a)$, then $\boldsymbol{x}(a) \succsim_a \boldsymbol{y}$ a.e.,
(E-2) $\int_A \boldsymbol{x}(a)d\mu \leq \int_A \omega(a)d\mu$.

The condition (E-1) implies that almost all consumers "maximizes his/her utility" under the budget constraint and the condition (E-2) implies that the resource condition is met.

We have a straightforward generalization of the first fundamental theorem of welfare economics, Proposition 5.1 given in Chap. 2.

Proposition 3.1. *Let* $(\boldsymbol{p}, \boldsymbol{x}(\cdot))$ *be a competitive equilibrium of an (exchange) economy* $\mathcal{E}: (A, \mathcal{A}, \mu) \to (\mathcal{P} \times \mathbb{R}^\ell, \mathcal{B}(\mathcal{P} \times \mathbb{R}^\ell))$ *such that* $\mathcal{E}(a) \subset \mathcal{P}_{lns} \times \mathbb{R}^\ell$ *a.e. Then the allocation* $\boldsymbol{x}(\cdot)$ *is Pareto optimal.*

Proof. On the contrary, suppose that $\boldsymbol{x}$ is not Pareto optimal. Then, there exists a feasible allocation $\boldsymbol{y}$, which is an integrable map of A to $\mathbb{R}^\ell$ such that:

$$\boldsymbol{y} \in X_a \quad \text{a.e.,} \quad \text{and} \quad \int_A \boldsymbol{y}(a) d\mu \leq \int_A \omega(a) d\mu,$$

and it satisfies:

$$\boldsymbol{y}(a) \succsim_a \boldsymbol{x}(a) \text{ a.e.} \quad \text{and} \quad \mu(\{a \in A | \boldsymbol{x}(a) \prec_a \boldsymbol{y}(a)\}) > 0.$$

Then, it follows that $\boldsymbol{p}\boldsymbol{y}(a) \geq \boldsymbol{p}\omega(a)$ a.e. Since the inequality follows trivially for $a \in A$ with $\boldsymbol{p}\omega_a \leq \inf \boldsymbol{p}X_a$, suppose that $\boldsymbol{p}\omega_a > \inf \boldsymbol{p}X_a$ and $\boldsymbol{p}\boldsymbol{y}(a) < \boldsymbol{p}\omega_a$ on a set B of positive measure. Then, by the local nonsatiation (LNS), there exists a consumption vector $\boldsymbol{z}(a) \in X_a$ close enough to $\boldsymbol{y}(a)$ such that $\boldsymbol{y}(a) \prec_a \boldsymbol{z}(a)$, and $\boldsymbol{p}\boldsymbol{z}(a) < \boldsymbol{p}\omega(a)$ on B. This contradicts the condition (E-1) from Definition 3.1. Furthermore, for $a \in A$ such that $\boldsymbol{x}(a) \prec_a \boldsymbol{y}(a)$, we have $\boldsymbol{p}\boldsymbol{y}(a) > \boldsymbol{p}\omega(a)$. Integrating these inequalities over A, we have $\boldsymbol{p} \int_A \boldsymbol{y}(a) d\mu > \boldsymbol{p} \int_A \omega(a) d\mu$. On the other hand, by the feasibility condition (E-2) of Definition 3.1 and $\boldsymbol{p} \geq \boldsymbol{0}$, it follows that $\boldsymbol{p} \int_A \boldsymbol{y}(a) d\mu \leq \boldsymbol{p} \int_A \omega(a) d\mu$, which is a contradiction. □

Recall that $\mathcal{P}$ is the set of all irreflexive, transitive, and continuous preference-consumption set pairs $(X, \prec)$. Consider the exchange economy:

$$\mathcal{E}: (A, \mathcal{A}, \mu) \to (\mathcal{P} \times \mathbb{R}^\ell, \mathcal{B}(\mathcal{P} \times \mathbb{R}^\ell)), \quad \mathcal{E}(a) = ((X_a, \prec_a), \omega_a).$$

Let X be a closed and convex subset of $\mathbb{R}^\ell$, which is bounded from below. In the following theorem, we assume that (almost) all consumers have the same consumption set X.

Theorem 3.1. *Let* $\mathcal{E}: (A, \mathcal{A}, \mu) \to (\mathcal{P} \times \mathbb{R}^\ell, \mathcal{B}(\mathcal{P} \times \mathbb{R}^\ell))$ *be an exchange economy such that* $X_a = X$ *a.e.,* $\omega(a) \in \mathbb{R}^\ell_+$ *a.e. and* $\int_A \omega(a) d\mu \gg 0$. *Moreover, we assume the minimum income assumption:*

(MI) $\boldsymbol{p}\omega(a) > \inf \boldsymbol{p}X$ *for almost all* $a \in A$.

Then, there exists a competitive equilibrium $(\boldsymbol{p}, (\boldsymbol{x}(a)))$ *for* $\mathcal{E}$.

Before proving this theorem, we note that the convexity of preferences (CV) is not assumed. As we pointed out in Section 3.1, this is the first aggregation effect in the large markets. We can illustrate this by the following simple example.

Example 3.1. Consider a sequence of finite-exchange economies $\mathcal{E}(m)$ with two commodities and m consumers, $A(m) = \{1 \cdots m\}, m = 1 \ldots.$

For each m, all consumers $a \in A(m)$ are assumed to have the same preference relation represented by convex (not concave!) utility function:

$$u(x, y) = x^2 + y^2 \quad \text{for } \boldsymbol{x} = (x, y) \in \mathbb{R}^2_+$$

defined on the consumption set $X = \mathbb{R}^2_+$. Also, all consumers have the same endowment vector $\omega = (1, 1)$ (see Fig. 3.1). Since the indifference curves are either circular or concave to the origin, for every non-negative price vector $\boldsymbol{p} = (p, q)$, the consumer maximizes his/her utility with vectors on the x-axis or on the y-axis. More specifically, from the budget equation:

$$px + qy = p + q,$$

the consumer demands either $\boldsymbol{x} = (\frac{p+q}{p}, 0)$ or $\boldsymbol{x} = (0, \frac{p+q}{q})$. Therefore, there does not exist any equilibrium for the one-consumer economy $\mathcal{E}(1)$ with $A(1) = \{1\}$.

However, there exists an equilibrium with the equilibrium price vector $\boldsymbol{p} = (1/2, 1/2) \in \Delta$ for the two-persons economy $\mathcal{E}(2)$ with $A(2) = \{1, 2\}$. An equilibrium allocation is given by $\{\boldsymbol{x}_1, \boldsymbol{x}_2\} = \{(2, 0), (0, 2)\}$. The market condition $(1/2)(\boldsymbol{x}_1 + \boldsymbol{x}_2) = (1/2)((2, 0) + (0, 2)) = (1, 1)$ is certainly met.

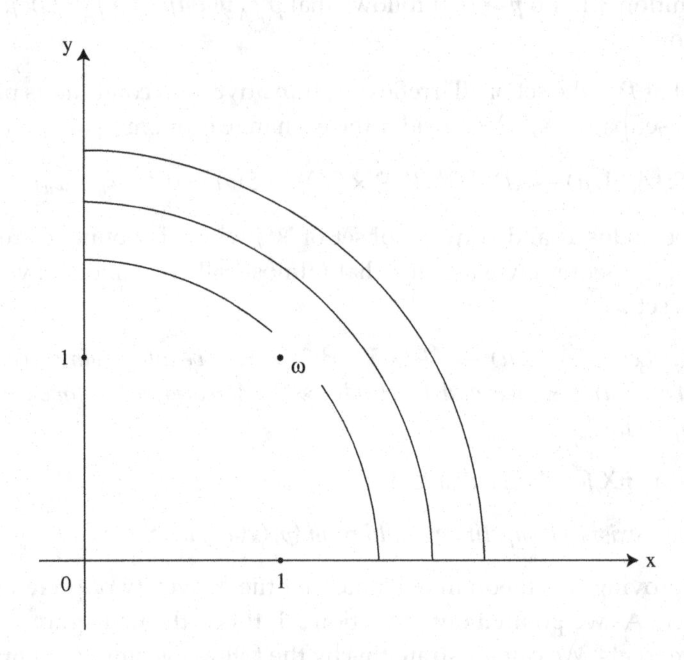

Figure 3.1. Example 3.1.

The reason for the existence of such an equilibrium given in Example 3.1 is of course, as the number of the consumers m increases, the number of the values in the mean demand correspondence $\Phi(\boldsymbol{p}) = (1/m)\sum_{a=1}^{m}\phi(a,\boldsymbol{p})$ increases (see Figs. 3.2(a) and 3.2(b)).

We expect that in the limit of $m \to +\infty$, the segment $co\{(2,0),(0,2)\}$ is full of the values of the mean demand, and indeed it is, thanks to Theorem G10.

Theorem G10 of Appendix G. *Let ϕ be a correspondence of an atomless measurable space $(\Omega, \mathcal{A})$ to $\mathbb{R}^\ell$. Then, the integral $\int_\Omega \phi(\omega)d\mu$ is a convex subset of $\mathbb{R}^\ell$, where the integration of the correspondence ϕ is defined by:*

$$\int_A \phi(a)d\mu = \left\{\int_\Omega f(a)d\mu \in \mathbb{R}^\ell \middle| f\colon A \to \mathbb{R}^\ell \text{ is integrable, } f(a) \in \phi(a) \text{ a.e.}\right\}.$$

Note that in the above example, $\phi(a) = \{(2,0),(0,2)\}$ for all $a \in A(m) = \{1\cdots m\}$, hence:

$$\int_A \phi(a)d\mu = \left\{(1/m)\sum_{a=1}^{m} f(a) \middle| f(a) = (2,0),(0,2)\right\}.$$

Our first task is to make sure that the mean-demand correspondence is nonempty. This can be achieved by using Proposition 3.2.

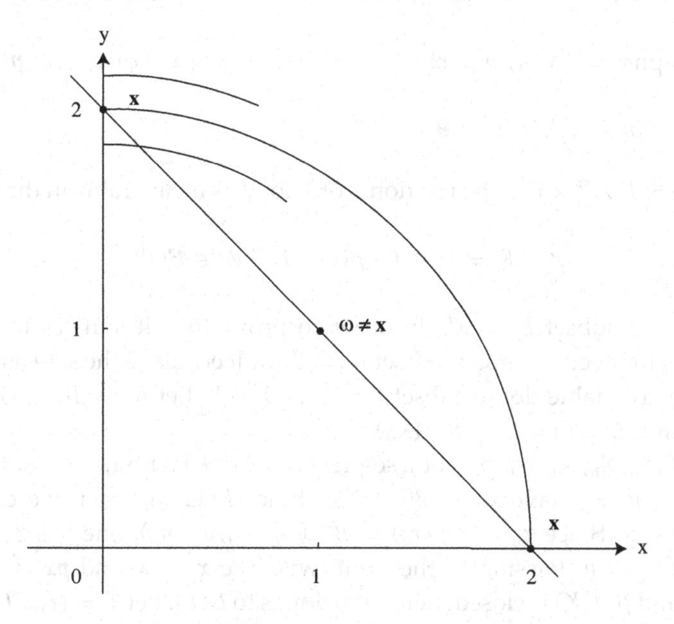

Figure 3.2(a). One-Consumer Case.

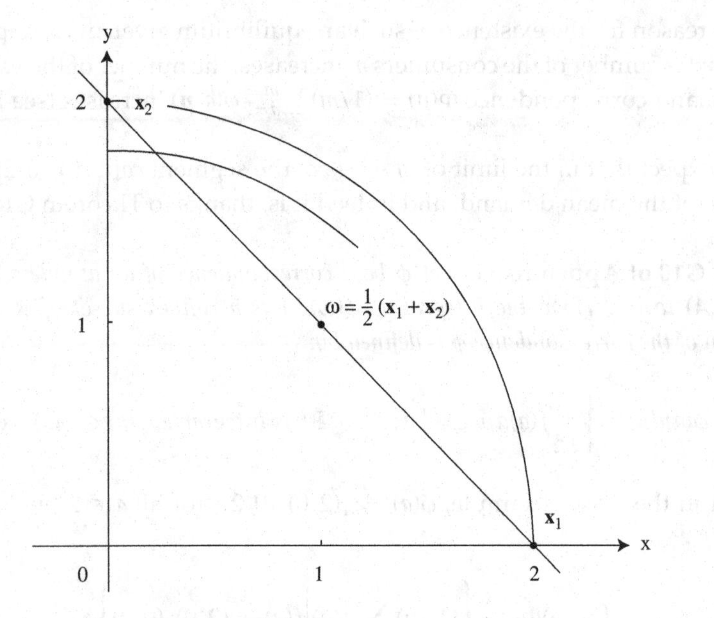

Figure 3.2(b). Two-Consumer Case.

Proposition 3.2. *Let $\beta(X, \prec, w, \boldsymbol{p}) = \{\boldsymbol{x} \in X \mid \boldsymbol{px} \leq w\}$ be the budget relation and $\phi(X, \prec, w, \boldsymbol{p}, \boldsymbol{x}) = \{\boldsymbol{x} \in X \mid \boldsymbol{x} \in \beta(X, \prec, w, \boldsymbol{p})$, and $\boldsymbol{x} \succsim \boldsymbol{y}$ whenever $\boldsymbol{y} \in \beta(X, \prec, w, \boldsymbol{p})\}$ be the demand relation. Then, the graph of the demand relation which is given by:*

$$\text{Graph}\phi = \{(X, \prec, w, \boldsymbol{p}, \boldsymbol{x}) \in \mathcal{P} \times \mathbb{R} \times \mathbb{R}^\ell \times \mathbb{R}^\ell \mid \boldsymbol{x} \in \phi(X, \prec, w, \boldsymbol{p})\}$$

is a Borel subset in $\mathcal{P} \times \mathbb{R} \times \mathbb{R}^\ell \times \mathbb{R}^\ell$.

Proof. Let $T = \mathcal{P} \times \mathbb{R} \times \mathbb{R}^\ell$. The relation β of T to $\mathbb{R}^\ell$ is measurable in the sense that:

$$\beta^{-1}(F) = \{\tau \in T \mid \beta(\tau) \cap F \neq \emptyset\} \in \mathcal{B}(T)$$

for every closed subset $F \subset \mathbb{R}^\ell$. In order to prove this, it suffices to show that $\beta^{-1}(K) \in \mathcal{B}(\tau)$ for every compact subset K of $\mathbb{R}^\ell$. Indeed, since the set F is separable, there exists a countable dense subset $\{x_0, x_1, \ldots\}$ of F. Let $K_n = B(x_n, 1) \cap F$. Then $F = \cup_{n=0}^{\infty} K_n$ and $\beta^{-1}(F) = \cup_{n=0}^{\infty} \beta^{-1}(K_n)$.

Let $\{\tau_n\} = \{(X_n, \prec_n, w_n, \boldsymbol{p}_n)\}$ be a sequence in $\beta^{-1}(K)$ with $\tau_n \to \tau = (X, \prec, w, \boldsymbol{p})$. Then for each n, we have $\boldsymbol{x}_n \in \beta(\tau_n) \cap K$. Since K is compact, we can assume that $\boldsymbol{x}_n \to \boldsymbol{x} \in K$. Since $\boldsymbol{x}_n \in \beta^{-1}(\tau_n) = \beta^{-1}(X_n, \prec_n, w_n, \boldsymbol{p}_n)$, one has $\boldsymbol{x}_n \in X_n$ and $\boldsymbol{p}_n \boldsymbol{x}_n \leq w_n$ for all n. Passing to the limit, we have $\boldsymbol{x} \in X$ and $\boldsymbol{px} \leq w$. Hence, $\boldsymbol{x} \in \beta(\tau) \cap K$ and $\beta^{-1}(K)$ is closed, hence it belongs to $\mathcal{B}(T)$. Let $\hat{T} = \{\tau \in T | \beta(\tau) \neq \emptyset\}$. Obviously, $\hat{T} \in \mathcal{B}(T)$.

By Proposition G7 of Appendix G, there exists a sequence $\{f_n\}_{n\in\mathbb{N}}$ of measurable functions of $\hat{T}$ to $\mathbb{R}^\ell$ such that:

$$\beta(\tau) = \text{closure}\{f_n(\tau) \mid n = 0, 1, \ldots\}$$

for every $\tau \in \hat{T}$. We define:

$$\phi_n(\tau) = \{\boldsymbol{x} \in \beta(\tau) \mid \boldsymbol{x} \succsim f_n(\tau)\}, \quad n \in \mathbb{N}.$$

We will show that the graph G_n of the relation ϕ_n is a Borel subset of $T \times \mathbb{R}^\ell$. By Theorem 2.1, the set:

$$\{(X, \prec, \boldsymbol{x}, \boldsymbol{y}) \in \mathcal{P} \times \mathbb{R}^\ell \times \mathbb{R}^\ell \mid \boldsymbol{x}, \boldsymbol{y} \in X \text{ and } \boldsymbol{y} \succsim \boldsymbol{x}\}$$

is closed. In a similar argument which proved the closedness of $\beta^{-1}(K)$, it is easy to show that the set:

$$G - \{(\tau, \boldsymbol{x}, \boldsymbol{y}) \in \hat{T} \times \mathbb{R}^\ell \times \mathbb{R}^\ell \mid \boldsymbol{x}, \boldsymbol{y} \in \beta(\tau) \text{ and } \boldsymbol{y} \succsim_\tau \boldsymbol{x}\}$$

is closed, where the preference $\succsim_\tau$ corresponds to $\tau \in \hat{T}$, or $\tau = (X, \prec_\tau, w, \boldsymbol{p})$.

Let g_n be the mapping of $\hat{T} \times \mathbb{R}^\ell$ to $\hat{T} \times \mathbb{R}^\ell \times \mathbb{R}^\ell$ is defined by:

$$(\tau, \boldsymbol{x}) \mapsto g_n(\tau, \boldsymbol{x}) = (\tau, f_n(\tau), \boldsymbol{x}).$$

Then, we have $G_n = g_n^{-1}(G)$. Since $f_n(\cdot)$ is a measurable mapping of $\hat{T}$ to $\mathbb{R}^\ell$, the mapping g_n is Borel measurable by Propositions G1 and G2 of Appendix G; hence $G_n \in \mathcal{B}(\hat{T} \times \mathbb{R}^\ell) \subset \mathcal{B}(T \times \mathbb{R}^\ell)$.

Finally, we show that $\phi(\tau) = \cap_{n=0}^{\infty}\phi_n(\tau)$. Obviously, $\phi(\tau) \subset \cap_{n=0}^{\infty}\phi_n(\tau)$. On the other hand, let $\boldsymbol{x} \in \cap_{n=0}^{\infty}\phi_n(\tau)$ but $\boldsymbol{x} \notin \phi(\tau)$. Then, there exists a vector $\boldsymbol{y} \in \beta(\tau)$ with $\boldsymbol{x} \prec_\tau \boldsymbol{y}$. Since the set $\{\boldsymbol{z} \in \beta(\tau) | \boldsymbol{x} \prec_\tau \boldsymbol{z}\}$ is open relative to $\beta(\tau)$, there exists an integer N such that $\boldsymbol{x} \prec f_N(\tau)$, or $\boldsymbol{x} \notin \phi_N(\tau)$, a contradiction. □

Proof of Theorem 3.1. For the initial endowment vector $\omega(a) = (\omega^t(a))$, let B be a non-negative number defined by:

$$B = \int_A \omega^1(a) d\mu + \cdots \int_A \omega^\ell(a) d\mu.$$

Since $\int_A \omega(a) d\mu \gg \boldsymbol{0}$, $B > 0$. We define the set $A_k \subset A$ by:

$$A_k = \{a \in A \mid \omega(a) \leq k(B \cdots B)\}.$$

Then, the set A_k is measurable by the measurability of the map $\omega: A \to \mathbb{R}^\ell$, and by the definition of B, we have $A_1 \neq \emptyset$. For if not, there exists $t \in \{1 \cdots \ell\}$ such that $\int_A \omega^t(a)d\mu > \int_A Bd\mu = B = \sum_{t=1}^{\ell} \int_A \omega^t(a)d\mu \geq \int_A \omega^t(a)d\mu$, a contradiction. Since $A_1 \subset A_2 \subset \cdots \subset A_k \subset \cdots$, we have $A_k \neq \emptyset$ for all $k \geq 1$. We define the sub-σ-field $\mathcal{A}_k$ of $\mathcal{A}$ by:

$$\mathcal{A}_k = \{C \cap A_k \mid C \in \mathcal{A}\},$$

and let μ_k be the restriction of μ to $(A_k, \mathcal{A}_k)$. Since the measure space $(A, \mathcal{A}, \mu)$ is atomless, so are the measure spaces $(A_k, \mathcal{A}_k, \mu_k), k = 1, 2, \ldots$. Now, for each positive integer k, let:

$$X_k = \{\boldsymbol{x} \in X \mid \boldsymbol{x} \leq k(B \cdots B)\},$$

and for each $a \in A_k$, we define the restricted budget sets and demand sets as,

$$\beta_k(a, \boldsymbol{p}) = \{\boldsymbol{x} \in X_k \mid \boldsymbol{p}\boldsymbol{x} \leq \boldsymbol{p}\omega(a)\},$$

and

$$\phi_k(a, \boldsymbol{p}) = \{\boldsymbol{x} \in \beta_k(a, \boldsymbol{p}) \mid \boldsymbol{x} \succsim_a \boldsymbol{z} \text{ for every } \boldsymbol{z} \in \beta_k(a, \boldsymbol{p})\},$$

respectively. Finally, the restricted excess mean demand is defined by:

$$\zeta_k(\boldsymbol{p}) = \int_{A_k} \phi_k(\boldsymbol{p})d\mu - \int_{A_k} \omega(a)d\mu.$$

We will show that the correspondence ζ_k of Δ to $Z_k \equiv \int_{A_k} X_k d\mu_k - \{\int_{A_k} \omega(a)d\mu_k\}$ satisfies the conditions, which are needed to apply the Gale–Nikaido lemma (Theorem D3 of Appendix D),

(i) $\zeta_k: \Delta \to Z_k$ is an upper hemi-continuous correspondence from Δ to a compact and convex set Z_k;
(ii) for every $\boldsymbol{p} \in \Delta$, $\zeta_k(\boldsymbol{p})$ is a non-empty convex set; and
(iii) for every $\boldsymbol{p} \in \Delta, \boldsymbol{p}\zeta_k(\boldsymbol{p}) \leq 0$.

First, note that the set $\int_{A_k} X_k d\mu_k$ is compact and convex by Theorem G10 of Appendix G (Liapunov's theorem) and Corollary G1 of Appendix G. Hence, so is the set Z_k. In order to show that $\zeta_k(\boldsymbol{p}) \neq \emptyset$, it is sufficient from Theorem G9 (measurable selection theorem) of Appendix G to show that:

$$\phi_k(a, \boldsymbol{p}) \neq \emptyset \text{ a.e.} \quad \text{for every } \boldsymbol{p} \in \Delta,$$
$$\text{Graph}\, \phi_k = \{(a, \boldsymbol{x}) \in A_k \times X_k \mid \boldsymbol{x} \in \phi_k(a, \boldsymbol{p})\} \in \mathcal{A}_k \times \mathcal{B}(X_k).$$

For almost all $a \in A_k$, it follows from the minimum income condition (MI) that $\beta(a,\boldsymbol{p}) \neq \emptyset$. Since the set $\beta_k(a,\boldsymbol{p})$ is compact and the preference $\prec_a$ is continuous, we have $\phi_k(a,\boldsymbol{p}) \neq \emptyset$ by Proposition 3.1 in Chap. 2. By Proposition 3.2, the graph of the demand correspondence ϕ_k is measurable. This proves $\int_{A_k} \phi_k(a,\boldsymbol{p})d\mu_k \neq \emptyset$, and hence $\zeta_k(\boldsymbol{p}) \neq \emptyset$.

By Theorem G10 of Appendix G, we have $\int_{A_k} \phi_k(a,\boldsymbol{p})d\mu_k$ is a convex set, so is $\zeta_k(\boldsymbol{p})$. From the definition of the demand set, $\boldsymbol{p}\phi_k(a,\boldsymbol{p}) \leq \boldsymbol{p}\omega(a)$ a.e. in A_k, therefore integrating over A_k, one sees that:

$$\boldsymbol{p}\zeta_k(\boldsymbol{p}) = \boldsymbol{p}\left(\int_{A_k} \phi_k(a,\boldsymbol{p}) - \omega(a)d\mu_k\right) \leq 0.$$

Finally, by Proposition 3.3 in Chap. 2, the correspondence $\phi_k(a,\boldsymbol{p})$ is closed for almost all $a \in A_k$, so is $\int_{A_k} \phi(a,\boldsymbol{p})d\mu_k$ by Corollary G2 of Appendix G. Therefore, the correspondence $\boldsymbol{p} \mapsto \zeta_k(\boldsymbol{p}) \subset Z_k$ is upper hemi-continuous at every $\boldsymbol{p} \in \Delta$.

We now apply the Gale–Nikaido lemma to the correspondence $\zeta_k(\boldsymbol{p})$ and obtain the price vector $\boldsymbol{p}_k \in \Delta$ and an integrable function f_k of A_k to $\mathbb{R}^\ell$ such that:

$$f_k(a) \in \phi(a,\boldsymbol{p}_k) \text{ a.e. in } A_k \quad \text{and} \quad \int_{A_k} f_k(a)d\mu_k \leq \int_{A_k} \omega(a)d\mu_k.$$

We define a sequence of measurable functions $\{g_k(a)\}_{k\in\mathbb{N}}$ of A to $\mathbb{R}^\ell$ by:

$$g_k(a) = \begin{cases} f_k(a) & \text{if } a \in A_k, \\ \omega(a) & \text{otherwise.} \end{cases}$$

Then, by the construction, it follows that:

$$g_k(a) \in \phi(a,\boldsymbol{p}_k) \quad \text{a.e. in } A_k, \text{ and } \int_A g_k(a)d\mu \leq \int_A \omega(a)d\mu.$$

Without loss of generality, we can assume that the sequence $\{\boldsymbol{p}_k\}$ is convergent, say, $\lim_{k\to\infty} \boldsymbol{p}_k = \boldsymbol{p} \in \Delta$. Let $\boldsymbol{b} = (b^t)$ be a lower bound of the consumption set X. Since the function $g_k(a)$ is bounded from below by the vector $(\gamma^t) = (\min\{0, b^t\})$ a.e. in A for all k, the sequence $\{\int_A g_k(a)d\mu\}$ is bounded from below. Consequently, it is bounded, since it is bounded from above by the vector $\int_A \omega(a)d\mu - (\gamma^t)$. Then, it follows from Fatou's lemma in ℓ-dimensions (Theorem G11 of Appendix G) that there exists an integrable function f of A to $\mathbb{R}^\ell$ such that $f(a)$ is a cluster point of the sequence $\{g_k(a)\}$ and $\int_A f(a)d\mu \leq \int_A \omega(a)d\mu$. Let $k(a)$ be a positive vector with $k(a) \geq \|\omega(a)/B\|$. Then, one sees that:

$$0 \leq \omega^t(a) \leq \|\omega(a)\| \leq k(a)B, \quad t = 1 \cdots \ell,$$

hence $a \in A_k$ for $k \geq k(a)$. Then, it follows that:

$$k > k(a) \text{ implies that } g_k(a) \in \phi_k(a, \boldsymbol{p}_k).$$

From this, we can show that:

$$f(a) \in \phi(a, \boldsymbol{p}) \quad \text{a.e. in } A.$$

Indeed, by the definition of $g_k(a)$, one sees that $f(a) \in \beta(a, \boldsymbol{p})$ a.e., since $f(a)$ is a cluster point of $\{g_k(a)\}$. If $\boldsymbol{x} \in X$ and $\boldsymbol{px} < \boldsymbol{p}\omega(a)$, then $\boldsymbol{p}_k\boldsymbol{x} < \boldsymbol{p}_k\omega(a)$ or all k sufficiently large. Hence, by the continuity of the preferences (CT), one has $f(a) \succsim_a \boldsymbol{x}$ a.e. in A. If $\boldsymbol{x} \in X$ and $\boldsymbol{px} = \boldsymbol{p}\omega(a)$, then by the minimum income condition (MI), we can take a sequence $\boldsymbol{x}_n$ converging to $\boldsymbol{x}$ and $\boldsymbol{px}_n < \boldsymbol{px}$, $n \in \mathbb{N}$. For each n, $f(a) \succsim_a \boldsymbol{x}_n$. Again by the continuity (CT), we have $f(a) \succsim_a \boldsymbol{x}$ a.e. in A. This proves the condition (E-1). The condition (E-2) has been already proved. Therefore, $(\boldsymbol{p}, f(a))$ is a competitive equilibrium. □

From a technical point of view, in the above proof, the role of the Fatou's lemma in ℓ-dimensions is apparent. However, under the assumption of the monotone preferences, we have a simpler proof of the theorem thanks to Theorem D4 of Appendix D, which is a strengthened version of the Gale–Nikaido lemma. The following concept will also be useful in later discussions.

Definition 3.3. The quasi-demand relation is defined by:

$$\tilde{\phi}(X, \prec, w, \boldsymbol{p}) = \begin{cases} \phi(X, \prec, w, \boldsymbol{p}) & \text{if } \inf \boldsymbol{p}X < w, \\ \{\boldsymbol{x} \in X \mid \boldsymbol{px} = \inf \boldsymbol{p}X\} & \text{if } \inf \boldsymbol{p}X \geq w. \end{cases}$$

A pleasant property of the quasi-demand relation is that it is a nonempty, convex-valued and closed (hence measurable) relation, if the consumption set is convex.

Proposition 3.3. *Let* $T = \mathcal{P} \times \mathbb{R} \times \mathbb{R}^\ell$ *and* $T_c = \{\tau = (X, \prec, w, \boldsymbol{p}) \in T \mid X \text{ is convex}\}$. *Then for each* $\tau \in T_c$, *the set* $\tilde{\phi}(\tau)$ *is nonempty and convex subset of* $\mathbb{R}^\ell$, *and the graph of the quasi-demand relation*:

$$\{(X, \prec, w, \boldsymbol{p}, \boldsymbol{x}) \in T_c \times \mathbb{R}^\ell \mid \boldsymbol{x} \in \tilde{\phi}(x, \prec, w, \boldsymbol{p})\}$$

is closed.

Proof. By definition, it is clear that $\tilde{\phi}(\tau)$ is nonempty and convex for each $\tau \in T_c$. Let $\tau_n = (X_n, \prec_n, w_n, \boldsymbol{p}_n)$ be a sequence converging to $\tau = (X, \prec, w, \boldsymbol{p})$, $\tau_n, \tau \in T_c$ for all n and $\boldsymbol{x}_n \to \boldsymbol{x}$, where $\boldsymbol{x}_n \in \tilde{\phi}(\tau_n)$ for all n. If $\inf \boldsymbol{p}X < w$, then $\inf \boldsymbol{p}_n X_n < w_n$ for all n large enough. Hence, by definition, $\tilde{\phi}(\tau_n) = \phi(\tau_n)$ and we can show that $\boldsymbol{x} \in \tilde{\phi}(\tau)$ in the same way as the proof of Proposition 3.3 in Chap. 2. If $\inf \boldsymbol{p}X \geq w$, then we have to show only that $\boldsymbol{px} = \inf \boldsymbol{p}X$. Since $\boldsymbol{x} \in X$, it is obvious that $\boldsymbol{px} \geq \inf \boldsymbol{p}X$. But, it is impossible that $\boldsymbol{px} > \inf \boldsymbol{p}X$, since if it was the case, $\boldsymbol{p}_n\boldsymbol{x}_n > \inf \boldsymbol{p}_n X_n$, for all n

sufficiently large. Hence, $\boldsymbol{x}_n \in \phi(\tau_n) \subset \beta(\tau_n)$ and $w_n \geq \boldsymbol{p}_n\boldsymbol{x}_n > \inf \boldsymbol{p}_n X_n = w_n$ for all n large enough. In this limit, we have $w \geq \boldsymbol{px} \geq \inf \boldsymbol{p}_X \geq w$. □

Theorem 3.2. *Let* $\mathcal{E}: (A, \mathcal{A}, \mu) \to (\mathcal{P} \times \mathbb{R}^\ell, \mathcal{B}(\mathcal{P} \times \mathbb{R}^\ell))$ *be an exchange economy such that* $\mathcal{E}(a) \subset \mathcal{P}_{mo} \times \mathbb{R}^\ell_+$, $X_a = \mathbb{R}^\ell_+$ *a.e., and* $\int_A \omega(a) d\mu \gg \mathbf{0}$. *Then, there exists a competitive equilibrium* $(\boldsymbol{p}, (\boldsymbol{x}(a)))$ *for* $\mathcal{E}$.

Proof. Let $\tilde{\phi}(a, \boldsymbol{p})$ be the quasi-demand relation and we define the quasi-mean demand by:

$$\tilde{\zeta}(\boldsymbol{p}) = \int_A \tilde{\phi}(a, \boldsymbol{p}) d\mu - \int_A \omega(a) d\mu,$$

for every $\boldsymbol{p} \gg \mathbf{0}$.

First, we shall show that there exists a price vector $\boldsymbol{p}^*$ such that $\tilde{\zeta}(\boldsymbol{p}^*) = \mathbf{0}$. The existence of such a price vector is a consequence of Theorem D4 of Appendix D, and we want to verify that:

(i) for every strictly positive vector $\boldsymbol{p} \gg \mathbf{0}$, it follows that $\boldsymbol{p}\tilde{\zeta}(\boldsymbol{p}) \leq \mathbf{0}$;
(ii) the correspondence $\tilde{\zeta}$ is compact and convex-valued, bounded from below, and upper hemi-continuous; and
(iii) for every sequence $\{\boldsymbol{p}_n\}$ in interior Δ converging to

$$\boldsymbol{p} \in \text{bdry}\Delta = \{\boldsymbol{p} = (p^t) \in \Delta \mid p^t = 0 \text{ for some } t\},$$

it follows that:

$$\inf \left\{ \sum_{t=1}^{\ell} z^t \,\middle|\, \boldsymbol{z} = (z^t) \in \tilde{\zeta}(\boldsymbol{p}_n) \right\} > 0 \quad \text{for } n \text{ large enough.}$$

Since preferences are monotone, it is obvious that $\boldsymbol{px} = \boldsymbol{p}\omega(a)$ for every $\boldsymbol{x} \in \tilde{\phi}(a, \boldsymbol{p})$ when $\inf \boldsymbol{p}X < \boldsymbol{p}\omega(a)$. If $\inf \boldsymbol{p}X \geq \boldsymbol{p}\omega(a)$, then since $\inf \boldsymbol{p}X = \inf \boldsymbol{p}\mathbb{R}^\ell_+ = 0$, it follows that $0 \leq \boldsymbol{px} \leq \boldsymbol{p}\omega(a) = 0$. Hence, property (i) is proved.

Let $\bar{\boldsymbol{p}} \gg \mathbf{0}$. Then, there exists a neighborhood U of $\bar{\boldsymbol{p}}$ consisting of strictly positive vectors. For fixed $a \in A$, the correspondence $\tilde{\phi}(a, \cdot)$ is closed at $\bar{\boldsymbol{p}}$ by Proposition 3.3. Moreover, setting $h(a) = (1/\pi)(\sum_{t=1}^{\ell} \omega^t(a))$, where $\pi = \min\{p^t | \boldsymbol{p} = (p^t) \in U\}$, we obtain an integrable function such that $\|\tilde{\phi}(a, \boldsymbol{p})\| \leq h(a)$ for every $\boldsymbol{p} \in U$ a.e. Then, by Corollary G2 of Theorem G12 of Appendix G, the correspondence $\tilde{\Phi}(\boldsymbol{p}) = \int_A \tilde{\phi}(a, \boldsymbol{p}) d\mu$ is closed at $\bar{\boldsymbol{p}}$. Since the correspondence $\tilde{\Phi}(\boldsymbol{p})$ is bounded on the neighborhood U of $\bar{\boldsymbol{p}}$, it follows that $\tilde{\Phi}(\boldsymbol{p})$ is compact-valued and upper hemi-continuous at $\bar{\boldsymbol{p}}$. Therefore, the property (ii) is verified. In order to prove property (iii), we will show that:

$$\inf\{\|\boldsymbol{x}\| | \boldsymbol{x} \in \tilde{\Phi}(\boldsymbol{p}_n)\} \to \infty \quad \text{as } n \to \infty$$

for every sequence $\{p_n\}$ of strictly positive vectors converging to $p \in \text{bdry}\Delta$. If not, then there exists a bounded set $B \subset \mathbb{R}^\ell$ such that $\tilde{\Phi}(p_n) \cap B \neq \emptyset$ for infinitely many n's. Since $\int_A \omega(a)d\mu \gg \mathbf{0}$, we have $\inf p\mathbb{R}^\ell_+ = 0 < p\omega(a)$ on a subset of A with positive measure. Hence, we can assume that $\phi(a,p) \cap B \neq \emptyset$ for infinitely many ns on the non-null set. Let $x_n(a) \in \phi(a,p_n) \cap B$. There is a converging subsequence, say $x_n(a) \to x(a)$. Since $\inf p\mathbb{R}^\ell_+ < p\omega(a)$, it follows from Proposition 3.3 that $x \in \phi(a,p)$, which is impossible. Indeed, this set is empty since $\prec_a \in \mathcal{P}_{mo}$ and the budget set $\beta(a,p)$ is unbounded for $p \in \text{bdry}\Delta$.

By Theorem D4 of Appendix D, there exists a vector $p^* \gg \mathbf{0}$ such that $\mathbf{0} \in \tilde{\zeta}(p)$. Then, there exists a measurable map $x(a) \in \tilde{\phi}(a,p^*)$ such that $\int_A x(a)d\mu = \int_A \omega(a)d\mu$. Hence, the condition (E-2) is satisfied. For $a \in A$ such that $\inf p^*\mathbb{R}^\ell_+ < p^*\omega(a)$, one has $\tilde{\phi}(a,p^*) = \phi(a,p)$, hence the condition (E-1) holds good for such an $a \in A$. For $a \in A$ such that $0 = \inf p^*\mathbb{R}^\ell_+ \geq p^*\omega(a)$, it follows from $p^* \gg \mathbf{0}$ that the budget set is a singleton $\beta(a,p^*) = \{\mathbf{0}\}$. Therefore, the condition (E-1) trivially holds good for such an a. □

Note that the minimum income condition (MI) in Theorem 3.1, is replaced by the survival condition $\omega(a) \in X$ a.e. in Theorem 3.2.

As discussed in the classical finite (or simple)-exchange economies in Chap. 2, we can prove the second fundamental theorem of welfare economics, or the converse of Proposition 3.1.

Definition 3.4. An allocation f for the economy $\mathcal{E}: A \to \mathcal{P} \times \mathbb{R}^\ell$ is called an equilibrium relative to a price vector $p \in \mathbb{R}^\ell, p \neq \mathbf{0}$ if:

(R-1) $f(a) \succsim_a x$ for every $x \in X_a$ with $px \leq pf(a)$ a.e. in A,
(E-2) $\int_A f(a)d\mu \leq \int_A \omega(a)d\mu$.

The allocation f is called a quasi-equilibrium relative to a price vector $p \in \mathbb{R}^\ell$, $p \neq \mathbf{0}$ if the condition (R-1) is replaced by:

(Q-1) $f(a) \succsim_a x$ for every $x \in X_a$ with $px < pf(a)$ a.e. in A.

Recall that $\mathcal{P}^*$ is the set of all preference relations with negative transitivity and $\mathcal{P}_{mo}$ the set of all local monotone preference relations. Let $\mathcal{P}^*_{mo} = \mathcal{P}^* \cap \mathcal{P}_{mo}$. The second fundamental theorem of welfare economics now reads:

Theorem 3.3. *Let $\mathcal{E}: A \to \mathcal{P} \times \mathbb{R}^\ell$ be an economy which satisfies $\mathcal{E}(a) \subset \mathcal{P}^*_{mo}$ and X_a is convex a.e. in A. Then, every Pareto-optimal allocation is a quasi-equilibrium relative to some price vector $p(\neq \mathbf{0}) \in \mathbb{R}^\ell_+$.*

Proof. Let $f: A \to \mathbb{R}^\ell$ be a Pareto-optimal allocation. For every $a \in A$, we define:

$$P(a) = \{x \in X_a | f(a) \prec_a x\}.$$

As shown in the proof of Proposition 3.2, one can show that the correspondence $P(a)$ has the measurable graph. By Theorem G10 of Appendix G, $\int_A P(a)d\mu$ is a convex subset of $\mathbb{R}^\ell$. Obviously, $\int_A \omega(a)d\mu \notin \int_A P(a)d\mu$, since the allocation f is Pareto-optimal. Then, we can apply the separation hyperplane theorem (Theorem A1) of Appendix A and obtain a vector $\boldsymbol{p} \in \mathbb{R}^\ell, \boldsymbol{p} \neq \boldsymbol{0}$ such that:

$$\boldsymbol{p}\int_A \omega(a)d\mu \leq \inf\left\{\boldsymbol{px}\middle|\boldsymbol{x} \in \int_A P(a)d\mu\right\}.$$

We have $\boldsymbol{p} \geq \boldsymbol{0}$ by the monotonicity (MT) of $\prec_a$. Since f is feasible, or $\int_A f(a)d\mu \leq \int_A \omega(a)d\mu$, we have $\boldsymbol{p}\int_A f(a)d\mu \leq \boldsymbol{p}\int_A \omega(a)d\mu$. Therefore, it follows from Proposition G9 of Appendix G that:

$$\boldsymbol{p}\int_A f(a)d\mu \leq \inf \boldsymbol{p}\int_A P(a)d\mu = \int_A \inf \boldsymbol{p}P(a)d\mu.$$

Since $\boldsymbol{p}f(a) \geq \inf \boldsymbol{p}P(a)$ a.e. in A, we see that $\boldsymbol{p}f(a) = \inf \boldsymbol{p}P(a)$ a.e. in A. Consequently, we have:

$$\text{if there exists } \boldsymbol{x} \in X_a \text{ with } \boldsymbol{px} < \boldsymbol{p}f(a), \text{ then } f(a) \precsim_a \boldsymbol{x} \text{ a.e. in } A.$$

Hence, the condition (Q-1) is established. □

As in the case of the classical economies, the quasi-equilibrium is reduced to the equilibrium relative to $\boldsymbol{p}$, if the situation $\boldsymbol{p}f(a) = \inf \boldsymbol{p}X_a$ is excluded.

Corollary. *Under the assumptions of Theorem 3.3, if the Pareto-optimal allocation $f(a)$ satisfies $\boldsymbol{p}f(a) > \inf \boldsymbol{p}X_a$, then it is an equilibrium relative to $\boldsymbol{p}$.*

Proof. Let $\boldsymbol{x} \in X_a$ be such that $\boldsymbol{px} = \boldsymbol{p}f(a)$ and suppose that $\inf \boldsymbol{p}X_a < \boldsymbol{p}f(a)$. Since X_a is convex, we obtain a sequence $\{\boldsymbol{x}_n\}$ such that $\boldsymbol{px}_n < \boldsymbol{p}f(a)$ and $\boldsymbol{x}_n \to \boldsymbol{x}$. By the continuity of preferences, we have $f(a) \precsim_a \boldsymbol{x}$. □

3.4. THE EQUIVALENCE OF THE CORE AND EQUILIBRIA

When the set of consumers A is finite, every subset C of A is called a coalition. For an atomless measure space of consumers $(A, \mathcal{A}, \mu)$, every measurable subset $C \in \mathcal{A}$ is called a coalition. If it is μ-measure zero, or $\mu(C) = 0$, the coalition C is called null. The null coalitions are considered to have no influence on the market. The empty coalition $C = \emptyset$ is an example of the null coalition.

The core $\mathcal{C}$ of an economy $\mathcal{E}: A \to \mathcal{P} \times \mathbb{R}^\ell$ is now defined in a natural way.

Definition 4.1. An allocation $f: A \to \mathbb{R}^\ell, f(a) \in X_a$ a.e. in A, is blocked by a coalition $C \in \mathcal{A}$, if there exists an allocation $g: A \to \mathbb{R}^\ell, g(a) \in X_a$ a.e. in A such that:

$$f(a) \prec_a g(a) \quad \text{for every } a \in C$$

and

$$\int_C g(a)d\mu \leq \int_C \omega(a)d\mu.$$

The set of feasible allocations which are not blocked by any non-null coalition is called the core and denoted by $\mathcal{C}$.

In Section 2.6, we proved the core limit theorem (Theorem 6.1), which says that the core approaches the equilibrium when the population of the economy becomes proportionally large. In the economies with an atomless measure space of consumers, we can prove the next core equivalence theorem.

Theorem 4.1. *Let $\mathcal{E}: A \to \mathcal{P} \times \mathbb{R}^\ell$ be an atomless exchange economy such that $\mathcal{E}(a) \subset \mathcal{P}_{mo} \times \mathbb{R}^\ell_+, X_a = \mathbb{R}^\ell_+$ a.e. in A, and the mean endowment is strictly positive or $\int_A \omega(a)d\mu \gg 0$. Then, the core coincides with the set of equilibrium allocation, namely that if $\boldsymbol{x} \in \mathcal{C}$, then there exists a price vector $\boldsymbol{p}$ such that $(\boldsymbol{p}, \boldsymbol{x})$ is a competitive equilibrium.*

Proof. Let $\mathcal{W}$ be the set of competitive allocations,

$$\mathcal{W} = \{\boldsymbol{x}: A \to \mathbb{R}^\ell \mid (\boldsymbol{p}, \boldsymbol{x}) \text{ is a competitive equilibrium for some } \boldsymbol{p} \in \mathbb{R}^\ell_+\}$$

and $\boldsymbol{x} \in \mathcal{W}$. Then, there exists a price vector $\boldsymbol{p} \in \mathbb{R}^\ell_+$ such that $(\boldsymbol{p}, \boldsymbol{x})$ is a competitive equilibrium for $\mathcal{E}$. Suppose that $\boldsymbol{x} \notin \mathcal{C}$, hence the allocation $\boldsymbol{x}$ is blocked by a non-null coalition C through an allocation $\boldsymbol{y}$ such that:

$$\boldsymbol{x}(a) \prec_a \boldsymbol{y}(a) \quad \text{for every } a \in C$$

and

$$\int_C \boldsymbol{y}(a)d\mu \leq \int_C \omega(a)d\mu.$$

Then, by the condition (E-1) of Definition 3.1, we have $\boldsymbol{p}\boldsymbol{y}(a) > \boldsymbol{p}\omega(a)$ a.e. in C. Hence, $\boldsymbol{p}\int_C \boldsymbol{y}(a)d\mu = \int_C \boldsymbol{p}\boldsymbol{y}(a) > \int_C \boldsymbol{p}\omega(a)d\mu = \boldsymbol{p}\int_C \omega(a)d\mu$. On the other hand, since $\boldsymbol{y}$ is feasible in C, we have $\boldsymbol{p}\int_C \boldsymbol{y}(a)d\mu \leq \boldsymbol{p}\int_C \omega(a)d\mu$, a contradiction. Therefore, $\boldsymbol{x} \in$ C and this proves $\mathcal{W} \subset \mathcal{C}$.

In order to prove $\mathcal{C} \subset \mathcal{W}$, let $\boldsymbol{x} \in \mathcal{C}$ and we define:

$$P(a) = \{\boldsymbol{y} \in X_a \mid \boldsymbol{x} \prec_a \boldsymbol{y}\}, \quad Q(a) = P(a) - \{\omega(a)\}.$$

We now claim:

Lemma. *There exists a measurable set U with $\mu(U) = \mu(A) = 1$ such that* $\mathbf{0} \notin$ interior$\{co(\cup_{a\in U} Q(a))\}$.

Proof. For each $\boldsymbol{y} \in \mathbb{R}^\ell$, let $Q^{-1}(\boldsymbol{y}) = \{a \in A \mid \boldsymbol{y} \in Q(a)\} = \{a \in A \mid \boldsymbol{x}(a) \prec_a \boldsymbol{y} + \omega(a)\}$. Since the map $\mathrm{Proj}_{\mathcal{P}} \circ \mathcal{E}: a \mapsto \prec_a$ is measurable, where $\mathrm{Proj}_{\mathcal{P}}$ is the projection map $(\prec_a, \omega(a)) \mapsto \prec_a$, it follows that $Q^{-1}(\boldsymbol{y})$ is measurable for each $\boldsymbol{y} \in \mathbb{R}^\ell$.

Let N be the set of all rational points $\boldsymbol{q}$ of $\mathbb{R}^\ell$ (the points with rational coordinates) such that $\mu(Q^{-1}(\boldsymbol{q})) = 0$. Obviously, N is a countable set. We define $U = A \setminus \cup_{\boldsymbol{q}\in N} Q^{-1}(\boldsymbol{q})$. Then, U is of full measure; $\mu(U) = 1$.

Suppose that $\mathbf{0} \in$ interior$\{co(\cup_{a\in U} Q(a))\}$. Then, there exists a point $\boldsymbol{x} \gg \mathbf{0}$ such that $-\boldsymbol{x} \in co(\cup_{a\in U} Q(a))$. Then, we have $a_1 \cdots a_k \in U$ and $\boldsymbol{x}_i \in Q(a_i), \theta_i \in \mathbb{R}$, $i = 1 \cdots k$ such that:

$$\mathbf{0} \gg -\boldsymbol{x} = \sum_{i=1}^{k} \theta_i \boldsymbol{x}_i, \quad \sum_{i=1}^{k} \theta_i = 1, \quad \theta_i \geq 0, \ i = 1 \cdots k.$$

By the continuity (CT) of preferences, we can find rational points $\boldsymbol{q}_i \in Q(a_i)$ which are sufficiently close to $\boldsymbol{x}_i$, and positive rational points τ_i sufficiently close to θ_i and $\sum_{i=1}^{k} \tau_i = 1$, so that we still have $\sum_{i=1}^{k} \tau_i \boldsymbol{q}_i \ll \mathbf{0}$. Let $-\boldsymbol{q} = \sum_{i=1}^{k} \tau_i \boldsymbol{q}_i$, and take an arbitrary consumer a_0 in U. Since $\boldsymbol{q} \gg \mathbf{0}$, we see $\boldsymbol{x}(a_0) \ll \alpha \boldsymbol{q} + \omega(a_0)$ for sufficiently large α. By the monotonicity (MT) of preferences, $\boldsymbol{x}(a_0) \prec_{a_0} \alpha\boldsymbol{q} + \omega(a_0)$, or $\alpha \boldsymbol{q} \in Q(a_0)$. We now define $\boldsymbol{q}_0 = \alpha \boldsymbol{q}$, $\alpha_0 = 1/(\alpha + 1)$, $\alpha_i = \alpha\tau_i/(\alpha + 1)$, $i = 1 \cdots k$. Since $\alpha, \tau_i > 0$, it follows that $\alpha_i > 0$ for all $i = 0 \cdots k$ and $\sum_{i=0}^{k} \alpha_i = 1$. Moreover, we have:

$$\sum_{i=0}^{k} \alpha_i \boldsymbol{q}_i = \left(\frac{\alpha}{\alpha+1}\right)\boldsymbol{q} + \left(\frac{\alpha}{\alpha+1}\right)\sum_{i=1}^{k} \tau_i \boldsymbol{q}_i = \mathbf{0},$$

and $\boldsymbol{q}_i \in Q(a_i), i = 0 \cdots k$. Then, $a_i \in Q^{-1}(\boldsymbol{q}_i)$ and $\boldsymbol{q}_i \notin N$, since $a_i \in U$. Hence, $Q^{-1}(\boldsymbol{q}_i)$ is of positive measure for all $i = 0 \cdots k$. Therefore, we can find disjoint subsets C_i of $Q^{-1}(\boldsymbol{q}_i)$ such that $\mu(C_i) = \delta\alpha_i$ for $\delta > 0$ sufficiently small. We define a coalition C by $C = \cup_{i=0}^{k} C_i$, and an allocation $\boldsymbol{y}$ by:

$$\boldsymbol{y}(a) = \begin{cases} \boldsymbol{q}_i + \omega(a) & \text{if } a \in C_i, \\ \omega(a) & \text{otherwise.} \end{cases}$$

We have to check that $\boldsymbol{y}(a) \in X_a$ for all $a \in A$. For $a \notin C_i$, this is obvious. For $a \in C_i \subset Q^{-1}(\boldsymbol{q}_i)$, it follows from $\boldsymbol{q}_i \in Q(a)$. Since, we have:

$$\int_C \boldsymbol{y}(a)d\mu = \sum_{i=0}^{k} \delta\alpha_i \boldsymbol{q}_i + \int_C \omega(a)d\mu = \mathbf{0} + \int_C \omega(a)d\mu,$$

the allocation $\boldsymbol{y}$ is feasible for the coalition C. It is obviously feasible for the grand coalition A, since $\boldsymbol{y}(a) = \omega(a)$ for $a \notin C$ by definition. Finally, since $C_i \subset Q^{-1}(\boldsymbol{q}_i)$, it follows that $\boldsymbol{x}(a) \prec_a \boldsymbol{q}_i + \omega(a)(=\boldsymbol{y}(a))$ for $a \in C$. Hence, the coalition C blocks the allocation $\boldsymbol{x}$, which was assumed to be in the core, a contradiction. This proves the lemma.

In the following, all consumers a will be an element of U, hence we will often omit the phrase "for $a \in U$". By the separation hyperplane theorem (Theorem A1) of Appendix A, there exists a vector $\boldsymbol{p} \neq \boldsymbol{0}$ such that $\boldsymbol{py} \geq 0$ for all $\boldsymbol{y} \in co(\cup_{a \in U} Q(a))$. Hence, $\boldsymbol{py} \geq 0$ for every $\boldsymbol{y} \in Q(a)$ if $a \in U$, or

$$(*) \quad \boldsymbol{py} \geq \boldsymbol{p}\omega(a) \quad \text{for every } \boldsymbol{y} \in P(a) \quad \text{when } a \in U.$$

By the monotonicity (MT) of preferences, $P(a) + \mathbb{R}^{\ell}_{+} \subset P(a)$, hence $\boldsymbol{p} \geq \boldsymbol{0}$. We shall show that $(\boldsymbol{p}, \boldsymbol{x})$ is a competitive equilibrium for $\mathcal{E}$, or $\boldsymbol{x} \in \mathcal{W}$.

By the monotonicity (MT) of preferences, we have a bundle $\boldsymbol{z}$, which is arbitrarily close to $\boldsymbol{x}(a)$ and $\boldsymbol{x}(a) \prec_a \boldsymbol{z}$. Therefore, it follows that $\boldsymbol{x}(a) \in$ closure $P(a)$. Hence, by $(*)$, we have $\boldsymbol{px}(a) \geq \boldsymbol{p}\omega(a)$. If $\boldsymbol{px}(a) > \boldsymbol{p}\omega(a)$ on a set of positive measure, then $\int_A \boldsymbol{px}(a) d\mu > \int_a \boldsymbol{p}\omega(a) d\mu$, contradicting the feasibility of the allocation $\boldsymbol{x}$. Therefore, $\boldsymbol{px}(a) \leq \boldsymbol{p}\omega(a)$ a.e., namely that $\boldsymbol{x}(a)$ is in the budget set of $a \in A$ a.e.

Finally, we show that the condition (E-1) of Definition 3.1, or:

$$(**) \quad \boldsymbol{pz} > \boldsymbol{p}\omega(a) \quad \text{for every } \boldsymbol{z} \in P(a).$$

In order to show this, first we will prove that $\boldsymbol{p} \gg \boldsymbol{0}$. Otherwise, at least one coordinate is 0, say $p^1 = 0$. Since $\boldsymbol{p} \neq \boldsymbol{0}$, at least one coordinate of $\boldsymbol{p}$ does not vanish. For example, let $p^2 > 0$. Since $\int_A \omega(a) \gg \boldsymbol{0}$ and $\boldsymbol{x} \in \mathcal{C}$, there exists a non-null coalition on which $\boldsymbol{x}(a) \gg \boldsymbol{0}$, hence $x^2(a) > 0$ on a set of positive measure. By the monotonicity (MT), we see that:

$$\boldsymbol{x}(a) \prec_a \boldsymbol{x}(a) + (1, 0 \cdots 0).$$

By the continuity (CT) of preferences, it follows that:

$$\boldsymbol{x}(a) \prec_a \boldsymbol{x}(a) + (1, -\delta, 0 \cdots 0),$$

for $\delta > 0$ sufficiently small. Then by $(*)$,

$$\begin{aligned} \boldsymbol{p}\omega(a) &\leq \boldsymbol{px}(a) + \boldsymbol{p}(1, -\delta, 0 \cdots 0) \\ &= \boldsymbol{px}(a) + p^1 - \delta p^2 \\ &= \boldsymbol{px}(a) - \delta p^2 < \boldsymbol{px}(a), \end{aligned}$$

a contradiction, since $\boldsymbol{x}(a)$ is in the budget set. This proves that $\boldsymbol{p} \gg \boldsymbol{0}$.

Let $\boldsymbol{y} \in P(a)$. First suppose that $\omega(a) > \boldsymbol{0}$. Then, $\boldsymbol{p}\omega(a) > 0$, since $\boldsymbol{p} \gg \boldsymbol{0}$. Hence by $(*)$, $\boldsymbol{py} > 0$, so that $\boldsymbol{y} \neq \boldsymbol{0}$. Therefore at least one coordinate is greater than 0, say $y^1 > 0$. By the continuity (CT), we have:

$$\boldsymbol{y} - (\delta, 0 \cdots 0) \in P(a)$$

for $\delta > 0$ sufficiently small. Then by (*),

$$\begin{aligned} \boldsymbol{p}\omega(a) &\leq \boldsymbol{p}(\boldsymbol{y} - (-\delta, 0 \cdots 0) \\ &= \boldsymbol{py} - \delta p^1 < \boldsymbol{py}, \end{aligned}$$

which proves (**). If $\omega(a) = \mathbf{0}$ and $\boldsymbol{y} > \mathbf{0}$, then $\boldsymbol{py} > 0 = \boldsymbol{p}\omega(a)$, proving (**). Finally, suppose that $\omega(a) = \mathbf{0}$ and $\boldsymbol{y} = \mathbf{0}$. Since $\boldsymbol{y} \in P(a)$, this implies that $\boldsymbol{x}(a) \prec_a \mathbf{0} = \omega(a)$. If the set on which this case is null, we may ignore it. If it is of positive measure, then the non-null coalition would block $\boldsymbol{x}$ thorough the endowment $\omega(a)$, contradicting $\boldsymbol{x} \in \mathcal{C}$. Since the feasibility condition (E-2) has already been established, the proof of the theorem is completed. □

3.5. EXCHANGE ECONOMIES WITH A NON-CONVEX CONSUMPTION SET

We have assumed that the consumption sets of consumers are convex so far. In the present and in the next sections, we will discuss economies with a non-convex consumption set. We will do so not simply as a mathematical generalization, but also with significant economic motivations.

The first example of non-convex consumption sets is the case of indivisible commodities.

In Fig. 3.3, the commodity x is assumed to be perfectly divisible. Hence, its quantity of consumption can be any value of real numbers, as it has been so far. However, the commodity y is an indivisible commodity, so that it is consumed only in the integer units, and its coordinate axis is represented by $\mathbb{N}$, the set of natural numbers. Therefore, the consumption set X is given by $X = \mathbb{R}^{\ell}_{+} \times \mathbb{N}$.

The second example is given by the case of mutually exclusive commodities.

In Fig. 3.4, the commodity x is a good which is consumed at Tokyo and the commodity y is consumed at New York at the same time. Usually, any individual consumer can consume a commodity either at Tokyo or at New York, but not at both places simultaneously; the consumption set X which represents this situation is something like:

$$X = \{(x, 0) \mid x \in \mathbb{R}\} \cup \{(0, y) \mid y \in \mathbb{R}\},$$

which is obviously not convex.

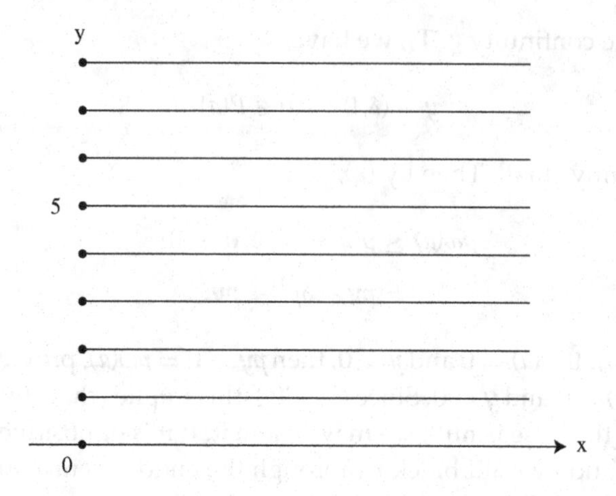

Figure 3.3. Indivisible Commodities.

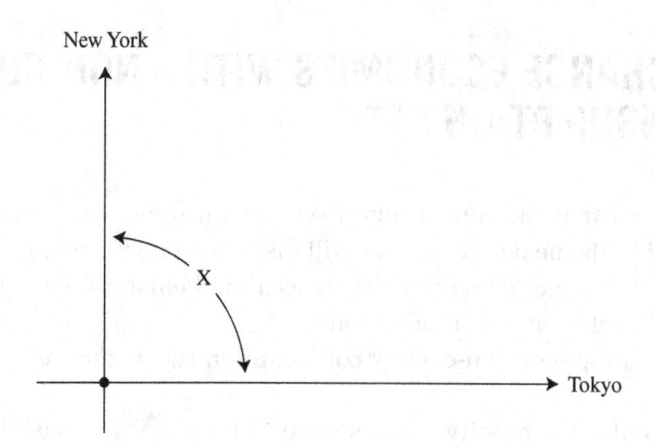

Figure 3.4. Mutually Exclusive Goods.

Recall that the commodities are distinguished not only by their physical properties but also by their consumption characteristics. Some of the characteristics of commodities, say locations or dates, will induce this kind of situation; the consumption of a commodity by an individual consumer necessarily excludes the possibility of consumption by the same individual of another commodity.

These examples indicate that the economic models in which the consumption sets are assumed to be convex are not adequate for some of their applications. However, a technical problem arises from the models with non-convex consumption sets; the demand correspondences generally exhibit discontinuous behaviors as the following example shows.

Example 5.1. Consider a non-convex consumption set X which is defined by:

$$X = \{(x,y) \in \mathbb{R}^2 \mid x \geq 1,\ \ y \geq 0\} \cup \{(x,y) \in \mathbb{R}^2 \mid x \geq 0,\ y \geq 3\}.$$

The set of consumers A is given by the unit interval $A = [0,1]$, and all consumers $a \in [0,1]$ have the same preference, which is represented by the convex (not concave!) utility function:

$$u(x,y) = x^2 + y^2, \quad (x,y) \in X.$$

Suppose that the endowment vector $\omega(a)$ is the same for every consumer $a \in [0,1]$ and given by $\omega(a) = (1,1)$ for $a \in [0,1]$ (see Fig. 3.5).

Since the indifference curves are circle-centered at the origin, for every price vector $\boldsymbol{p} = (p,q)$, a consumer maximizes his/her utility at the consumption bundle on the x-axis or the y-axis, or $\boldsymbol{x} = (x,0)$ or $= (0,y)$. In the following, we normalize the price vector as $\boldsymbol{p} = (p, 1-p),\ 0 \leq p \leq 1$. Hence, the budget equation of the consumer is given by:

$$px + (1-p)y = 1.$$

From the budget equation, we see that the demand of the consumer $a \in [0,1]$ is determined by:

$$\phi(a,p) = \begin{cases} \left(\dfrac{1}{p}, 0\right) & \text{for } 0 \leq p < 2/3, \\[2ex] \left(0, \dfrac{1}{p-1}\right) & \text{for } 2/3 \leq p \leq 1, \end{cases}$$

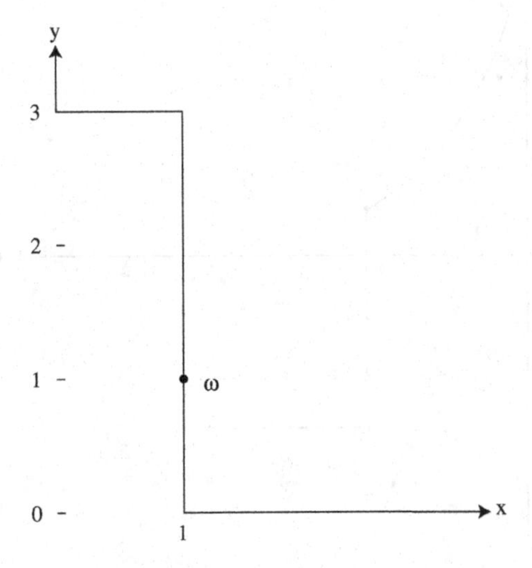

Figure 3.5. Example 5.1.

which is discontinuous at $p = 2/3$ for every $a \in [0,1]$. Since all consumers have the same demand function, the mean excess demand function can be simply calculated as:

$$\zeta(p) = \int_0^1 \phi(a,p)d\mu - \int_0^1 \omega(a)d\mu$$
$$= \begin{cases} \left(\dfrac{1}{p} - 1, -1\right) & \text{for } 0 \le p < 2/3, \\ \left(-1, \dfrac{1}{p-1} - 1\right) & \text{for } 2/3 \le p \le 1, \end{cases}$$

and the mean demand function is also discontinuous at $p = 2/3$, and it stays away from zero. Therefore, the economy does not have an equilibrium. Figure 3.6 illustrates the graph of the x-coordinate of the mean demand function $\zeta(p)$.

The reason of the discontinuous behavior of $\zeta(p)$ is of course that the consumers on a set of positive measure (actually, whole space $[0,1]$ in this example) behaves discontinuously at a particular price. Therefore, we expect that if for each price vector, the set of the discontinuous consumers is of measure zero, the mean

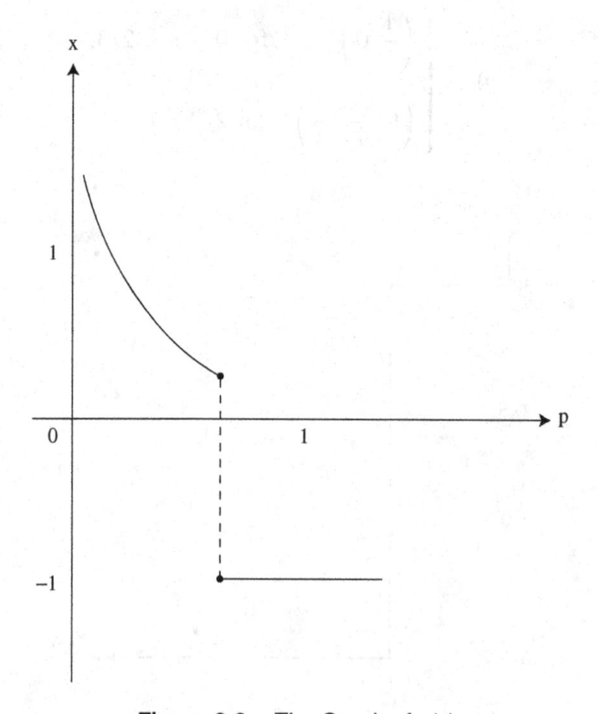

Figure 3.6. The Graph of $\zeta(p)$.

demand $\Phi(\boldsymbol{p}) = \int_A \phi(a,\boldsymbol{p})d\mu$ will be continuous, and so is the mean demand correspondence $\zeta(\boldsymbol{p}) = \Phi(\boldsymbol{p}) - \int_A \omega(a)d\mu$. This is, indeed the case, as the following example shows.

Example 5.2. For each consumer, $a \in [0,1]$, the consumption set X and the utility function is the same as given in Example 5.1:

$$X = \{(x,y) \in \mathbb{R}^2 \mid x \geq 1,\ y \geq 0\} \cup \{(x,y) \in \mathbb{R}^2 \mid x \geq 0,\ y \geq 3\},$$
$$u(x,y) = x^2 + y^2, \quad (x,y) \in X.$$

The endowment vector $\omega(a)$, however, is not constant but it depends on a in such a way that:

$$\omega(a) = (1,a) \quad \text{for } a \in [0,1].$$

The budget equation of the consumer a is now given by:

$$px + (1-p)y = p + (1-p)a.$$

The discontinuity of the demand of a arises at $\boldsymbol{x} = (0,3)$. From the budget equation, setting $\boldsymbol{x} = (0,3)$, the price at which the demand being discontinuous is calculated as:

$$3(1-p) = p + (1-p)a, \quad 3-a = (4-a)p, \quad p = \frac{3-a}{4-a}$$

which depends on $a \in [0,1]$. The demand function of a is then determined by:

$$\phi(a,p) = \begin{cases} \left(1 + \dfrac{(1-p)a}{p}, 0\right) & \text{for } 0 \leq p < \dfrac{3-a}{4-a}, \\ \left(0, \dfrac{p}{1-p} + a\right) & \text{for } \dfrac{3-a}{4-a} \leq p \leq 1. \end{cases}$$

Now, we compute the mean excess demand function for, say, x-commodity. Since the discontinuity of a appears at $p = 3 - a/4 - a$, for given p, the consumer who "jumps" at the price p is given by:

$$3(1-p) = p + (1-p)a, \quad a = \frac{3-4p}{1-p}.$$

From this, the mean excess demand function for x can be calculated as:

$$\zeta(p) = \int_0^{\frac{3-4p}{1-p}} \left(1 + \frac{(1-p)a}{p}\right) d\mu - \int_0^1 d\mu = \left(a + \frac{1-p}{2p}a^2\right)_0^{\frac{3-4p}{1-p}} - 1$$

$$= \frac{3-4p}{1-p} + \frac{(3-4p)^2}{2p(1-p)} - 1 = \frac{1}{2p(1-p)}\{2p(3-4p) + (3-4p)^2 - 2p(1-p)\}$$

$$= \frac{1}{2p(1-p)}(10p^2 - 20p + 9),$$

which is continuous for $0 < p < 1$, and the equilibrium price for which $\zeta(p) = 0$ is given by:

$$10p^2 - 20p + 9 = 0, \quad p = \frac{10 - \sqrt{10}}{10}.$$

From this observation, we are naturally led to:

Definition 5.1. The endowment distribution of the economy $\mathcal{E}: A \to \mathcal{P} \times \mathbb{R}^\ell, \mathcal{E}(a) = (\prec_a, \omega(a))$ is said to be dispersed at $\boldsymbol{p} \in \mathbb{R}^\ell_+, \boldsymbol{p} \neq \mathbf{0}$ if and only if the measure ν_p on A defined by:

$$\nu_p(B) = \mu(\{a \in A \mid \boldsymbol{p}\omega(a) \in B\})$$

for every Borel set $B \in \mathcal{B}(\mathbb{R})$ does not give positive value to any particular value $b \in \mathbb{R}$, or:

$$\nu_p(\{b\}) = 0$$

for every $b \in \mathbb{R}$. The endowment distribution is said to be dispersed if it is dispersed at every $\boldsymbol{p} \in \mathbb{R}^\ell_+, \boldsymbol{p} \neq \mathbf{0}$.

Note that in Example 5.3, the endowment distribution is dispersed at every $\boldsymbol{p} = (p, q) \in \mathbb{R}^2_+$, but $\boldsymbol{p} = (p, 0)$.

When the consumption set is not convex, the monotonicity (MT) of preferences is too strong, since it implicitly assumes that $X_a + \mathbb{R}^\ell_+ \subset X_a$. Instead, we assume the weak desirability.

(WD) (Weak desirability) for every $\boldsymbol{x} = (x^t) \in X_a$ and $t = 1 \cdots \ell$, there exists a vector $\boldsymbol{y} \in X_a$ such that $y^t > x^t$ and $y^s \leq x^s$ for $s \neq t$, and $\boldsymbol{x} \prec_a \boldsymbol{y}$.

Let $\mathcal{P}_{wd}$ denote the set of all preference relations, which satisfy the weak desirability, and $\mathcal{P}_{wd,lns} = \mathcal{P}_{wd} \cap \mathcal{P}_{lns}$, where $\mathcal{P}_{lns}$ is the set of all preferences which satisfy the local nonsatiation (LNS). In the remainder of this section, we assume that all consumers in the economy have the same consumption set $X \subset \mathbb{R}^\ell$, which is closed and bounded from below. We do not assume that X to be convex. The regularizing effect for the mean demand by a large number of consumers now reads as follows.

Theorem 5.1. *Let $\mathcal{E}: A \to \mathcal{P} \times \mathbb{R}^\ell$ be an economy such that $\mathcal{E}(a) \subset \mathcal{P}_{wd,Ins} \times X$ a.e. Suppose that the endowment distribution is dispersed. Then, we have:*

(i) *for every $\boldsymbol{p} \in$ interior Δ, there exists a μ-null set $A_p \subset A$ such that for all $a \in A \backslash A_p$, the individual demand correspondence $\phi(a, \boldsymbol{p})$ of interior Δ to $\mathbb{R}^\ell$ is upper hemi-continuous at $\boldsymbol{p}$ and*

(ii) *the mean demand correspondence $\Phi(\boldsymbol{p})$: interior $\Delta \to \mathbb{R}^\ell$, $\Phi(\boldsymbol{p}) = \int_A \phi(a, \boldsymbol{p}) d\mu$ is compact, convex-valued and upper hemi-continuous.*

Proof. Let $\boldsymbol{p} \neq \boldsymbol{0}$ in $\mathbb{R}^\ell_+$ be given. The open half-ball centered at $\boldsymbol{x} \in \mathbb{R}^\ell$ with radius $\delta > 0$ is defined by:

$$HB_{\boldsymbol{p}}(\boldsymbol{x}, \delta) = \{\boldsymbol{z} \in \mathbb{R}^\ell \mid \|\boldsymbol{x} - \boldsymbol{z}\| < \delta, \boldsymbol{pz} < \boldsymbol{px}\}.$$

We define the subset $NC_{\boldsymbol{p}}$ of X and $CW_{\boldsymbol{p}}$ of $\mathbb{R}$ by:

$$NC_{\boldsymbol{p}} = \{\boldsymbol{x} \in X \mid HB_{\boldsymbol{p}}(\boldsymbol{x}, \delta) \cap X = \emptyset \text{ for some } \delta > 0\},$$

and

$$CW_{\boldsymbol{p}} = \{w \in \mathbb{R} \mid w = \boldsymbol{px} \text{ for some } \boldsymbol{x} \in NC_{\boldsymbol{p}}\},$$

respectively. □

We call a point $\boldsymbol{z}$ in $HB_{\boldsymbol{p}}(\boldsymbol{x}, \delta) \cap X$, the local cheaper point of $\boldsymbol{x}$. $NC_{\boldsymbol{p}}$ is the set of consumption vectors which have no local cheaper points. In Example 5.2, the vector $(0, 3)$ has no local cheaper points. Then, the consumer a such that $\boldsymbol{p}\omega(a) \in CW_{\boldsymbol{p}}$ possibly behaves discontinuously. The wealth level w in $CW_{\boldsymbol{p}}$ is called the critical wealth level (see Fig. 3.7). The fundamental lemma is the following.

Lemma 5.1. *The set $CW_{\boldsymbol{p}}$ is a countable set for every $\boldsymbol{p} \in \mathbb{R}^\ell$ with $\boldsymbol{p} \neq \boldsymbol{0}$.*

Proof. Suppose that $CW_{\boldsymbol{p}}$ is an uncountable set $CW_{\boldsymbol{p}} = \{w_\alpha\}$. Then, there are uncountably many points $\{\boldsymbol{x}_\alpha\}$ in $NC_{\boldsymbol{p}}$ such that $\boldsymbol{px}_\alpha = w_\alpha$ for each $w_\alpha \in CW_{\boldsymbol{p}}$. Denote $C = \{\boldsymbol{x}_\alpha\}$ and let $\epsilon > 0$ be a fixed positive number. Then, the space $\mathbb{R}^\ell$ can be covered by countably many closed balls with radius ϵ. Since C is uncountable subset of $\mathbb{R}^\ell$, at least one of these balls, say B_0, must contain uncountably many points in C. Set $C_0 = C \cap B_0$.

We now claim that there exists a point $\boldsymbol{x}$ in C_0 and a sequence $\{\boldsymbol{x}_n\}$ converging to $\boldsymbol{x}$ such that $\boldsymbol{px}_n < \boldsymbol{px}$ for all $n \in \mathbb{N}$. Suppose not. Then, for every $\boldsymbol{x} \in C_0 \subset NC_{\boldsymbol{p}}$, there exists a positive number $\delta_{\boldsymbol{x}}$ such that $HB_{\boldsymbol{p}}(\boldsymbol{x}, \delta_{\boldsymbol{x}}) \cap C_0 = \emptyset$. For each positive integer $n \geq 1$, we define a subset C_n of C_0 by:

$$C_n = \{\boldsymbol{x} \in C_0 \mid HB_{\boldsymbol{p}}(\boldsymbol{x}, 1/n) \cap C_0 = \emptyset\}.$$

Then, $C_0 = \cup_{n=1}^\infty C_n$. Since C_0 is uncountable, at least one of the C_ns, say C_N contains uncountable many points of C_0. Therefore, one can take a sequence $\{\boldsymbol{z}_n\}$ in C_N such

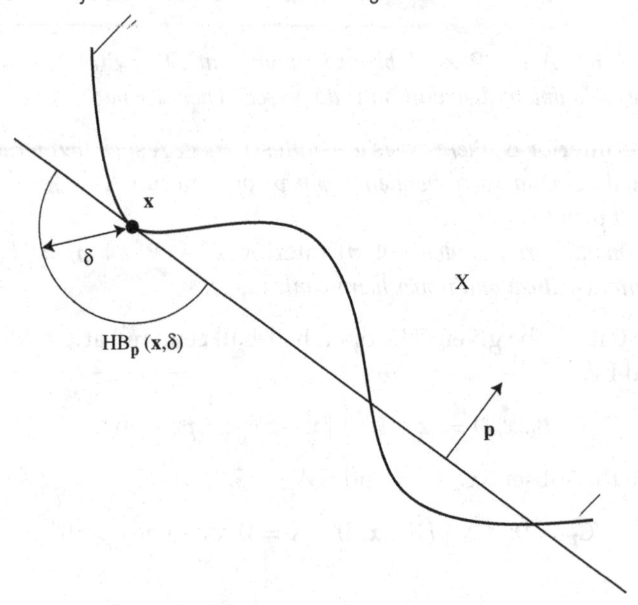

Figure 3.7. $x \in NC_p$.

that $pz_n \neq pz_m$ for $n \neq m$. Since the set C_N is bounded, we can assume that for sufficiently large j and k with $j \neq k$, one has $\|z_j - z_k\| < 1/2N$ and $pz_j \neq pz_k$, say $pz_j < pz_k$. Hence, one obtains $z_j \in HB_p(z_k, 1/N)$. This contradicts the fact that both z_j and z_k are distinct elements of the set C_N. Therefore, our claim is verified.

Take a point $x \in C_0$ and a sequence $\{x_n\}$ in C_0, which converges to x and $px_n < px$ for each n. Since $x \in NC_p$, there is a positive number δ such that $HB_p(x, \delta) \cap X = \emptyset$. But for n sufficiently large, $x_n \in HB_p(x, \delta)$, is a contradiction. This proves Lemma 5.1.

We now continue the proof of Theorem 5.1. For given $p \neq \mathbf{0}$ in $\mathbb{R}^\ell_+$, we define a set $A_p \subset A$ by:

$$A_p = \{a \in A \mid p\omega(a) \in CW_p\}.$$

Then by Lemma 5.1, CW_p is a countable set $\{w_0, w_1 \ldots\}$. Hence, $\mu(A_p) = \sum_{n=0}^{\infty} \mu(\{a \in A \mid p\omega(a) = w_n\}) = 0$, since the endowment distribution is dispersed. We will show that the relation $\{(q, x) \in \mathbb{R}^\ell \times \mathbb{R}^\ell \mid x \in \phi(a, q)\}$ is closed at p for each $a \in A \backslash A_p$. Let $(p_n . x_n)$ be a sequence such that $p_n \to p$, $x_n \to x$, and $x_n \in \phi(a, p_n)$ for all $n \in \mathbb{N}$. For each $n \in \mathbb{N}$, $p_n x_n \leq p_n \omega(a)$. Passing to the limit, we have $px \leq p\omega(a)$. For $z \in X$ such that $pz < p\omega(a)$, one has $p_n z < p_n \omega(a)$ for all n large enough. Thus, $x_n \succsim_a z$ for all n large enough. By the continuity (CT) of $\succsim_a$, it follows that $x \succsim_a z$. For $z \in X$ with $pz = p\omega(a)$, there exists a sequence $\{z_n\}$ such that $z_n \to z$ and $pz_n < p\omega(a)$, since $a \notin A_p$. For every $n \in \mathbb{N}$, we have $x \succsim_a z_n$. Hence, again by the continuity of $\succsim_a$, one sees that $x \succsim_a z$. Therefore, the relation $\phi(a, q)$ is closed at p.

Let $\boldsymbol{p} \in$ interior Δ be given. Since $\omega(a) \in X$, the budget set $\beta(a,\boldsymbol{p}) = \{\boldsymbol{x} \in X \mid \boldsymbol{px} \leq \boldsymbol{p}\omega(a)\}$ is nonempty and compact. By Proposition 3.1 in Chap. 2, we have $\phi(a,\boldsymbol{p}) \neq \emptyset$. Let V be a compact neighborhood of $\boldsymbol{p} \gg \boldsymbol{0}$ with strictly positive vectors and we define $\pi > 0$ by $\pi = \min\{q^t \mid \boldsymbol{q} = (q^t) \in V,\ t = 1 \cdots \ell\}$. Then, $\phi^t(a,\boldsymbol{p}) \leq (1/\pi)\sum_{t=1}^{\ell} |\omega^t(a)|$ for all $\boldsymbol{q} \in V$, $t = 1 \cdots \ell$. Since the consumption set X is bounded from below, there exists a compact set $K \subset \mathbb{R}^\ell$ such that $\phi(a,\boldsymbol{q}) \in K$ for all $\boldsymbol{q} \in V$. Since $\phi(a,\cdot)$ is closed at $\boldsymbol{p} \in V$, it is upper hemi-continuous. This proves the part (i) of Theorem 5.1.

By Proposition 3.2, the correspondence $\phi(a,\boldsymbol{p})$ has a measurable graph. Recall that there exists a lower bound of X, say $b \in \mathbb{R}$, or $(b \cdots b) \leq \boldsymbol{x}$ for all $\boldsymbol{x} \in X$. We define a function $h: A \to \mathbb{R}^\ell_+$ by $h(a) = \max\{|b|, \sum_{t=1}^{\ell} |\omega^t(a)|\}$. Then, the function $h(a)$ is obviously integrable and $|\phi(a,\boldsymbol{p})| \leq h(a)$ a.e. in A. Therefore, by Theorems G9, G10, and Corollaries G1 and G2 of Appendix G, the mean demand correspondence $\int_A \phi(a,\boldsymbol{p})d\mu$ is nonempty, compact, and convex-valued and upper hemi-continuous. □

Now we can present the existence of equilibrium without the convexity assumptions.

Theorem 5.2. *Let $\mathcal{E}: A \to \mathcal{P} \times \mathbb{R}^\ell$ be an economy such that $\mathcal{E}(a) \subset \mathcal{P}_{wd,lns} \times X$ a.e. Suppose that the endowment distribution is dispersed. Then, there exists a competitive equilibrium for $\mathcal{E}$.*

Proof. Consider the mean demand excess demand correspondence from interior Δ to $\mathbb{R}^\ell$,

$$\zeta(\boldsymbol{p}) = \int_{A_k} \phi_k(a,\boldsymbol{p})d\mu - \int_A \omega(a)d\mu$$

for every $\boldsymbol{p} \in \Delta$. We will show that the correspondence ζ satisfies the conditions which are needed to apply the Gale–Nikaido lemma (Theorem D4) of Appendix D,

(i) for every $\boldsymbol{p} \in$ interior Δ, and $\boldsymbol{z} \in \zeta(\boldsymbol{p})$, $\boldsymbol{pz} = 0$;
(ii) the correspondence $\zeta(\cdot)$ is compact and convex-valued, bounded from below, and upper hemi-continuous; and
(iii) for every sequence $\{\boldsymbol{p}_n\}$ in interior Δ converging to $\boldsymbol{p} \in$ bdry $\Delta = \{\boldsymbol{p} = (p^t) \in \Delta \mid p^t = 0$ for some $t\}$, it follows that:

$$\inf\left\{\sum_{t=1}^{\ell} z^t \,\middle|\, \boldsymbol{z} = (z^t) \in \zeta(\boldsymbol{p})\right\} > 0 \quad \text{for } n \text{ large enough.}$$

Since preferences are locally nonsatiated, we have $\boldsymbol{px} = \boldsymbol{p}\omega(a)$ for each $\boldsymbol{x} \in \phi(a,\boldsymbol{p})$. Hence, the property (i) follows from Proposition G9 of Appendix G. We have already shown the property (ii) in Theorem 5.1(b). In order to prove property (iii), let $(\boldsymbol{p}_n)$

be a sequence in interior Δ converging to $\boldsymbol{p} = (p^t)$ with $p^s = 0$ for some s. It suffices to show that:

$$\inf \left\{ \sum_{t=1}^{\ell} x^t \,\middle|\, \boldsymbol{x} = (x^t) \in \phi(a, \boldsymbol{p}_n) \right\} \to \infty \quad \text{as } n \to \infty$$

for every $a \in A \backslash A_{\boldsymbol{p}}$, since $\mu(A_{\boldsymbol{p}}) = 0$. Suppose not. Then, there exists a bounded set B in $\mathbb{R}^{\ell}$ such that $\phi(a, \boldsymbol{p}_n) \cap B \neq \emptyset$ for infinitely many ns. Take $\boldsymbol{x}_n \in \phi(a, \boldsymbol{p}_n) \cap B$ for each n. We can take a converging subsequence, still denoted by $\{\boldsymbol{x}_n\}$ with $\boldsymbol{x}_n \to \boldsymbol{x}$. In the proof of Theorem 5.1(b), we showed that the relation $\phi(a, \cdot)$ is closed at $\boldsymbol{p}$, since $a \notin A_{\boldsymbol{p}}$. Hence, $\boldsymbol{x} \in \phi(a, \boldsymbol{p})$. By the weak desirability (WD), there exists $\boldsymbol{y} \in X$ such that $y^s > x^s$, $y^t \leq x^t$ for $t \neq s$, and $\boldsymbol{x} \prec_a \boldsymbol{y}$, since $p^s = 0$, $\boldsymbol{p}\boldsymbol{y} \leq \boldsymbol{p}\omega(a)$, a contradiction. □

3.6. PRODUCTION ECONOMIES WITH A NON-CONVEX CONSUMPTION SET

In this section, we introduce the possibility of production into the economy. For each $a \in A$, let $Y(a) \subset \mathbb{R}^{\ell}$ be a production possibility set for the consumer $a \in A$. The element $\boldsymbol{y}$ in $Y(a)$ is called the production vector.

For a production vector $\boldsymbol{y} \in Y(a)$, the commodity t is an output when $y^t > 0$, and it is an input when $y^t < 0$. Hence, for a given price vector $\boldsymbol{p} \in \mathbb{R}^{\ell}_{+}$, the value $\pi = \boldsymbol{p}\boldsymbol{y}$ is the profit of a production vector $\boldsymbol{y}$ evaluated by the price system $\boldsymbol{p}$. Each agent $a \in A$ will be assumed to choose a production vector $\boldsymbol{y}$, which maximizes the profit $\boldsymbol{p}\boldsymbol{y}$ in its own production set $Y(a)$, given the price system $\boldsymbol{p}$. Therefore, each agent in the economy is assumed to behave competitively both as a consumer and a producer. This seems to be natural in the equilibrium model of a large market.

For simplicity, we will assume that the production sets $Y(a)$ are compact and convex subset of $\mathbb{R}^{\ell}$ and contain the origin, $\boldsymbol{0} \in Y(a)$ for all a, which means that no production activities are possible. Let $\mathcal{Y}$ be the family of all nonempty, compact, and convex subsets of $\mathbb{R}^{\ell}$, which contain $\boldsymbol{0}$. Since $\mathbb{R}^{\ell}$ is separable, the space $(\mathcal{K}, \delta)$ of all non-empty compact subsets of $\mathbb{R}^{\ell}$ with the Hausdor's distance metric δ is separable by Proposition F1 of Appendix F. Hence, so is the set $(\mathcal{Y}, \delta)$ by Proposition C2 of Appendix C. We assume that the correspondence

$$Y\colon (A, \mathcal{A}, \mu) \to (\mathcal{Y}, \mathcal{B}(\mathcal{Y})), \quad a \mapsto Y(a)$$

is (Borel)-measurable. A pair of measurable mappings $(\mathcal{E}, Y)$, where $\mathcal{E}\colon A \to \mathcal{P} \times \mathbb{R}^{\ell}$ is an exchange economy, is called a coalition production economy, or simply an economy. The term "coalition" comes from that in the economy $(\mathcal{E}, Y)$, each coalition has a possibility to access the production set $\int_C Y(a)d\mu$, that is, the production correspondence $Y: A \to \mathcal{Y}$ on the set of agents A induces the production correspondence on the set of coalitions $\mathcal{A}$,

$$\mathbf{Y}: \mathcal{A} \to \mathcal{Y}, \ C \mapsto \mathbf{Y}(C) = \int_C Y(a)d\mu.$$

Recall that a measurable map $f: A \to \mathbb{R}^\ell$ such that $f(a) \in X_a$ a.e. is called an allocation. An allocation of the production economy $(\mathcal{E}, Y)$ is now defined by a pair (f, g) of maps, where f is an allocation of the exchange economy $\mathcal{E}$, and $g: A \to \mathbb{R}^\ell$ is a measurable map with $g(a) \in Y(a)$ a.e. in A, which is called a production plan. An allocation (f, g) of the production economy $(\mathcal{E}, Y)$ is said to be feasible, if and only if:

$$\int_A f(a)d\mu \leq \int_A \omega(a)d\mu + \int_A g(a)d\mu.$$

Then, we are naturally led to the definition of the competitive equilibrium given below.

Definition 6.1. Let $(\mathcal{E}, Y)$ be a production economy. A triple consisting of a price vector $\boldsymbol{p} \in \mathbb{R}^\ell_+$ and a feasible allocation $f: A \to \mathbb{R}^\ell$, and a production plan $g: A \to \mathbb{R}^\ell$, $(\boldsymbol{p}, f(\cdot), g(\cdot))$ is called a competitive equilibrium of the economy $(\mathcal{E}, Y)$ if the following conditions hold good.

(E-1) $\boldsymbol{p}f(a) \leq \boldsymbol{p}\omega(a) + \boldsymbol{p}g(a)$ and $f(a) \succsim_a \boldsymbol{x}$, whenever $\boldsymbol{px} \leq \boldsymbol{p}\omega(a) + \boldsymbol{p}g(a)$ a.e.;
(E-2) $\boldsymbol{p}g(a) = \max\{\boldsymbol{py} \mid \boldsymbol{y} \in Y(a)\}$ a.e.; and
(E-3) $\int_A f(a)d\mu \leq \int_A \omega(a)d\mu + \int_A g(a)d\mu$.

Note that in the condition (E-1), the budget set of the consumer $a \in A$ is now given by:

$$\beta(a, \boldsymbol{p}) = \{\boldsymbol{x} \in X_a \mid \boldsymbol{px} \leq \boldsymbol{p}\omega(a) + \max \boldsymbol{p}Y(a)\}.$$

In other words, the income of $a \in A$ consists of the market value of the endowment and the profit which is obtained by his/her production activity. The condition (E-2) means that the agent a maximizes its profit within its own production possibility set, given the price system $\boldsymbol{p}$. The condition (E-3) is the usual market feasibility condition.

As we saw already, the income level of each consumer in the production economy is determined by the value of the endowment vector and the profit value earned by his/her production activity. Therefore, even if the endowment distribution is dispersed in the sense of Definition 5.1, the distribution of the profit may counteract the dispersed endowments, and the income level of the consumer could concentrate on some particular level, so that the equilibrium may cease to exist. The following example illustrates this situation.

Example 6.1. For each consumer $a \in [0,1]$, the consumption set X and the utility function are the same as in Examples 5.1 and 5.2.

$$X = \{(x,y) \in \mathbb{R}^2 \mid x \geq 1, y \geq 0\} \cup \{(x,y) \in \mathbb{R}^2 \mid x \geq 0, y \geq 3\},$$

$$u(x,y) = x^2 + y^2, \quad (x,y) \in X.$$

The endowment vector $\omega(a)$ is given as in Example 5.2:

$$\omega(a) = (1,a) \quad \text{for } a \in [0,1].$$

Therefore, the endowment distribution is dispersed. Suppose that the consumer a has its production possibility set as:

$$Y(a) = co\{(0,0),(0,1-a)\}, \quad a \in [0,1]$$

where as usual coS denotes the convex hull of the set S. Note that $Y(a)$ produces $1-a$ amount of the commodity y from nothing (free production). Hence, for every price system $(p, 1-p)$, $0 \leq p \leq 1$, the point $(0, 1-a)$ maximizes the profit in $Y(a)$. Therefore, the budget equation of the consumer $a \in [0,1]$ is now given by:

$$px + (1-p)y = p + (1-p)a + (1-p)(1-a) = 1$$

which does not depend on a, and the situation comes back to that of Example 5.1. The mean excess demand function is calculated here, as:

$$\zeta(p) = \begin{cases} \left(\dfrac{1}{p} - 1, -1\right) & \text{for } 0 \leq p < 2/3, \\ \left(-1, \dfrac{1}{p-1} - 1\right) & \text{for } 2/3 \leq p \leq 1, \end{cases}$$

noting that $\omega(a) + g(a) = (1,a) + (0,1-a) = (1,1)$ for every $a \in A$.

This economy has a dispersed endowment distribution, but does not have any equilibria.

Example 6.1 shows that we have to impose some additional conditions on the distribution of production sets in order to obtain the existence of equilibria. The conditions of the next theorem is one of the simplest among the possible ones.

Theorem 6.1. *Let $(\mathcal{E}, Y)$ be a production economy such that the exchange economy $\mathcal{E}: A \to \mathcal{P} \times \mathbb{R}^\ell$ satisfies that $\mathcal{E}(a) \subset \mathcal{P}_{wd,lns} \times X$ a.e., and the endowment distribution is dispersed. We also assume that the production correspondence $Y: A \to \mathbb{R}^\ell$ is simple, or there exists a finite partition of A, $\{A_1 \cdots A_n\}$ with $A_i \in \mathcal{A}$, $A_i \cap A_j = \emptyset$ and production sets $\{Y_1 \cdots Y_n\} \subset \mathcal{Y}$ such that $Y(a) = Y_i$ on A_i, $i = 1 \cdots n$. Then, we have:*

(i) *for every $\boldsymbol{p} \in$ interior Δ, there exists a μ-null set $A_p \subset A$ such that for all $a \in A \backslash A_p$, the individual demand correspondence $\phi(a, \boldsymbol{p})$ of interior Δ to $\mathbb{R}^\ell$ is upper hemi-continuous at $\boldsymbol{p}$ and*

(ii) *the mean demand correspondence* $\Phi(\boldsymbol{p})$: interior $\Delta \to \mathbb{R}^\ell$, $\Phi(\boldsymbol{p}) = \int_A \phi(a,\boldsymbol{p})d\mu$ *is compact, convex-valued, and upper hemi-continuous.*

Proof. Let $\boldsymbol{p} \neq \mathbf{0}$ in $\mathbb{R}^\ell_+$ be given. Let the sets $HB_{\boldsymbol{p}}(\boldsymbol{x},\delta)$, $NC_{\boldsymbol{p}}$, and $CW_{\boldsymbol{p}}$ be defined as in Theorem 5.1:

$$HB_{\boldsymbol{p}}(\boldsymbol{x},\delta) = \{\boldsymbol{z} \in \mathbb{R}^\ell \mid \|\boldsymbol{x} - \boldsymbol{z}\| < \delta,\ \boldsymbol{pz} < \boldsymbol{px}\},$$
$$NC_{\boldsymbol{p}} = \{\boldsymbol{x} \in X \mid HB_{\boldsymbol{p}}(\boldsymbol{x},\delta) \cap X = \emptyset \text{ for some } \delta > 0\},$$
$$CW_{\boldsymbol{p}} = \{w \in \mathbb{R} \mid w = \boldsymbol{px} \text{ for some } \boldsymbol{x} \in NC_{\boldsymbol{p}}\},$$

respectively. We now define:

$$A_{\boldsymbol{p}} = \{a \in A \mid \boldsymbol{p}\omega(a) + \max \boldsymbol{p}Y(a) \in CW_{\boldsymbol{p}}\}.$$

By Lemma 5.1, the set $CW_{\boldsymbol{p}}$ is a countable set, $\{w_1, w_2, \ldots\}$. Since the range of the map $Y: A \to \mathcal{Y}$ is a finite set $\{Y_1 \cdots Y_n\}$, it follows that:

$$\mu(A_{\boldsymbol{p}}) = \mu(\{a \in A \mid \boldsymbol{p}\omega(a) + \max \boldsymbol{p}Y(a) \in CW_{\boldsymbol{p}}\})$$
$$\leq \sum_{j=1}^{\infty}\sum_{k=1}^{n} \mu(\{a \in A \mid \boldsymbol{p}\omega(a) = w_j - \max \boldsymbol{p}Y_k\}) = 0,$$

since the endowment distribution is dispersed. As shown in the proof of Theorem 5.1(i), we can show that the relation $\phi(a,\boldsymbol{p})$ is closed at $\boldsymbol{p} \in \mathbb{R}^\ell_+, \boldsymbol{p} \neq \mathbf{0}$ for each $a \in A \setminus A_{\boldsymbol{p}}$.

Let $\boldsymbol{p} \in$ interior Δ be given. Since $\omega(a) \in X$ a.e. and $Y(a)$ is a compact subset of $\mathbb{R}^\ell$ with $\mathbf{0} \in Y(a)$, the budget set:

$$\beta(a,\boldsymbol{p}) = \{\boldsymbol{x} \in X \mid \boldsymbol{px} \leq \boldsymbol{p}\omega(a) + \max \boldsymbol{p}Y(a)\}$$

is nonempty and compact. Hence, by Proposition 3.1 given in Chap. 2, we have $\phi(a,\boldsymbol{p}) \neq \emptyset$. Let V be a compact neighborhood of $\boldsymbol{p}$ with strictly positive vectors. Let $Y_1 \cdots Y_n$ be the values of the simple correspondence $Y: A \to \mathcal{Y}$ and we define $\hat{w} = \max\{\boldsymbol{p}\omega(a) + \boldsymbol{py} \mid \boldsymbol{p} \in V, \boldsymbol{y} \in \cup_{i=1}^n Y_i\}$. Since each Y_i is compact, we have $\hat{w} < +\infty$. Then, it follows that $\boldsymbol{p}\phi(a,\boldsymbol{p}) \leq \hat{w}$ for all $\boldsymbol{p} \in V$. Since the consumption set X is bounded from below, we can take a compact set $K \subset \mathbb{R}^\ell$ such that $\phi(a,\boldsymbol{p}) \in K$ for all $\boldsymbol{p} \in V$. Since $\phi(a,\cdot)$ is closed at $\boldsymbol{p} \in V$, it is upper hemi-continuous.

Part (ii) can be proved exactly as in the same way as that of Theorem 5.1(ii). □

Once we have obtained the upper hemi-continuity of the mean demand correspondence $\Phi(\boldsymbol{p}) = \int_A \phi(a,\boldsymbol{p})d\mu$ thanks to the regularizing effect of the aggregation, the existence of equilibrium can be immediately deduced.

Theorem 6.2. *Let $(\mathcal{E}, Y): A \to \mathcal{P} \times X \times \mathcal{Y}$ be a production economy such that $\mathcal{E}(a) \subset \mathcal{P}_{wd,lns} \times X$ a.e. Suppose that the endowment distribution is dispersed and the production correspondence Y is simple. Then, there exists a competitive equilibrium for $(\mathcal{E}, Y)$.*

Proof. We define the mean supply correspondence $\Psi(\boldsymbol{p})$ of interior Δ to $\mathbb{R}^\ell$ by:

$$\Psi(\boldsymbol{p}) = \left\{ \boldsymbol{y} \in \int_A Y(a)d\mu \,\middle|\, \boldsymbol{py} = \max \boldsymbol{p} \int_A Y(a)d\mu \right\},$$

and consider the mean- and excess-demand correspondences from interior Δ to $\mathbb{R}^\ell$,

$$\zeta(\boldsymbol{p}) = \Phi(a, \boldsymbol{p})d\mu - \Psi(\boldsymbol{p}) - \int_A \omega(a)d\mu,$$

for every $\boldsymbol{p} \in \Delta$. We will show that the correspondence Z satisfies the conditions, which are needed to apply the Gale–Nikaido lemma (Theorem D4) of Appendix D,

(i) for every $\boldsymbol{p} \in$ interior Δ, and $\boldsymbol{z} \in \zeta(\boldsymbol{p})$, $\boldsymbol{pz} = 0$;
(ii) the correspondence $\zeta(\cdot)$ is compact and convex-valued, bounded from below and upper hemi-continuous; and
(iii) for every sequence $\{\boldsymbol{p}_n\}$ in interior Δ converging to $\boldsymbol{p} \in$ bdry $\Delta = \{\boldsymbol{p} = (p^t) \in \Delta \mid p^t = 0 \text{ for some } t\}$, it follows that:

$$\inf \left\{ \sum_{t=1}^{\ell} z^t \,\middle|\, \boldsymbol{z} = (z^t) \in \zeta(\boldsymbol{p}) \right\} > 0 \quad \text{for } n \text{ large enough.}$$

The property (i) follows from Proposition G9 of Appendix G and the local non-satiation (LNS) of preferences. By Theorem 6.1(ii), the mean-demand correspondence $\Phi(\boldsymbol{p})$ is upper hemi-continuous. Since the correspondence Y is compact-valued and simple, it is integrably bounded. By the Liapunov's theorem (Theorem G10) of Appendix G, $\int_A Y(a)d\mu$ is a convex set. Hence, $\Psi(\boldsymbol{p})$ is compact and convex-valued, and upper hemi-continuous. Therefore, so is $\zeta(\boldsymbol{p})$. Clearly, $\zeta(\boldsymbol{p})$ is bounded from below, since X is bounded from below and $\int_A Y(a)d\mu$ is compact.

In order to prove property (iii), let $\{\boldsymbol{p}_n\}$ be a sequence in interior Δ converging to $\boldsymbol{p} \in$ bdry Δ. Since $\int_A Y(a)d\mu$ is compact, it suffices to show that:

$$\inf \left\{ \sum_{t=1}^{\ell} x^t \,\middle|\, \boldsymbol{x} = (x^t) \in \phi(a, \boldsymbol{p}_n) \right\} \to +\infty \quad \text{as } n \to +\infty$$

for all $a \in A \backslash A_{\boldsymbol{p}}$. However, this can be shown as exactly the same way as in the proof of Theorem 5.2. Then, we can apply Theorem D4 of Appendix D and the proof is complete. □

The condition that the production correspondence Y is simple in Theorems 6.1 and 6.2 is very restrictive. However, we can apply these results to a generic analysis of the space of production economies. In Section 2.8, we examined the regularity of equilibria for exchange economies and observed that the regular economies are open and dense subset of the space of all exchange economies. In the following, we will pursue the similar analysis on a topological space of economies. Example 6.1 showed that the dispersedness endowment distribution is not sufficient for the existence of equilibria. However, in Theorem 6.3, we will see that the set of production economies with equilibria is a dense subset of the space of production economies with dispersed endowment distributions. Unfortunately, the set is not open, as Example 6.2 shows.

In order to present these results, we have to describe the topology endowed with the set of economies. We do so in terms of sequences.

Let $(A_n, \mathcal{A}_n, \mu_n)_{n\in\mathbb{N}}$ be a sequence of probability spaces. We say that a sequence of exchange economies as:

$$\mathcal{E}_n\colon (A_n, \mathcal{A}_n, \mu_n) \to \mathcal{P} \times \mathbb{R}^\ell, \quad a \mapsto (\prec_{a,n}, \omega_n(a))$$

converges to an exchange economy:

$$\mathcal{E}\colon (A, \mathcal{A}, \mu) \to \mathcal{P} \times \mathbb{R}^\ell, \quad a \mapsto (\prec_a, \omega(a)),$$

if and only if:

(i) the sequence $(\mathcal{E}_n)$ converges to $\mathcal{E}$ in distribution, or the sequence of measures $\nu_n \equiv \mu_n \circ \mathcal{E}_n^{-1}$ converges weakly to the measure $\nu \equiv \mu \circ \mathcal{E}^{-1}$ and
(ii) the mean endowment converges, or $\int_{A_n} \omega_n(a) d\mu_n \to \int_A \omega(a) d\mu$.

Similarly, we say that a sequence of production economies:

$$(\mathcal{E}_n, Y_n)\colon (A_n, \mathcal{A}_n, \mu_n) \to \mathcal{P} \times \mathbb{R}^\ell \times \mathcal{Y}, \quad a \mapsto (\prec_{a,n}, \omega_n(a), Y_n(a))$$

converges to a production economy:

$$(\mathcal{E}, Y)\colon (A, \mathcal{A}, \mu) \to \mathcal{P} \times \mathbb{R}^\ell \times \mathcal{Y}, \quad a \mapsto (\prec_a, \omega(a), Y(a)),$$

when the condition (i) is replaced by (i′), the sequence $(\mathcal{E}_n, Y_n)$ converges to $(\mathcal{E}, Y)$ in distribution, or the sequence of measures $\nu_n \equiv \mu_n \circ (\mathcal{E}_n, Y_n)^{-1}$ converges weakly to the measure $\nu \equiv \mu \circ (\mathcal{E}, Y)^{-1}$.

In the following theorem, we will discuss the case in which $(A_n, \mathcal{A}_n, \mu_n) = (A, \mathcal{A}, \mu)$ for all $n \in \mathbb{N}$. We now state the main result of this section.

Theorem 6.3. *Let $(\mathcal{E}, Y)\colon A \to \mathcal{P} \times X \times \mathcal{Y}$ be a production economy such that $\mathcal{E}(a) \subset \mathcal{P}_{wd,lns} \times X$ a.e. Suppose that the endowment distribution is dispersed. Then, there exists a sequence of production correspondences $Y\colon A \to \mathcal{Y}$ such that:*

(i) *for each n, the production economy $(\mathcal{E}, Y_n)$ has an equilibrium and*
(ii) *the sequence $(\mathcal{E}, Y_n)$ converges to $(\mathcal{E}, Y)$ in distribution.*

Note that the statement (ii) implies that the sequence of production economies converges to the production economy $(\mathcal{E}, Y)$, since the exchange economy (consumption sector) $\mathcal{E}$ does not change; hence, the mean endowments of the sequence economies are constant.

Proof of Theorem 6.3. By the theorem of Rådström (Theorem F1) of Appendix F, the space $(\mathcal{K}_c, \delta)$ is embedded as a convex cone with the vertex at the origin to a real-normed linear space L by an isomorphic embedding map ξ_1. Note that $\xi_1(\{\mathbf{0}\}) = \mathbf{0}$. According to the theorem of completion (Theorem H1) of Appendix H, there exists an isometric isomorphism ξ_2 between L and a dense linear subspace of a Banach space B. Since $\mathcal{Y} \subset \mathcal{F}_0$ is a separable metric space, the set $\{Y(a) \in \mathcal{Y} \mid a \in A\}$ is separable by Proposition C2; hence, the set $\{\xi_2 \circ \xi_1 \circ Y(a) \in \mathcal{Y} \mid a \in A\}$ is also a separable subset of B. Since the maps ξ_1 and ξ_2 are both continuous and the correspondence $Y: A \to \mathcal{Y}$ is $\mathcal{A}$-measurable, it follows that for any continuous linear functional on B, the map $f \circ \xi_2 \circ \xi_1 \circ Y(\cdot)$ is $\mathcal{A}$-measurable. According to Petti's theorem (Theorem G14) of Appendix G, there exists a sequence $(\tilde{Y}_n)_{n \in \mathbb{N}}$ of simple functions of A into B with $\|\tilde{Y}_n(a) - \xi_2 \circ \xi_1 \circ Y(a)\| \to 0$ a.e. For each n, we have $\{\tilde{Y}_n(a) \in B \mid a \in A\} \subset \{\xi_2 \circ \xi_1 \circ Y(a) \in B \mid a \in A\} \cup \{\mathbf{0}\}$. Therefore, we have a sequence of simple functions $(Y_n(\cdot))_{n \in \mathbb{N}}$ of A to $\mathcal{Y}$ with $\delta(Y_n(a), Y(a)) \to 0$ as $n \to \infty$, where δ is the Hausdor's distance. Let d be a metric on $\mathcal{P} \times \mathbb{R}^\ell \times \mathcal{Y}$ defined by:

$$d((\prec, \omega, Y), (\prec', \omega', Y')) = \hat{\delta}(\prec, \prec') + \|\omega - \omega'\| + \delta(Y, Y'),$$

where $\hat{\delta}$ is the metric of the topology of closed convergence τ_c. Then, it follows that $d((\mathcal{E}(a), Y_n(a)), (\mathcal{E}(a), Y(a)) = \delta(Y_n(a), Y(a)) \to 0$ a.e. The map $\mathcal{E}$ is $\mathcal{B}(\mathcal{P} \times \mathbb{R}^\ell)$-measurable and the mappings Y, Y_n, are $\mathcal{B}(\mathcal{Y})$-measurable. Hence, the mappings $(\mathcal{E}, Y)$ and $(\mathcal{E}, Y_n)$ are $\mathcal{B}(\mathcal{P} \times \mathbb{R}^\ell) \times \mathcal{B}(\mathcal{Y}) = \mathcal{B}(\mathcal{P} \times \mathbb{R}^\ell \times \mathcal{Y})$-measurable by Propositions G1 and G4 of Appendix G, since the sets $\mathcal{P} \times \mathbb{R}^\ell$ and $\mathcal{Y}$ are both separable metric spaces. Then, by Theorems G5 and G6 of Appendix G, we see that $(\mathcal{E}, Y_n) \to (\mathcal{E}, Y)$ in distribution, and for each n, the economy $(\mathcal{E}_n, Y)$ has an equilibrium according to Theorem 6.2. □

The following example shows that the set of production economies with an equilibrium is not open.

Example 6.2. Consider the exchange economy given in Example 5.2. For each consumer, $a \in [0, 1]$, the consumption set X and the utility function is:

$$X = \{(x, y) \in \mathbb{R}^2 \mid x \geq 1, y \geq 0\} \cup \{(x, y) \in \mathbb{R}^2 \mid x \geq 0, y \geq 3\},$$

$$u(x, y) = x^2 + y^2, \quad (x, y) \in X,$$

and the endowment vector $\omega(a)$ is:

$$\omega(a) = (1, a) \quad \text{for } a \in [0, 1].$$

This exchange economy (with a dispersed endowment distribution) can be seen as a production economy, if we give a production correspondence:

$$Y(a) = \{(0,0)\}, \quad a \in [0,1].$$

We know that the excess-demand function for x-commodity is:

$$\zeta(p) = \frac{1}{2p(1-p)}(10p^2 - 20p + 9),$$

where we normalized the price vector $\boldsymbol{p} = (p, 1-p)$. The equilibrium price of the commodity x is:

$$10p^2 - 20p + 9 = 0, \ \ p = \frac{10 - \sqrt{10}}{10}.$$

We can approximate this "production" economy $(\mathcal{E}, Y)$ by a sequence of production economies without equilibrium $(\mathcal{E}, Y_n)$ in the following way. Note that this shows that $(\mathcal{E}, Y)$ is not an interior point of the space of production economies with an equilibrium. Therefore, it is not an open subset of the economies with dispersed endowment distributions.

Let $(\epsilon_n)_{n\in\mathbb{N}}$ be a sequence of positive numbers decreasing to zero. The demand function of the consumer $a \in [0,1]$ has been already calculated as:

$$\phi(a,p) = \begin{cases} \left(1 + \dfrac{(1-p)a}{p}, 0\right) & \text{for } 0 \le p < \dfrac{3-a}{4-a}, \\ \left(0, \dfrac{p}{1-p} + a\right) & \text{for } \dfrac{3-a}{4-a} \le p \le 1. \end{cases}$$

For given p, we can identify the consumer a who "jumps" at the price $p = (3-a)/(4-a)$, and solving for a, $a = (3-4p)/(1-p)$. Then, we calculate:

$$\begin{aligned} \zeta^x(p) &= \int_{\epsilon_n}^1 (\phi^x(a,\boldsymbol{p}) - \omega^x(a))\, da = \frac{1-p}{p}\int_{\epsilon_n}^{3-4p/1-p} a da - \int_{3-4p/1-p}^1 da \\ &= \frac{1-p}{2p}\left\{\left(\frac{3-4p}{1-p}\right)^2 - \epsilon_n^2\right\} - \left(1 - \frac{3-4p}{1-p}\right) \\ &= \frac{1}{2p(1-p)}\left((10-\epsilon_n^2)p^2 - 2(10-\epsilon_n^2)p + 9 - \epsilon_n^2\right). \end{aligned}$$

Solving $(10-\epsilon_n^2)p^2 - 2(10-\epsilon_n^2)p + 9 - \epsilon_n^2 = 0$, we obtain:

$$p = \frac{10 - \epsilon_n^2 - \sqrt{10 - \epsilon_n^2}}{10 - \epsilon_n^2} \equiv p_n.$$

Note that $p_n \to p = \frac{10-\sqrt{10}}{10}$ as $n \to \infty$, since $\epsilon_n \to 0$.

For each n, we define the production correspondence $Y_n: A \to \mathcal{Y}$ by

$$Y(a) = \begin{cases} co\left\{(0,0), \left(0, \dfrac{3-4p_n}{1-p_n} - a\right)\right\} & \text{for } a \in [0, \epsilon_n], \\ \{(0,0)\} & \text{for } a \in (\epsilon_n, 1]. \end{cases}$$

Since $\epsilon_n \to 0$, it follows that $\delta(Y_n, Y) \to 0$ a.e. Hence, as in the proof of Theorem 6.3, we have from Theorems G5 and G6 of Appendix G that $(\mathcal{E}, Y_n) \to (\mathcal{E}, Y)$ in distribution. Furthermore, by the construction of Y_n, the excess demand for the x-commodity of the consumer $a \in [0, \epsilon_n)$ is:

$$\phi^x(a,p) - \omega^x(a) = \begin{cases} \left(\dfrac{1-p}{p}\right)\left(\dfrac{3-4p_n}{1-p_n}\right) & \text{for } 0 \le p < p_n, \\ -1 & \text{for } p_n \le p \le 1. \end{cases}$$

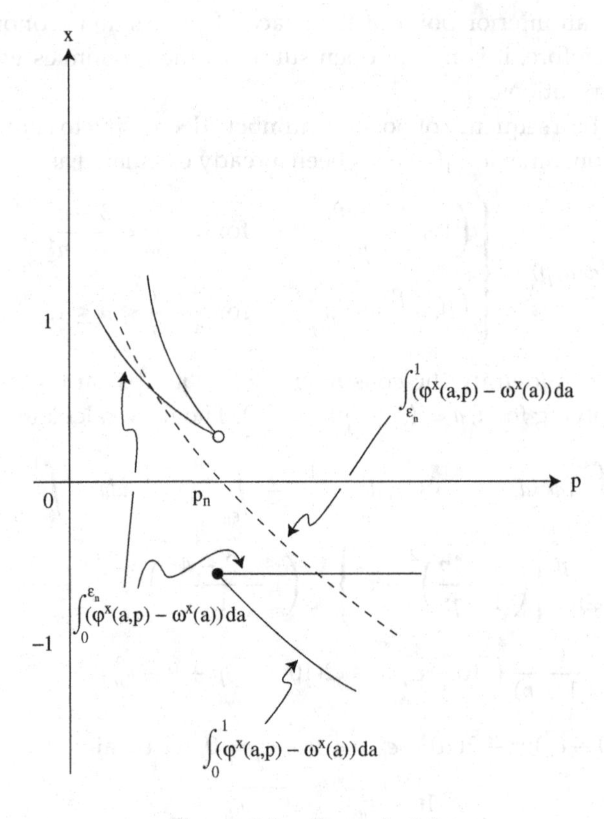

Figure 3.8. Example 6.2.

Therefore, the mean-demand correspondence for x is:

$$\begin{aligned}\zeta^x(p) &= \int_0^1 \phi^x(a,\boldsymbol{p}) - \omega^x(a)da \\ &= \int_0^{\epsilon_n} \phi^x(a,\boldsymbol{p}) - \omega^x(a)da + \int_{\epsilon_n}^1 \phi^x(a,\boldsymbol{p}) - \omega^x(a)da\end{aligned}$$

is discontinuous at $p = p_n$, hence the economy $(\mathcal{E}, Y_n)$ does not have an equilibrium (see Fig. 3.8).

3.7. ON THE LAW OF DEMAND

In this section, we present a remarkable observation on the law of demand proposed by Hildenbrand. Throughout this section, we assume that the consumer's income $w \in \mathbb{R}_+$ is independent of the market price vector $\boldsymbol{p} \in \mathbb{R}_+^\ell$. Hence, the budget set and the demand set of a consumer are given by:

$$\begin{aligned}\beta(\boldsymbol{p}, w) &= \{\boldsymbol{x} \in X \mid \boldsymbol{px} \le w\} \quad \text{and} \\ \phi(\boldsymbol{p}, w) &= \{\boldsymbol{x} \in X \mid \boldsymbol{px} \le w, \boldsymbol{x} \succsim \boldsymbol{z} \ \text{ whenever } \boldsymbol{pz} \le w\},\end{aligned}$$

respectively. Suppose that the distribution of the income w is given by the density function $\rho(w)$ over the unit interval $[0, 1]$. The market demand function $\Phi(\boldsymbol{p}) = (\Phi^t(\boldsymbol{p}))$ is then defined by:

$$\Phi(\boldsymbol{p}) = \int_0^1 \phi(\boldsymbol{p}, w)\rho(w)dw.$$

The law of demand states that the market demand function for each commodity t is decreasing in the price of the commodity t when the prices of the other commodities $s \neq t$ remain constant:

$$\frac{\partial \Phi^t(\boldsymbol{p})}{\partial p^t} \le 0.$$

Hildenbrand (1983) found that when the density function $\rho(w)$ is decreasing, then the law of demand holds good. This fact is not obvious when one sees the behavior of the individual demand function $\phi(\boldsymbol{p}, w)$, which is described by the Slutzky equation

(Theorem 4.2 in Chap. 2),

$$\frac{\partial}{\partial p^t}\phi^t(\boldsymbol{p},w) = \frac{\partial}{\partial p^t}\phi^t_{\boldsymbol{x}}(\boldsymbol{p}) - \phi(\boldsymbol{p},w)\frac{\partial}{\partial w}\phi^t(\boldsymbol{p},w).$$

By Theorem 4.1(ii) in Chap. 2, we know that the "substitution effect" is negative,

$$\frac{\partial}{\partial p^t}\phi^t_{\boldsymbol{x}}(\boldsymbol{p}) \le 0.$$

However, the sign of the "income effect" is undetermined:

$$\phi(\boldsymbol{p},w)\frac{\partial}{\partial w}\phi^t(\boldsymbol{p},w) \gtreqless 0.$$

Remarkably, however, we can determine the sign of the total income effect and it is negative, since:

$$\begin{aligned} -\int_0^1 \phi(\boldsymbol{p},w)\frac{\partial}{\partial w}\phi^t(\boldsymbol{p},w)\rho(w)dw &= -(1/2)\int_0^1 \frac{\partial}{\partial w}\left(\phi^t(a,\boldsymbol{p})\right)^2 \rho(w)dw \\ &= (1/2)\int_0^1 \left(\phi^t(\boldsymbol{p},w)\right)^2 \rho'(w)dw \le 0 \end{aligned}$$

under the assumptions that $\rho'(w) \le 0$, $\rho(1) = 0$ and $\phi^t(\boldsymbol{p},0) = 0$.

This simple observation can be generalized as follows. We say that the market demand function $\Phi(\boldsymbol{p})$ is monotone if for any price vectors $\boldsymbol{p}$ and $\boldsymbol{q}$,

$$(\boldsymbol{q}-\boldsymbol{p})(\Phi(\boldsymbol{q}) - \Phi(\boldsymbol{p})) \le 0.$$

It is said to be strictly monotone if for any price vectors $\boldsymbol{p}$ and $\boldsymbol{q}$ with $\boldsymbol{q} \ne \boldsymbol{p}$,

$$(\boldsymbol{q}-\boldsymbol{p})(\Phi(\boldsymbol{q}) - \Phi(\boldsymbol{p})) < 0.$$

The monotonicity of Φ clearly implies that the law of demand holds for all markets, $t = 1 \cdots \ell$.

When the individual demand function $\phi(a,\cdot)$ is continuously differentiable and the density function $\rho(w)$ of the wealth level w is continuous, then the mean-demand function $\Phi(\boldsymbol{p}) = \int_0^1 \phi(\boldsymbol{p},w)\rho(w)dw$ is continuously differentiable and it is easily verified by the Taylor's expansion that the mean demand $\Phi(\boldsymbol{p})$ is strictly monotone, if its Jacobian matrix is negative definite, or:

$$\boldsymbol{v}'\partial\Phi(\boldsymbol{p})\boldsymbol{v} = \sum_{t=1}^{\ell}\sum_{s=1}^{\ell} v^t v^s \partial_s \Phi^t(\boldsymbol{p}) < 0$$

for every $\boldsymbol{v} = (v^t) \ne \boldsymbol{0}$, where $\boldsymbol{v}'$ is the transpose of $\boldsymbol{v}$. The next theorem is proposed by Hildenbrand (1983).

Theorem 7.1. *Suppose that the individual demand function $\phi(\boldsymbol{p}, w)$ is continuously differentiable in $\boldsymbol{p}$ and the substitution matrix $S_{\boldsymbol{x}}(\boldsymbol{p}) = (\partial_s \phi^t_{\boldsymbol{x}}(\boldsymbol{p}))$ is of the rank $\ell - 1$. Then, if the density $\rho(w)$ is continuous and decreasing, then the function $\Phi(\boldsymbol{p})$ is negative definite.*

Proof. Let $S_{\boldsymbol{x}}(\boldsymbol{p}) = \left(\partial_s \phi^t_{\boldsymbol{x}}(\boldsymbol{p})\right)$ be the substitution matrix (see Section 2.4), and let $A(\boldsymbol{p}, w)$ be the matrix defined by $A(\boldsymbol{p}, w) = (a^s_t) = (\phi^s(\boldsymbol{p}, w)\partial_w \phi^t(\boldsymbol{p}, w))$, $s, t = \cdots \ell$. Then, the Slutzky equation (Theorem 4.2) can be written as:

$$\partial\phi(\boldsymbol{p}, w) = S_{\boldsymbol{x}}(\boldsymbol{p}) - A(\boldsymbol{p}, w).$$

Integrating over w, we have:

$$\partial\Phi(\boldsymbol{p}) = \int_0^1 \left(S_{\boldsymbol{x}}(\boldsymbol{p}) - A(\boldsymbol{p}, w)\right) \rho(w) dw.$$

By Theorem 4.1(ii), the substitution matrix is negative semi-definite. Since we assumed that rank $S_{\boldsymbol{x}}(\boldsymbol{p}) = \ell - 1$ and $\boldsymbol{p}' S_{\boldsymbol{x}}(\boldsymbol{p}) = \boldsymbol{0}$ by Proposition 4.4 in Chap. 2, it follows that:

$$\boldsymbol{v}' S_{\boldsymbol{x}}(\boldsymbol{p}) \boldsymbol{v} < 0$$

for every $\boldsymbol{v} \in \mathbb{R}^\ell$, which is not of the form $\boldsymbol{v} = \lambda \boldsymbol{p}$, $\lambda \neq 0$. Hence, the matrix $\int_0^1 S_{\boldsymbol{x}}(\boldsymbol{p})\rho(w)dw$ is also negative definite for all $\boldsymbol{v}$, which is not collinear with $\boldsymbol{p}$, or for all $\boldsymbol{v} \in \mathbb{R}^\ell$ which is not of the form $\boldsymbol{v} = \lambda \boldsymbol{p}$, $\lambda \neq 0$. Therefore, it remains to show that:

$$\boldsymbol{v}' \int_0^b A(\boldsymbol{p}, w)\rho(w)dw\, \boldsymbol{v} \geq 0$$

for every $\boldsymbol{v} \in \mathbb{R}^\ell$ with strict inequality for $\boldsymbol{p} = \lambda \boldsymbol{v}$, $\lambda \neq 0$. Now, we have:

$$\begin{aligned}
\boldsymbol{v}' \int_0^b A(\boldsymbol{p}, w)\rho(w)dw\boldsymbol{v} &= \sum_{s=1}^{\ell}\sum_{t=1}^{\ell} v^s v^t \int_0^b \phi^s(\boldsymbol{p}, w)\partial_w \phi^t(\boldsymbol{p}, w)\rho(w)dw \\
&= \int_0^b \left(\sum_{s=1}^{\ell} v^s \phi^s(\boldsymbol{p}, w)\right)\left(\sum_{t=1}^{\ell} v^t \partial_w \phi^t(\boldsymbol{p}, w)\right)\rho(w)dw \\
&= \int_0^b (\partial_w \phi(\boldsymbol{p}, w)\boldsymbol{v})(\phi(\boldsymbol{p}, w)\boldsymbol{v})\rho(w)dw \\
&= (1/2)\int_0^b \partial_w (\phi(\boldsymbol{p}, w)\boldsymbol{v})^2 \rho(w)dw.
\end{aligned}$$

Since ρ is decreasing, it follows from the intermediate value theorem of the elementary calculus that there exists $0 \leq \xi \leq b$ such that:

$$\int_0^b \partial_w(\phi(\boldsymbol{p},w)\boldsymbol{v})^2\rho(w)dw = \rho(0)\int_0^\xi \partial_w\left(\phi(\boldsymbol{p},w)\boldsymbol{v}\right)^2$$

$$= \rho(0)(\phi(\boldsymbol{p},\xi)\boldsymbol{v})^2 \geq 0.$$

If $\boldsymbol{v} = \lambda\boldsymbol{p}$, then $\phi(\boldsymbol{p},w)\boldsymbol{v} = \lambda w$. Hence, we have:

$$\boldsymbol{v}'\int_0^b A(\boldsymbol{p},w)\rho(w)dw\boldsymbol{v} = (1/2)\int_0^b \partial_w(\lambda w)^2\rho(w)dw > 0.$$

Therefore, $\Phi(\boldsymbol{p})$ is negative definite. □

3.8. NOTES

Section 3.2: A basic reference is Hildenbrand (1974), Chap. 1. Hildenbrand and Kirman (1986) is a readable introductory textbook.

Section 3.3: The idea of the proof of the equilibrium existence theorem which truncates the consumption set and invokes the Fatou's lemma in ℓ-dimension is originally proposed by Aumann (1966) and Shcmeidler (1969). We owe Yamazaki (1978) for the proof of Theorem 3.1. The proof of Theorem 3.3 is put forth by Hildenbrand (1974). For the case that the consumption sets can be different from each consumer, see Yamazaki (1981).

Section 3.4: Theorem 4.1 is due to Aumann (1964), and we follow this paper entirely. See also Hildenbrand (1974), Chap. 3 for the formulation of the limit theorem of the core in terms of the sequences of economies.

Section 3.5: Theorems 5.1, 5.2, and Lemma 5.1 are put forth by Yamazaki (1978). For the proof, we follow Suzuki (1995).

Section 3.6: Theorem 6.3 is put forth by Suzuki (1995).

Let $(\Omega, \mathcal{M})$ be a measurable space, and let μ be a (finite) measure on Ω. Let $(\Omega, \mathcal{M}_\mu, \bar{\mu})$ be the completion of $(\Omega, \mathcal{M}, \mu)$ (see Appendix G).

We define $\mathcal{M}_u = \cap\{\mathcal{M}_\mu | \mu$ is a finite measure on $(\Omega, \mathcal{M})\}$. The measurable space $(\Omega, \mathcal{M}_u)$ is called the universal completion of $(\Omega, \mathcal{M})$, and each set in $(\Omega, \mathcal{M}_u)$ is called universally measurable. For a topological space X, let $(X, \mathcal{B}_u(X))$ be the universal completion of the measurable space $(X, \mathcal{B}(X))$. Following Yamazaki (1986), we can prove the following theorem.

Theorem. The set $\mathcal{P}_{wd}$ is universally measurable.

In order to prove this theorem, we need some preparations.

Let Ω be a set. The concept of the sequences of subsets of Ω is generalized as follows. Let $\mathcal{N}_f$ be the set of all finite sequences of positive integers, $\mathcal{N}_f = \{(n_0 \cdots n_k) \mid$

$n_j \in \mathbb{N}, 0 \leq j \leq k, k \in \mathbb{N}\}$. Consider a map from $\mathcal{N}_f$ to 2^{Ω}, which maps a finite sequence to a subset of Ω,

$$(n_0 \cdots n_k) \mapsto B_{(n_0 \ldots n_k)} \subset \Omega.$$

For the image $\{B_{(n_0 \cdots n_k)}\}$ of this map, we can associate the set $B = \cup_{\boldsymbol{n}} \cap_{k \geq 0} B_{(n_0 \cdots n_k)}$, where the union is taken over all infinite sequences $\boldsymbol{n} = (n_0 \cdots n_k, n_{k+1} \cdots)$ of non-negative integers. We call this as the Suslin operation. Note that the countable unions and intersections are examples of the Suslin operation. Let $\mathcal{S}$ be a family of sets. We denote by $\mathcal{S}_s$ the family of all sets obtained from the sets of $\mathcal{S}$ by the Suslin operation. We call every element S of $\mathcal{S}_s$ a Suslin set. Then, we have:

Math. Theorem 5.1. Let $(\Omega, \mathcal{M})$ be a measurable space. Then, $\mathcal{M}_s \subset \mathcal{M}_u$.

Let X be a topological space. A subset A of X is called analytic set if there exists a complete separable metric space C and a continuous map $f: C \to X$ such that $A = f(C)$. Let $\mathcal{A}(X)$ be the set of all analytic sets of X. Then it follows that:

Math. Theorem 5.2. If X is a complete and separable metric space, then $\mathcal{B}(X) \subset \mathcal{A}(X)$, or a Borel set is an analytic set.

Math. Theorem 5.3. Let $\mathcal{F}(X)$ be the family of all closed subsets of a complete separable metric space X. Then, $\mathcal{A}(X) = \mathcal{F}(X)_s$.

We know that the inverse image of a Borel set by a measurable map is a Borel set. However, it may not be the case that the image of a Borel set by a measurable map is also a Borel set. For the analytic sets, however, the following proposition holds.

Math. Theorem 5.4. Let X, Y be complete and separable metric spaces, and $f: (X, \mathcal{B}(X)) \to (Y, \mathcal{B}(Y))$ be a Borel-measurable map. Then:

(i) for every $B \in \mathcal{A}(Y), f^{-1}(B) \in \mathcal{A}(X)$ and
(ii) for every $C \in \mathcal{A}(X), f(C) \in \mathcal{A}(Y)$.

Finally, we have:

Math. Theorem 5.5. Let $(\Omega, \mathcal{M})$ be a measurable space, X a complete and separable metric space. Let $pr_{\Omega}: \Omega \times X \to \Omega$ be a projection map. Then, for every $C \in (\Omega \times X)_s$, it follows that $pr_{\Omega}(C) \in \mathcal{M}_s$.

Then, we can prove the theorem as follows. For given $t\ (= 1 \cdots \ell)$ and $\boldsymbol{x} = (x^t) \in \mathbb{R}^{\ell}$, define the subsets of $\mathbb{R}^{\ell}$ by:

$$QN^t(\boldsymbol{x}) = \{\mathbf{z} \in \mathbb{R}^{\ell} \mid z^t > x^t, z^s \leq x^s \text{ for } s \neq t\}.$$

Then, setting for each $t = 1 \cdots \ell$, $\mathcal{Q}^t = \{(X, \prec) \in \mathcal{P} \mid$. There exists $\boldsymbol{x} \in X$ such that for each $\boldsymbol{z} \in QN^t(\boldsymbol{x}) \cap X, \boldsymbol{x} \succsim \mathbf{z}\}$, we have $\mathcal{P}_{wd} = \mathcal{P} \setminus \cup_{t=1}^{\ell} \mathcal{Q}^t$. Therefore, it suffices to show that each $\mathcal{Q}^t$ is universally measurable. Fix a t and for each non-negative integer $i \in \mathbb{N}$, define:

$$QN_i^t(\boldsymbol{x}) = \left\{ \mathbf{z} \in \mathbb{R}^\ell \middle| z^t \geq x^t + \frac{1}{i+1}, z^s \leq x^s \text{ for } s \neq t \right\},$$

$$\mathcal{G}_i = \{(X, \prec, \boldsymbol{x}) \in \mathcal{P} \times \mathbb{R}^\ell \mid QN_i^t(\boldsymbol{x}) \cap X \neq \emptyset\}.$$

They induce the correspondence $\xi_i \colon \mathcal{G}_i \to \mathbb{R}^\ell$ defined by $\xi_i(X, \prec, \boldsymbol{x}) = QN_i^t \cap X$. Since the correspondences $(X, \prec, \boldsymbol{x}) \to X$ and $(X, \prec, \boldsymbol{x}) \to QN_i^t(\boldsymbol{x})$ are closed, so is the correspondence ξ_i, hence it is measurable. By Proposition G7 of Appendix G, there exists a countable family of measurable mappings $\{f_{ij}\}_{j \in \mathbb{N}}$ of $\mathcal{G}_i \to \mathbb{R}^\ell$ such that:

$$\xi_i(X, \prec, \boldsymbol{x}) = cl\{f_{ij}(X, \prec, \boldsymbol{x}) \mid j \in \mathbb{N}\}.$$

Then, defining for $i, j \in \mathbb{N}$,

$$\mathcal{P}_{ij} = \{(X, \prec) \in \mathcal{P} \mid \boldsymbol{x} \succsim f_{ij}(X, \prec, \boldsymbol{x}) \text{ for some } \boldsymbol{x} \in X\},$$

we have $\mathcal{Q}^t = \cap_{i=0}^{\infty} \cap_{j=0}^{\infty} \mathcal{P}_{ij}$. Indeed, it is obvious that $\mathcal{Q}^t \subset \cap_{i=0}^{\infty} \cap_{j=0}^{\infty} \mathcal{P}_{ij}$. Suppose that there exists $(X, \prec)$ such that $(X, \prec) \in \cap_{i=0}^{\infty} \cap_{j=0}^{\infty} \mathcal{P}_{ij}$ and $(X, \prec) \notin \mathcal{Q}^t$. Then, for every $\boldsymbol{x} \in X$, there exists $\mathbf{z} \in QN_i^t(\boldsymbol{x}) \cap X$ such that $\boldsymbol{x} \prec \mathbf{z}$. Then we can take some $\epsilon > 0$ such that $\mathbf{z} \in QN_i^t(\boldsymbol{x}) \cap X$ and $\boldsymbol{x} \prec B(\mathbf{z}, \epsilon) \cap X$. Since, we have:

$$QN_i^t(\boldsymbol{x}) \cap X = cl\{f_{ij}(X, \prec, \boldsymbol{x}) \mid j \in \mathbb{N}\},$$

it follows that $\boldsymbol{x} \prec f_{ij}(X, \prec, \boldsymbol{x})$ for some j, contradicting that $(X, \prec) \in \mathcal{P}_{ij}$ for all i, j. Therefore, the proof is complete if we show that $\mathcal{P}_{ij}$ is universally measurable. Define a relation $F_{ij} \colon \mathcal{P} \to \mathbb{R}^\ell$ by:

$$F_{ij} = \{(X, \prec, \boldsymbol{x}) \in \mathcal{P} \times \mathbb{R}^\ell \mid \boldsymbol{x} \succsim f_{ij}(X, \prec, \boldsymbol{x})\},$$

and define a measurable mapping $g_{ij} \colon \mathcal{G}_i \to \mathcal{P} \times \mathbb{R}^\ell \times \mathbb{R}^\ell$ by:

$$g_{ij}(X, \prec, \boldsymbol{x}) = (X, \prec, f_{ij}(X, \prec, \boldsymbol{x}), \boldsymbol{x}).$$

Then, obviously, $F_{ij} = g_{ij}^{-1}(\{(X, \prec, \boldsymbol{y}, \boldsymbol{x}) \mid \boldsymbol{x} \succsim \boldsymbol{y}\})$. Since the set $\{(X, \prec, \boldsymbol{y}, \boldsymbol{x}) \mid \boldsymbol{x} \succsim \boldsymbol{y}\}$ is closed, F_{ij} is a Borel set in $\mathcal{P} \times \mathbb{R}^\ell$. Since $\mathcal{P}_{ij} = \{(X, \prec) \in \mathcal{P} \mid F_{ij} \neq \emptyset\}$, it follows from Math. Theorems 5.1, 5.2, 5.3, and 5.5, the set $\mathcal{P}_{ij}$ is a $\mathcal{B}(\mathcal{P})$-Suslin set, hence it is a universally measurable set.

Section 3.7: For Theorem 7.1, we follow entirely Hildenbrand (1983).

PRODUCTION ECONOMIES WITH INCREASING RETURNS

Chapter

4

4.1. CLASSICAL PRODUCTION ECONOMIES WITH COMPETITIVE FIRMS

In Section 3.6 of Chap. 3, we considered the model in which consumers can produce commodities according to their own production technologies. From now on, we will introduce into the markets the firms that make their production decisions independently on the consumers.

The number of firms will be assumed to be finite, say n, and each firm is indexed by $b(=1\cdots n)$. In the following, we sometimes write $B = \{1\cdots n\}$. The production technology of the firm b is represented by a closed set $Y_b \subset \mathbb{R}^\ell$. As usual, for $\boldsymbol{y} = (y^t) \in Y_b$, the commodity t is input commodity when $y^t < 0$, and it is output when $y^t > 0$. Throughout this book, we always assumed that:

(NFP) (No free production) $Y_b \cap \mathbb{R}^\ell_+ = \{\mathbf{0}\}$ and
(FD) (Free disposability) for every $\boldsymbol{y} \in Y_b, \boldsymbol{y} + \mathbb{R}^\ell_- \subset Y_b$.

Note that the no free production condition (NFP) implies the possibility of no production,

(NP) (Possibility of no production) $\mathbf{0} \in Y_b$.

We call Y_b, the production set of the firm b. The production set Y_b is said to be non-increasing returns to scale if it is convex. As a special case, it is constant returns to scale if it is a convex cone with vertex $\{\mathbf{0}\}$, or for every $\boldsymbol{y} \in Y_b$ and every $t \geq 0$, $t\boldsymbol{y} \in Y_b$. The set Y_b is said to be increasing returns if for every $\boldsymbol{y} \in Y_b$ and every $t \geq 1$, $t\boldsymbol{y} \in Y_b$. Figures 4.1(a)–4.1(c) represent the decreasing, constant, and increasing returns to scale production set, respectively.

In this chapter, we assume that there exist m consumers in the economy indexed by $a(=1\cdots m)$ and the consumption set $X_a \subset \mathbb{R}^\ell$ of the consumer a is convex, $a = 1\cdots m$. As usual, the consumer a is characterized by the preference

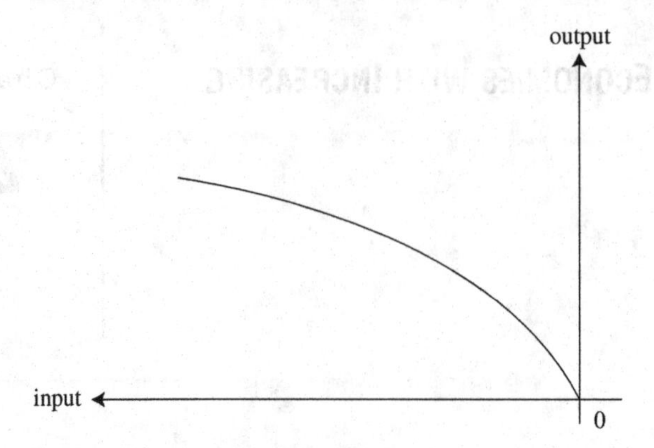

Figure 4.1(a). Decreasing Returns.

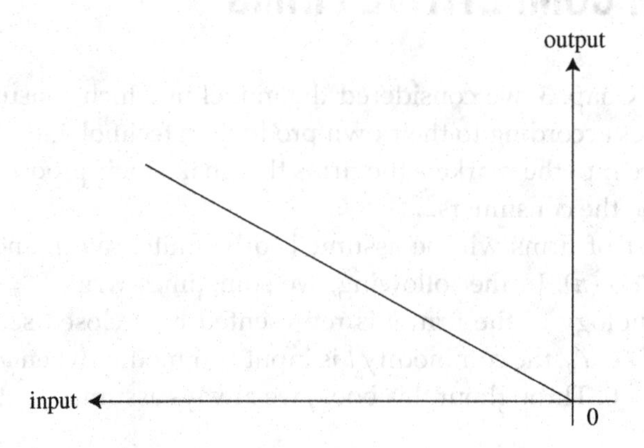

Figure 4.1(b). Constant Returns.

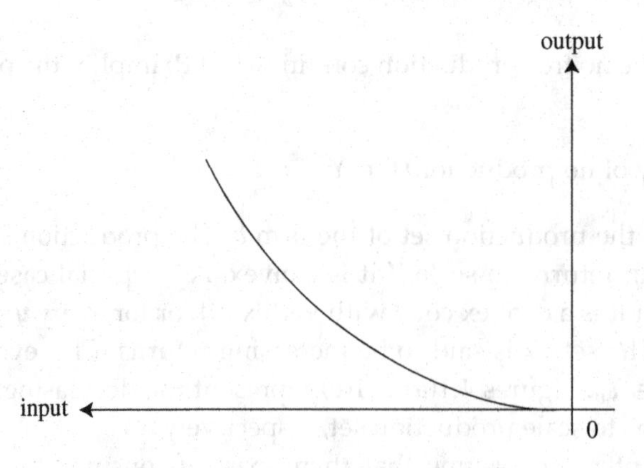

Figure 4.1(c). Increasing Returns.

relation $\prec_a \subset X_a \times X_a$, which is irreflexive and transitive, and the endowment vector $\omega_a \in \mathbb{R}^\ell$, $a = 1 \cdots m$. Let $\boldsymbol{x}_a \in X_a$ be a consumption vector of a and $\boldsymbol{y}_b \in Y_b$ a production vector of the firm b. Then the $(m+n)$-tuple of vectors $((\boldsymbol{x}_a), (\boldsymbol{y}_b))_{a \in A, b \in B}$ is called an allocation.

The allocation $((\boldsymbol{x}_a), (\boldsymbol{y}_b))$ is said to be feasible if and only if

$$\sum_{a=1}^{m} \boldsymbol{x}_a \leq \sum_{b=1}^{n} \boldsymbol{y}_b + \sum_{a=1}^{m} \omega_a.$$

In this book, we consider private ownership economies in which the consumer a owns the firm b with the fixed share $\theta_{ab} \geq 0$, $a = 1 \cdots m$, $b = 1 \cdots n$. Hence

$$\sum_{a=1}^{m} \theta_{ab} = 1$$

holds for each $b = 1 \cdots n$. The list $\mathcal{E}_Y = (X_a, \prec_a, \omega_a, \theta_{ab}, Y_b)_{a \in A, b \in B}$ is called the private ownership economy or simply a production economy.

The definition of the equilibrium of the production economy $\mathcal{E}_Y$ in which each firm behaves competitively reads as follows.

Definition 1.1. An $(m+n+1)$-tuple $(\boldsymbol{p}, (\boldsymbol{x}_a), (\boldsymbol{y}_b))$ of an allocation $(\boldsymbol{x}_a)_{a \in A}$, $(\boldsymbol{y}_b)_{b \in B}$ and a price vector $\boldsymbol{p} \in \mathbb{R}^\ell_+$ is said to consist of a competitive equilibrium or a Walras equilibrium if and only if

(E-1) $\boldsymbol{p}\boldsymbol{x}_a \leq \boldsymbol{p}\omega_a + \sum_{b=1}^{n} \theta_{ab}\boldsymbol{p}\boldsymbol{y}_b$ and $\boldsymbol{x}_a \succsim_a \boldsymbol{z}$ whenever $\boldsymbol{p}\boldsymbol{z} \leq \boldsymbol{p}\omega_a + \sum_{b=1}^{n} \theta_{ab}\boldsymbol{p}\boldsymbol{y}_b$, $a = 1 \cdots m$,
(E-2) $\boldsymbol{p}\boldsymbol{y}_b \geq \boldsymbol{p}\boldsymbol{y}$ for all $\boldsymbol{y} \in Y_b$, $b = 1 \cdots n$, and
(E-3) $\sum_{a=1}^{m} \boldsymbol{x}_a \leq \sum_{b=1}^{n} \boldsymbol{y}_b + \sum_{a=1}^{m} \omega_a$.

We now give a few remarks. In the condition (E-1), the income of the consumer a is the market value of his/her initial endowment plus the total value of the dividends $\theta_{ab}\boldsymbol{p}\boldsymbol{y}_b$, which are distributed from the firm $b = 1 \cdots n$. The condition (E-2) describes the profit maximization of the competitive firms who maximizes their profit value provided by the market price system $\boldsymbol{p}$ as given. Obviously, this condition is not generally compatible with the increasing returns to scale technologies. Generally speaking, the competitive equilibrium will prevail only in the market, where each firm operates under the decreasing or constant returns to scale technology. Simple observations of the actual market obviously show the limitation of this equilibrium concept. In Chap. 6, we discuss a more realistic equilibrium concept in which the firms have a technology with non-trivial setup cost and behave monopolistically. Finally, the condition (E-3) is the standard market condition which says that the equilibrium allocation has to be feasible.

A feasible allocation $((\boldsymbol{x}_a), (\boldsymbol{y}_b)) \in \prod_{a\in A} X_a \times \prod_{b\in B} Y_b$ is called Pareto optimal if and only if there exists no other feasible allocation $((\boldsymbol{x}'_a), (\boldsymbol{y}'_b)) \in \prod_{a\in A} X_a \times \prod_{b\in B} Y_b$ such that:

$$\boldsymbol{x}'_a \succsim_a \boldsymbol{x}_a \quad \text{for all } a \in A,$$

and

$$\boldsymbol{x}_a \prec \boldsymbol{x}'_a \quad \text{holds for at least one } a \in A.$$

The first and the second welfare theorems hold for the competitive equilibrium of the private ownership economy.

Proposition 1.1. *Let $(\boldsymbol{p}^*, (\boldsymbol{x}^*_a), (\boldsymbol{y}^*_b)) \in \mathbb{R}^\ell_+ \times \prod_{a\in A} X_a \times \prod_{b\in B} Y_b$ be a competitive equilibrium of an economy $\mathcal{E} = (X_a, \prec_a, \omega_a, \theta_{ab}, Y_b)_{a\in A, b\in B}$, which satisfies the local non-satiation of preferences (LNS) for all $a \in A$. Then, the allocation $((\boldsymbol{x}^*_a), (\boldsymbol{y}^*_b))$ is Pareto optimal.*

Proof. Suppose that $((\boldsymbol{x}^*_a), (\boldsymbol{y}^*_b))$ was not Pareto optimal. Then, there exists a feasible allocation $((\boldsymbol{x}_a), (\boldsymbol{y}_b)) \in \prod_{a\in A} X_a \times \prod_b Y_b$ such that:

$$\boldsymbol{x}_a \succsim_a \boldsymbol{x}^*_a \quad \text{for all } a \in A$$

and

$$\boldsymbol{x}^*_a \prec \boldsymbol{x}_a \quad \text{holds for at least one } a \in A.$$

Then, it follows that $\boldsymbol{p}^*\boldsymbol{x}_a \geq \boldsymbol{p}^*\omega_a + \boldsymbol{p}^* \sum_{b=1}^n \theta_{ab}\boldsymbol{y}^*_b$ for all $a \in A$. For if not, there is an a such that $\boldsymbol{p}^*\boldsymbol{x}_a < \boldsymbol{p}^*\omega_a + \boldsymbol{p}^* \sum_{b=1}^n \theta_{ab}\boldsymbol{y}^*_b$. By the local non-satiation (LNS), there exists a bundle $\boldsymbol{z} \in X_a$ which is close enough to $\boldsymbol{x}_a$ so that $\boldsymbol{p}^*\boldsymbol{z} < \boldsymbol{p}^*\omega_a + \boldsymbol{p}^* \sum_{b=1}^n \theta_{ab}\boldsymbol{y}^*_b$, and $\boldsymbol{x}_a \prec_a \boldsymbol{z}$, contradicting the condition (E-1). Furthermore, for a such that $\boldsymbol{x}^*_a \prec_a \boldsymbol{x}_a$, we have $\boldsymbol{p}^*\boldsymbol{x}_a > \boldsymbol{p}^*\omega_a + \boldsymbol{p}^* \sum_{b=1}^n \boldsymbol{y}_b$. Summing these inequalities over a, and using $\sum_{a=1}^m \theta_{ab} = 1$, one obtains $\sum_{a=1}^m \boldsymbol{p}^*\boldsymbol{x}_a > \sum_{a=1}^m \boldsymbol{p}^*\omega_a + \boldsymbol{p}^* \sum_{b=1}^n \boldsymbol{y}_b$. On the other hand, since the allocation $((\boldsymbol{x}_a), (\boldsymbol{y}_b))$ is feasible, we have $\sum_{a=1}^m \boldsymbol{x}_a \leq \sum_{a=1}^m \omega_a + \sum_{b=1}^n \boldsymbol{y}_b$, hence, $\sum_{a=1}^m \boldsymbol{p}^*\boldsymbol{x}_a \leq \sum_{a=1}^m \boldsymbol{p}^*\omega_a + \boldsymbol{p}^* \sum_{b=1}^n \boldsymbol{y}^*_b$. A contradiction. □

In order to state the second welfare theorem, we give:

Definition 1.2. Suppose a price vector $\boldsymbol{p} \in \mathbb{R}^\ell_+$ be given. An allocation $((\boldsymbol{x}_a), (\boldsymbol{y}_b))$ is said to be an equilibrium relative to the price vector $\boldsymbol{p}$ if and only if

(R-1) $\boldsymbol{x}_a \prec_a \boldsymbol{z}$ implies that $\boldsymbol{p}\boldsymbol{x}_a < \boldsymbol{p}\boldsymbol{z}$ for all $a \in A$,
(E-2) $\boldsymbol{p}\boldsymbol{y}_b \geq \boldsymbol{p}\boldsymbol{y}$ for all $\boldsymbol{y} \in Y_b, b = 1 \cdots n$, and
(E-3) $\sum_{a=1}^m \boldsymbol{x}_a \leq \sum_{b=1}^n \boldsymbol{y}_b + \sum_{a=1}^m \omega_a$.

As in the case of the exchange economies, if the condition (R-1) is replaced by:

(Q-1) $\boldsymbol{x}_a \prec_a \boldsymbol{z}$ implies that $\boldsymbol{p}\boldsymbol{x}_a \leq \boldsymbol{p}\boldsymbol{z}$ for all $a \in A$, the allocation $((\boldsymbol{x}_a), (\boldsymbol{y}_b))$ is said to be a quasi-equilibrium relative to the price vector $\boldsymbol{p}$. The second fundamental theorem of welfare economics reads as follows.

Theorem 1.1. *For every consumer $a \in A$, suppose that the preference relation satisfies the negative transitivity (NTR), the continuity (CT), the convexity (CV), and the monotonicity (MT). Assume that the production set Y_b is convex for all $b \in B$. Then, every Pareto-optimal allocation is a quasi-equilibrium.*

Proof. Let $((\boldsymbol{x}_a), (\boldsymbol{y}_b))$ be a Pareto-optimal allocation. Since it is a feasible allocation, the condition (E-3) is met. For each consumer a, we define the strictly preferred set $P_a(\boldsymbol{x})$ by:

$$P_a(\boldsymbol{x}) = \{\boldsymbol{z} \in X_a \mid \boldsymbol{x} \prec_a \boldsymbol{z}\}.$$

The set $P_a(\boldsymbol{x})$ is nonempty for all $\boldsymbol{x} \in X_a$ by the assumption (MT). As in the proof of Theorem 5.2 of Chap. 2, we can show that the set $P_a(\boldsymbol{x})$ is convex for all $\boldsymbol{x} \in X_a$.

Let $Q = \sum_{a=1}^m P_a(\boldsymbol{x}_a) - \sum_{b=1}^n Y_b$. Then, Q is convex as the sum of the convex sets. Since the allocation $((\boldsymbol{x}_a), (\boldsymbol{y}_b))$ is Pareto optimal, $\sum_{a=1}^n \omega_a \notin Q$. For if not, there exists a vector $\boldsymbol{z} \in Q$ or $\boldsymbol{z} = \sum_a \boldsymbol{x}'_a - \sum_b \boldsymbol{y}'_b = \sum_{a=1}^m \omega_a$ satisfying $\boldsymbol{x}'_a \in P_a(\boldsymbol{x}_a), \boldsymbol{y}'_b \in Y_b$, $a = 1 \cdots m, b = 1 \cdots n$, a contradiction. Hence, we can apply the separation theorem (Theorem A1) of Appendix A, we get a vector $\boldsymbol{p} \in \mathbb{R}^\ell$ with $\boldsymbol{p} \neq \boldsymbol{0}$ and:

$$\boldsymbol{p} \sum_{a=1}^m \omega_a \leq \boldsymbol{p} \sum_{a=1}^m \boldsymbol{z}$$

for every $\boldsymbol{z} \in Q$, or for every $\boldsymbol{z}$ of the form $\boldsymbol{z} = \sum_{a=1}^m \boldsymbol{x}'_a - \sum_{b=1}^n \boldsymbol{y}'_b$, $\boldsymbol{x}'_a \in P_a(\boldsymbol{x}_a)$, $\boldsymbol{y}'_b \in Y_b$, $a = 1 \cdots m$, $b = 1 \cdots n$. By the monotonicity (MT), it follows that $\boldsymbol{p} \geq \boldsymbol{0}$. For $c \neq a$, we can take $\boldsymbol{x}'_c$ arbitrarily close to $\boldsymbol{x}_c$. Hence, in the limit, setting $\boldsymbol{y}'_b = \boldsymbol{y}_b$, $b = 1 \cdots n$, we have:

$$\begin{aligned}
\boldsymbol{p} \sum_{a=1}^m \omega_a &\leq \boldsymbol{p} \sum_{a=1}^m \boldsymbol{x}'_a - \boldsymbol{p} \sum_{b=1}^n \boldsymbol{y}_b \leq \boldsymbol{p} \sum_{c \neq a} \boldsymbol{x}_c + \boldsymbol{p}\boldsymbol{x}'_a - \boldsymbol{p} \sum_{b=1}^n \boldsymbol{y}_b \\
&\leq \boldsymbol{p} \sum_{c=1}^m \boldsymbol{x}_c - \boldsymbol{p}\boldsymbol{x}_a + \boldsymbol{p}\boldsymbol{x}'_a + \boldsymbol{p} \sum_{a=1}^m \omega_a - \boldsymbol{p} \sum_{a=1}^m \boldsymbol{x}_a \\
&= \boldsymbol{p} \sum_{a=1}^m \omega_a - \boldsymbol{p}\boldsymbol{x}_a + \boldsymbol{p}\boldsymbol{x}'_a,
\end{aligned}$$

and from this, we obtain:

$$\boldsymbol{p}\boldsymbol{x}_a \leq \boldsymbol{p}\boldsymbol{x}'_a \quad \text{for every } \boldsymbol{x}'_a \in P_a(\boldsymbol{x}_a), \ a = 1 \cdots m.$$

Therefore, the condition (Q-1) is met. Next, we set $\boldsymbol{x}'_a = \boldsymbol{x}_a$ for all a and $\boldsymbol{y}'_c = \boldsymbol{y}_c$ for $c \neq b$ and obtain:

$$\boldsymbol{p}\sum_{a=1}^{m}\omega_a \leq \boldsymbol{p}\sum_{a=1}^{m}\boldsymbol{x}_a - \boldsymbol{p}\sum_{c\neq b}\boldsymbol{y}_c - \boldsymbol{p}\boldsymbol{y}'_b \leq \boldsymbol{p}\sum_{a=1}^{m}\boldsymbol{x}_a - \boldsymbol{p}\sum_{c=1}^{n}\boldsymbol{y}_c + \boldsymbol{p}\boldsymbol{y}_b - \boldsymbol{p}\boldsymbol{y}'_b$$

$$\leq \sum_{b=1}^{n}\boldsymbol{y}_b + \boldsymbol{p}\sum_{a=1}^{m}\omega_a - \boldsymbol{p}\sum_{c=1}^{m}\boldsymbol{y}_c + \boldsymbol{p}\boldsymbol{y}_b - \boldsymbol{p}\boldsymbol{y}'_b.$$

Therefore, we have:

$$\boldsymbol{p}\boldsymbol{y}'_b \leq \boldsymbol{p}\boldsymbol{y}_b \quad \text{for every } \boldsymbol{y}'_b \in Y_b, \ \ b = 1 \cdots n.$$

This proves the condition (E-2) and the theorem is established. □

As in the case of exchange economies, if the situation that $\boldsymbol{p}\boldsymbol{x}_a = \inf \boldsymbol{p}X_a$ is excluded, then we have an equilibrium rather than quasi-equilibrium.

Corollary. *Under the assumptions of Theorem 5.3 of Chap. 2, if the Pareto-optimal allocation* $((\boldsymbol{x}_a)_{a=1}^m, (\boldsymbol{y}_b)_{b=1}^n)$ *satisfies* $\boldsymbol{p}\boldsymbol{x}_a > \inf \boldsymbol{p}X_a$, *then it is an equilibrium relative to* $\boldsymbol{p}$.

The fundamental existence theorem of the competitive equilibrium now reads:

Theorem 1.2. *Suppose that an economy* $\mathcal{E} = (X_a, \prec_a, \omega_a, \theta_{ab}, Y_b)_{a\in A, b\in B}$ *satisfies the continuity (CT), the convexity (CV) and the minimum income condition (MI) for every* $a \in A$, *and for every* $b \in B$, *the following conditions hold.*

(NFP) *(No free production)* $Y_b \cap \mathbb{R}^\ell_+ = \{\boldsymbol{0}\}$;
(FD) *(Free disposability)* $\boldsymbol{y} + \mathbb{R}^\ell_- \subset Y_b$ *for every* $\boldsymbol{y} \in Y_b$; *and*
(PCV) *(Convex production sets) the set* Y_b *is convex.*

Finally, we assume that:

(BTP) *(Bounded total production) the set* $\{(\boldsymbol{y}_b) \in \prod_{b\in B} Y_b \mid \boldsymbol{z} \leq \sum_{b\in B}\boldsymbol{y}_b\}$ *is bounded for all* $\boldsymbol{z} \in \mathbb{R}^\ell$.

Then, there exists a competitive equilibrium $(\boldsymbol{p}^*, \boldsymbol{x}_1^* \cdots \boldsymbol{x}_m^*)$ for $\mathcal{E}$.

Proof. Take a positive number $\bar{k}$ such that $\| \sum_{a=1}^m \omega_a \| < \bar{k}$, and for each integer $k \geq \bar{k}$, set $\{\boldsymbol{x} \in \mathbb{R}^\ell \mid \ \|\boldsymbol{x}\| \leq k\boldsymbol{1}\}$, where $\boldsymbol{1} = (1 \cdots 1)$. As before, for each integer $k \geq \bar{k}$, we define the restricted consumption sets and the production sets,

$$\hat{X}_a(k) = X_a \cap K, \quad a = 1 \cdots m,$$
$$\hat{Y}_b(k) = Y_b \cap K, \quad b = 1 \cdots n.$$

The sets $\hat{X}_a(k)$ and $\hat{Y}_b(k)$ are compact and convex subsets of $\mathbb{R}^\ell$, $a = 1 \cdots m$, $b = 1 \ldots n$. For each $b(= 1 \cdots n)$, we define the (restricted) supply correspondence $\hat{\psi}_b$:

$$\hat{\psi}_b : \mathbb{R}^\ell_+ \to \hat{Y}_b(k), \quad \boldsymbol{p} \mapsto \hat{\psi}_b(\boldsymbol{p}) = \{\boldsymbol{y} \in \hat{Y}_b(k) \mid \boldsymbol{py} \geq \boldsymbol{p}\hat{Y}_b(k)\}$$

and the (restricted) profit function by:

$$\hat{\pi}_b : \mathbb{R}^\ell_+ \to \mathbb{R}_+, \quad \boldsymbol{p} \mapsto \boldsymbol{p}\hat{\psi}_b(\boldsymbol{p}),$$

respectively. Clearly, for all $t > 0$, we have $\hat{\psi}_b(t\boldsymbol{p}) = \hat{\psi}_b(\boldsymbol{p})$ and $\hat{\pi}_b(t\boldsymbol{p}) = t\hat{\pi}_b(\boldsymbol{p})$, $b = 1 \cdots n$. Therefore, without loss of generality, we can restrict the domain $\mathbb{R}^\ell_+$ to Δ, the unit simplex. Moreover, since $\boldsymbol{0} \in \hat{Y}_b(k)$, we have $\hat{\pi}_b(\boldsymbol{p}) \geq 0$ for all $\boldsymbol{p} \in \Delta$. Since, $\hat{Y}_b(k)$ is compact, it follows from the Berge's maximum theorem (Theorem D2) of Appendix D that the correspondence $\hat{\psi}_b$ is upper hemi-continuous and the function $\hat{\pi}_b$ is continuous. We now consider the budget correspondence $\hat{\beta}(a, \cdot) : \Delta \to \hat{X}_a(k)$ which is defined by:

$$\hat{\beta}(a, \boldsymbol{p}) = \left\{ \boldsymbol{x} \in \hat{X}_a(k) \;\middle|\; \boldsymbol{px} \leq \boldsymbol{p}\omega_a + \sum_{b=1}^{n} \theta_{ab}\hat{\pi}_b(\boldsymbol{p}) \right\}, \quad a = 1 \cdots m,$$

and the demand correspondence $\hat{\phi}(a, \cdot) : \Delta \to \hat{X}_a(k)$ defined by,

$$\hat{\phi}(a, \boldsymbol{p}) = \{\boldsymbol{x} \in \hat{\beta}(a, \boldsymbol{p}) \mid \boldsymbol{x} \succsim_a \boldsymbol{z} \text{ for all } \boldsymbol{z} \in \hat{\beta}(a, \boldsymbol{p})\}, \quad a = 1 \cdots m.$$

By the assumption of the minimum income (MI) and $\hat{\pi}_b(\boldsymbol{p}) \geq 0$, the budget set $\hat{\beta}_a(\boldsymbol{p})$ is nonempty for all $\boldsymbol{p} \in \Delta$. Obviously, it is compact and convex subset of $\mathbb{R}^\ell$. Since $\hat{\pi}_b(\boldsymbol{p})$ is continuous, we can easily show that the correspondence $\hat{\beta}_a$ is upper hemi-continuous. As in the proof of Proposition 3.2 of Section 2.3, we can prove that $\hat{\beta}_a$ is lower hemi-continuous, hence, it is continuous. Then, by the same way of the proof of Propositions 3.1 and 3.3 of Chap. 2, the correspondence $\hat{\phi}_a(\boldsymbol{p})$ is nonempty, compact-valued, and upper hemi-continuous. By the assumption of the convexity of preferences (CV), it is also convex-valued. Then, we define the aggregate excess-demand correspondence,

$$\hat{\zeta} : \Delta \to \mathbb{R}^\ell, \quad \hat{\zeta}(\boldsymbol{p}) = \sum_{a=1}^{m} \hat{\phi}(a, \boldsymbol{p}) - \sum_{a=1}^{m} \omega_a - \sum_{b=1}^{n} \hat{\psi}_b(\boldsymbol{p}).$$

Let $\hat{Z}(k) = \sum_{a=1}^{m} \hat{X}_a(k) - \sum_{a=1}^{m} \omega_a - \sum_{b=1}^{n} \hat{Y}_b(k)$. Then, the set $\hat{Z}(k)$ is also compact and convex, and $\zeta(\boldsymbol{p}) \subset \hat{Z}(k)$ for every $\boldsymbol{p} \in \Delta$. The correspondence $\hat{\zeta}$ is upper hemi-continuous and convex-valued and satisfies $\boldsymbol{p}\zeta(\boldsymbol{p}) \leq 0$ by the budget condition of each a. Applying Gale–Nikaido–Debreu lemma (Theorem D3) of Appendix D, we obtain $\boldsymbol{p}(k) \in \Delta$ such that:

$$\hat{\xi}(\boldsymbol{p}(k)) = \sum_{a=1}^{m} \hat{\phi}(a, \boldsymbol{p}(k)) - \sum_{a=1}^{m} \omega_a - \sum_{b=1}^{n} \hat{\psi}_b(\boldsymbol{p}(k)) \leq \mathbf{0}.$$

Take $\boldsymbol{x}_a(k) \in \hat{\phi}(\boldsymbol{p}(k))$, $a = 1 \cdots m$ and $\boldsymbol{y}_b(k) \in \hat{\psi}_b(\boldsymbol{p}(k))$, $b = 1 \cdots n$. Then, we have:

(E-1_k) $\boldsymbol{p}(k)\boldsymbol{x}_a(k) \leq \boldsymbol{p}(k)\omega_a + \sum_{b=1}^{n} \theta_{ab}\boldsymbol{p}(k)\boldsymbol{y}_b(k)$ and $\boldsymbol{x}_a(k) \succsim_a \mathbf{z}$ for every $\mathbf{z} \in \hat{X}_a(k)$ such that $\boldsymbol{p}(k)\mathbf{z} \leq \boldsymbol{p}(k)\omega_a + \sum_{b=1}^{n} \theta_{ab}\boldsymbol{p}(k)\boldsymbol{y}_b(k)$, $a = 1 \cdots m$,

(E-2_k) $\boldsymbol{p}(k)\boldsymbol{y}(k) \geq \boldsymbol{p}(k)\boldsymbol{y}$ for every $\boldsymbol{y} \in \hat{Y}_b(k)$, $b = 1 \cdots n$, and

(E-3_k) $\sum_{a=1}^{m} \boldsymbol{x}_a(k) \leq \sum_{a=1}^{m} \omega_a + \sum_{b=1}^{n} \boldsymbol{y}_b(k)$.

Since Δ is compact, we can assume that $\boldsymbol{p}(k) \to \boldsymbol{p}^* \in \Delta$. By the assumption of the bounded total production (BTP) and the consumption sets being bounded from below, the sequences $(\boldsymbol{y}_b(k))$ are bounded and the production sets being closed, we may assume that $\boldsymbol{y}_b(k) \to \boldsymbol{y}_b^* \in \hat{Y}_b(k)$, $b = 1 \cdots n$. Consequently, the sequences $(\boldsymbol{x}_a(k))$ are also bounded, we can assume that $\boldsymbol{x}_a(k) \to \boldsymbol{x}_a^* \in \hat{X}_a(k)$.

We claim that $(\boldsymbol{p}^*, \boldsymbol{x}_1^* \cdots \boldsymbol{x}_m^*)$ is a desired competitive equilibrium. Passing to the limit in the above market conditions (E-3_k), we obtain:

(E-3) $\sum_{a=1}^{m} \boldsymbol{x}_a^* \leq \sum_{a=1}^{m} \omega_a + \sum_{b=1}^{n} \boldsymbol{y}_b^*$.

First, we shall show that for all $b = 1 \cdots n$,

(E-2) $\boldsymbol{p}^*\boldsymbol{y}_b^* \geq \boldsymbol{p}^*\boldsymbol{y}$ for every $\boldsymbol{y} \in Y_b$.

Suppose to the contrary that for some b, there exists $\boldsymbol{y} \in Y_b$ such that $\boldsymbol{p}^*\boldsymbol{y}_b < \boldsymbol{p}^*\boldsymbol{y}$. For k large enough, $\boldsymbol{y} \in \hat{Y}_b(k)$. Since $\boldsymbol{p}(k) \to \boldsymbol{p}^*$ and $\boldsymbol{y}(k) \to \boldsymbol{y}_b^*$, it follows that $\boldsymbol{p}(k)\boldsymbol{y}_b(k) < \boldsymbol{p}(k)\boldsymbol{y}$, contradicting the condition (E-2_k).

Since $\boldsymbol{p}(k)\boldsymbol{x}_a(k) \leq \boldsymbol{p}(k)\omega_a + \sum_{b=1}^{n} \theta_{ab}\boldsymbol{p}(k)\boldsymbol{y}_b(k)$, passing to the limit, we have:

$$\boldsymbol{p}^*\boldsymbol{x}_a^* \leq \boldsymbol{p}^*\omega_a + \sum_{b=1}^{n} \theta_{ab}\boldsymbol{p}^*\boldsymbol{y}_b^*, \quad a = 1 \cdots m.$$

Suppose, there exists $\mathbf{z} \in X_a$ such that $\boldsymbol{p}^*\mathbf{z} \leq \boldsymbol{p}^*\omega_a + \sum_{b=1}^{n} \theta_{ab}\boldsymbol{p}^*\boldsymbol{y}_b^*$ and $\boldsymbol{x}_a^* \prec_a \mathbf{z}$. For k sufficiently large, we have $\mathbf{z} \in \hat{X}_a(k)$. By the continuity (CT) of the preference relation and the minimum income condition (MI), we can take $\mathbf{z}' \in \hat{X}_a(k)$ which is close enough to $\mathbf{z}$ such that $\boldsymbol{x}_a^* \prec_a \mathbf{z}'$ and $\boldsymbol{p}^*\mathbf{z}' < \boldsymbol{p}^*\omega_a + \sum_{b=1}^{n} \theta_{ab}\boldsymbol{p}^*\boldsymbol{y}_b^*$. Hence, for k large enough, we have $\boldsymbol{x}_a(k) \prec_a \mathbf{z}'$ and $\boldsymbol{p}(k)\mathbf{z}' < \boldsymbol{p}(k)\omega_a + \sum_{b=1}^{n} \theta_{ab}\boldsymbol{p}(k)\boldsymbol{y}_b(k)$, contradicting the condition (E-1_k). Therefore, the condition (E-1) is met. □

Note that the above proof shows that the assumption (BTP) is replaced by:

(BTP′) the set $\{(\boldsymbol{y}_b) \in \prod_{b=1}^{n} Y_b \mid \boldsymbol{b} - \sum_{a=1}^{m} \omega_a \leq \sum_{b=1}^{n} \boldsymbol{y}_b\}$ is bounded,

where $\boldsymbol{b} = (b^t)$ is defined by $b^t = \min\{b_a^t \mid a = 1 \cdots m\}$ and recall that $\boldsymbol{b}_a = (b_a^t)$ is the lower bound of the consumption set of the consumer a.

4.2. CORE OF AN ECONOMY WITH INCREASING RETURNS

In this section, we consider the core and equilibria of a production economy where the increasing returns prevail, or in other words the production sets of the producers are not assumed to be convex.

As discussed before, we assume that the commodity space is $\mathbb{R}^\ell$, and there exist m consumers indexed by $a(=1\cdots m)$. All consumers have the same consumption set which is assumed to be the non-negative orthant of $\mathbb{R}^\ell$, $X_a = \mathbb{R}^\ell_+$, $a = 1\cdots m$. The consumer a has the initial endowment vector $\omega_a \in \mathbb{R}^\ell$ and the preference relation $\prec_a \subset X_a \times X_a$ which is irreflexive and transitive binary relation on X_a. We assume:

(CT) (Continuity) the set $\prec = \{(\boldsymbol{x},\boldsymbol{z}) \in X_a \times X_a \mid \boldsymbol{x} \prec \boldsymbol{z}\}$ is open in $X_a \times X_a$,
(CV) (Convexity) for every $\boldsymbol{x} \in X_a$ the set $\{\boldsymbol{z} \in X_a | \boldsymbol{z} \succsim_a \boldsymbol{x}\}$ is convex,
(MT) (Monotonicity) if $\boldsymbol{x},\boldsymbol{z} \in X_a$ and $\boldsymbol{x} < \boldsymbol{z}$, then $\boldsymbol{x} \prec \boldsymbol{z}$.

For the production side of the economy, we assume that there exists a total (social) production possibility set $Y \subset \mathbb{R}^\ell$, which is a close subset of $\mathbb{R}^\ell$ available to all coalitions of the consumer $C \subset A = \{1\cdots m\}$. This formulation of a production economy is considered to be a special case of the coalition production economy discussed in Section 3.6 of Chap. 3. That is, the production correspondence is given by a constant:

$$Y(C) = Y \quad \text{for every } C \subset A.$$

The list $((\prec_a), Y)$ is called the coalition production economy and denoted by $\mathcal{E}_C$.

For the total production set Y, we assume that:

(NP) (Possibility of no production) $\mathbf{0} \in Y$,
(FD) (Free disposability) $\boldsymbol{y} + \mathbb{R}^\ell_- \subset Y$ for every $\boldsymbol{y} \in Y$,

An $(m+1)$-tuple of vectors $((\boldsymbol{x}_a), \boldsymbol{y})$ is called an allocation if $\boldsymbol{x}_a \in X_a$, $a = 1\cdots m$, and $\boldsymbol{y} \in Y$. The allocation $((\boldsymbol{x}_a), \boldsymbol{y})$ is said to be feasible if:

$$\sum_{a=1}^{m} \boldsymbol{x}_a \leq \sum_{a=1}^{m} \omega_a + \boldsymbol{y}.$$

The core of the coalition production economy $\mathcal{E}$ is naturally defined as follows.

Let $C \subset A$ be a coalition. A feasible allocation $((\boldsymbol{x}_a), \boldsymbol{y})$ is said to be blocked by the coalition C if and only if there exists an allocation $((\boldsymbol{x}'_a), \boldsymbol{y}') \in \prod_{a=1}^{m} X_a \times Y$

such that:

$$x'_a \succsim_a x_a \quad \text{for all } a \in C,$$

$$x_a \prec_a x'_a \quad \text{for some } a \in C,$$

and

$$\sum_{a \in C} x'_a \leq \sum_{a \in C} \omega_a + y'.$$

Definition 2.1. A feasible allocation is called a core allocation if no coalition can block it. The set of all core allocations is called the core.

In this section, we divide the commodities $t = 1 \cdots \ell$ into two categories. The first category, consisting of commodities $t = 1 \cdots k$, does not enter into the preference of consumers and is used only as the input for producers. We call these as the producer commodities. The remaining commodities $k+1 \cdots \ell$ will be in the second category. They are called the consumer commodities. Note that the consumer commodities may also enter into production as the production factor such as labor if leisure is involved in the preferences of consumers.

On account of these remarks, we will make the following additional assumptions on preferences and the production set.

(PC) (Producer commodities) for all $\boldsymbol{y} \in Y$, it follows that $y^t \leq 0$ for $t = 1 \cdots k$,
(CC) (Consumer commodities) for all $\boldsymbol{x}, \boldsymbol{z} \in X_a$ if $x^t = z^t$ for all $t = k+1 \cdots \ell$, then, $\boldsymbol{x} \sim_a \boldsymbol{z}$, $a = 1 \cdots m$.

Let $\Lambda = \{(y^1 \cdots y^\ell) \in \mathbb{R}^\ell \mid y^1 \geq 0, \ldots y^k \geq 0\}$ be a cone with the vertex at the origin in which the first k-coordinates are restricted to be nonnegative. Then the assumption (PC) is equivalent to:

$$Y \subset -\Lambda.$$

Scarf (1986) introduced the concept of distributive production sets which contains the increasing returns to scale production sets.

Definition 2.2. Let Y be a set in $\mathbb{R}^\ell$ with $\mathbb{R}^\ell_-(= -\mathbb{R}^\ell_+) \subset Y \subset -\Lambda$. We say that Y is a distributive set if for any finite number of vectors $\boldsymbol{y}_i \in Y$ and any nonnegative $\alpha_i \geq 0$, the vector $\boldsymbol{y} = \sum_i \alpha_i \boldsymbol{y}_i$ is also in Y, if $\boldsymbol{y}$ satisfies the condition that $\boldsymbol{y}_i - \boldsymbol{y} \in \Lambda$ for all i.

In other words, a set Y is distributive if and only if every non-negative weighted sum of the vectors in Y will be in Y if it uses more of the producer commodities than any of the original plans. Note that if all commodities in the economy are consumer commodities, then $\Lambda = \mathbb{R}^\ell$ and a distributive set is a convex cone with vertex at the origin.

Example 2.1. Let $\ell = 2$ and assume that the first commodity is a producer commodity and the second one is a consumer commodity. Let:

$$f: \mathbb{R}_+ \to \mathbb{R}_+, \quad y^1 \mapsto y^2 = f(y^1)$$

be a production function which is continuous and monotonically increasing. We define the production set as:

$$Y = \{(-y^1, y^2) \in \mathbb{R}_- \times \mathbb{R}_+ \mid y^1 \geq 0,\ y^2 \leq f(y^1)\}$$

in which the consumer good is produced from the producer good by the production function f.

It is obvious from Fig. 4.2 that the set Y is distributive if and only if the function f is non-decreasing returns to scale, or $\lambda f(y) \leq f(\lambda y)$ for all $\lambda \geq 1$.

In the general case in which several commodities in each category are involved, distributive sets exhibit non-decreasing returns to scale. For if $\boldsymbol{y} \in Y$ and $\lambda \geq 1$, then the first k-components of the producer commodities in $\boldsymbol{y} - \lambda\boldsymbol{y}$ will be nonnegative, hence, $\lambda\boldsymbol{y} \in Y$.

The next theorem, which is a version of the separating hyperplane theorem for the distributive sets, is useful in later discussions.

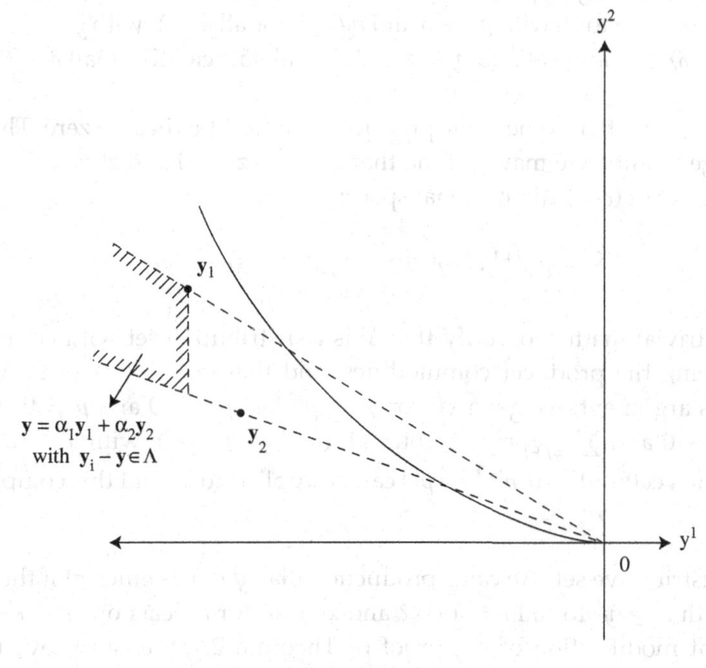

Figure 4.2. Example 2.1.

Theorem 2.1. *Let Y be a closed distributive set and $\mathbf{z} \notin Y$. Then, there exists a non-negative vector $\mathbf{p} \geq \mathbf{0}$ such that $\mathbf{pz} > 0$ and $\mathbf{py} \leq 0$ for all $\mathbf{y} \in Y \cap (\Lambda + \{\mathbf{z}\})$. Moreover, if any of the first k-components of $\mathbf{z}$ are zero, or $z^t = 0$ for some $t = 1 \cdots k$, then we can take the vector $\mathbf{p}$ as $p^t = 0$.*

Proof. We shall first prove the theorem in the case where $z^t < 0$ for $t = 1 \cdots k$. We define the set $T \subset \mathbb{R}^\ell$ by:

$$T = \left\{ \sum_{i=1}^{k} \alpha_i \mathbf{y}_i \middle| \alpha_i \geq 0, \ \mathbf{y}_i \in Y, \text{ and } \mathbf{y}_i - \mathbf{z} \in \Lambda, k = 1, 2, \ldots \right\}.$$

T is the smallest convex cone with vertex zero, containing the points of $Y \cap (\Lambda + \mathbf{z})$. Note that the cone T satisfies the free disposability assumption (FD), since it is additive, or $\mathbf{y}_1, \mathbf{y}_2 \in T$ implies that $\mathbf{y}_1 + \mathbf{y}_2 \in T$, and contains $(-\delta \cdots -\delta)$ for δ sufficiently small $\delta > 0$. Therefore, the closure of T does not contain $\mathbf{z}$. Suppose $\mathbf{z} \in$ closure T. Then, there exists a sequence $(\mathbf{z}_n)_{n \in \mathbb{N}}$ such that $\mathbf{z}_n \in T$ for all $n \in \mathbb{N}$ and $\mathbf{z}_n \to \mathbf{z}$. Using FD of T, we can assume without loss of generality that $\mathbf{z}_n \leq \mathbf{z}$ for all n. But, we can write $\mathbf{z}_n = \sum_i \alpha_i \mathbf{y}_i$ and $\mathbf{y}_i - \mathbf{z} \in \Lambda$ and this implies that $\mathbf{y}_i - \mathbf{z}_n \in \Lambda$. By the definition of the distributive set, we see that $\mathbf{z}_n \in Y$, and since Y is closed, it follows that $\mathbf{z} \in Y$, a contradiction.

By the separating hyperplane theorem (Theorem A1) of Appendix A, there exists a non-zero vector $\mathbf{p}$ with $\mathbf{pz} > 0$ and $\mathbf{py} \leq 0$ for all $\mathbf{y} \in Y$ with $\mathbf{y} - \mathbf{z} \in \Lambda$. Since $\mathbf{y} = (-\delta \cdots -\delta) \in Y$ and satisfies $\mathbf{y} - \mathbf{z} \in \Lambda$ for all sufficiently small $\delta > 0$, we see that $\mathbf{p} \geq \mathbf{0}$.

Next, suppose that some of the producer commodities in $\mathbf{z}$ are zero. Then without loss of generality, we may assume that $z^1 = \cdots z^j = 0$ and $z^{j+1} < 0, \ldots z^k < 0$. Consider the set in $(\ell - j)$-dimensional space:

$$\hat{Y} = \{(y^{j+1} \cdots y^\ell) | (0 \cdots 0, y^{j+1} \cdots y^\ell) \in Y\}.$$

It is a trivial matter to verify that $\hat{Y}$ is a distributive set with commodities $j + 1 \cdots k$ being the producer commodities, and that $(z^{j+1} \cdots z^\ell) \notin \hat{Y}$. Applying the previous argument, we get a vector $\hat{\mathbf{p}} = (p^{j+1} \cdots p^\ell) \geq \mathbf{0}$ and $\mathbf{p} \neq \mathbf{0}$ such that $\sum_{t=j+1}^{\ell} p^t z^t > 0$ and $\sum_{t=j+1}^{\ell} p^t y^t \leq 0$ for all $(y^{j+1} \cdots y^\ell) \in \hat{Y}$ with $y^t \geq z^t$ for $t = j + 1 \cdots k$. The vector $(0 \cdots 0, p^{j+1} \cdots p^\ell)$ can be applied to Y, and this completes the proof. □

Let Y be a distributive set. We call a production plan $\mathbf{y} \in Y$ is efficient if there exists no $\mathbf{z} \in Y$ with $z^t \geq y^t$ for all $t = 1 \cdots \ell$ and $z^t > y^t$ for at least one $t = k + 1 \cdots \ell$. With a slight modification of the proof of Theorem 2.1, we can prove the next proposition.

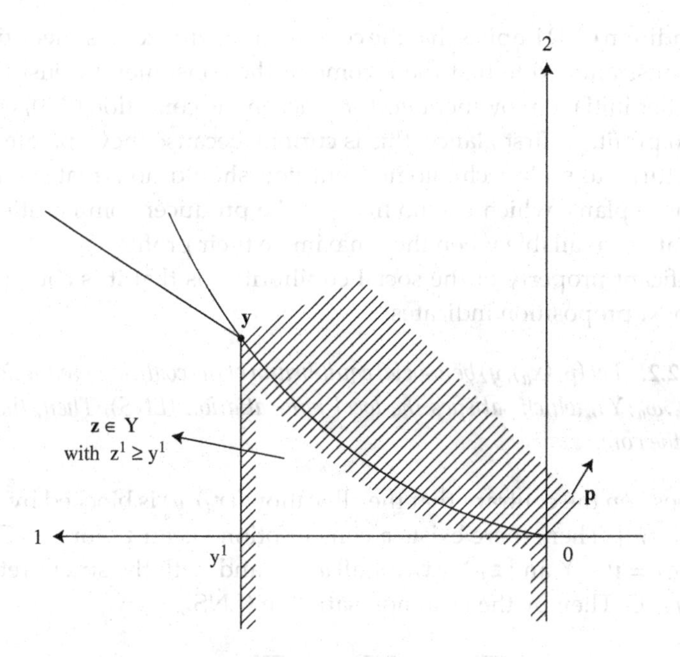

Figure 4.3. Proposition 2.1.

Proposition 2.1. *Let Y be a distributive set and $\boldsymbol{y} \in Y$ be an efficient production plan. Then, there exists a non-zero vector $\boldsymbol{p} \geq \boldsymbol{0}$ such that $\boldsymbol{py} = 0$ and $\boldsymbol{pz} \leq 0$ for all $\boldsymbol{z} \in Y$ with $z^t \geq y^t$ for all $t = 1 \cdots k$.*

The situation of Proposition 2.1 is illustrated by Fig. 4.3, where $\ell = 2$ and $k = 1$.

We now consider a coalition production economy $\mathcal{E}_C = (X_a, \prec_a, \omega_a, Y)$, where $X_a = \mathbb{R}_+^\ell$ and the preference relation $\prec_a$ satisfies the continuity (CT), the convexity (CV) and the monotonicity (MT). Moreover, we assume that the condition (CC) which says that preferences of the consumers are only concerned with the consumer commodities, $t = k+1 \cdots \ell$. The endowment vector ω_a will be assumed to be strictly positive, or:

(PE) (Positive endowment) $\omega_a \gg \boldsymbol{0}$ for all $a = 1 \cdots m$.

The production set Y is assumed to be distributive and satisfies (PC), or $Y \subset -\Lambda$.

Definition 2.3. An allocation $((\boldsymbol{x}_a)_{a\in A}, \boldsymbol{y})$ is called a social equilibrium if there exists a price vector $\boldsymbol{p} \geq \boldsymbol{0}$ such that:

(S-1) $\boldsymbol{px}_a \leq \boldsymbol{p}\omega_a$ and $\boldsymbol{x}_a \succsim_a \boldsymbol{x}$ whenever $\boldsymbol{px} \leq \boldsymbol{p}\omega_a$, $a = 1 \cdots m$,

(S-2) $\boldsymbol{py} = 0$ and $\boldsymbol{pz} \leq 0$ for all $\boldsymbol{z} = (z^t) \in Y$ such that $z^t \geq -\sum_{a=1}^m \omega_a^t$ for all $t = 1 \cdots k$, and

(E-3) $\sum_{a\in A} \boldsymbol{x}_a \leq \sum_{a\in A} \omega_a + \boldsymbol{y}$.

The condition (S-1) implies that the consumer maximizes his/her utility under the budget constraint. Note that the income of the consumer a is just the market value of his/her initial endowment vector, since in the condition (S-2), the producers yield zero profit. At first glance, this is curious because they operate under the increasing returns to scale technologies, but you should note that the firms consider only those plans, which use no more of the producer commodities than the quantities that are available when they maximize their profits.

A significant property of the social equilibrium is that it is contained in the core, as the next proposition indicates.

Proposition 2.2. *Let $(\boldsymbol{p}, (\boldsymbol{x}_a), \boldsymbol{y})$ be a social equilibrium of the coalition production economy $\mathcal{E}_C = (X_a, \prec_a, \omega_a, Y)$, which satisfies the local non-satiation (LNS). Then, the allocation $((\boldsymbol{x}_a), \boldsymbol{y})$ is in the core.*

Proof. Suppose on the contrary that the allocation $((\boldsymbol{x}_a), \boldsymbol{y})$ is blocked by a coalition $C \subset A = \{1 \cdots m\}$. Then, there exists a consumption vector $\boldsymbol{z}_a$ for $a \in C$ such that $\sum_{a \in C} (\boldsymbol{z}_a - \omega_a) = \hat{\boldsymbol{y}} \in Y$ and $\boldsymbol{z}_a \succsim_a \boldsymbol{x}_a$ for all $a \in C$ and with the strict preference for at least one $a \in C$. Then by the local non-satiation (LNS),

$$\sum_{a \in C} \boldsymbol{p}\boldsymbol{z}_a > \sum_{a \in C} \boldsymbol{p}\boldsymbol{x}_a \geq \sum_{a \in C} \boldsymbol{p}\omega_a,$$

hence, $\boldsymbol{p}\hat{\boldsymbol{y}} = \boldsymbol{p} \sum_{a \in C} (\boldsymbol{z}_a - \omega_a) > 0$. But $\hat{\boldsymbol{y}} \in Y$ and $\hat{\boldsymbol{y}} \geq -\sum_{a \in C} \omega_a \geq -\sum_{a \in A} \omega_a$, which contradicts the condition (S-2). □

The next theorem, which is the main result of this section, claims that the social equilibrium exists with the distributive production set; hence, by Proposition 2.2, the core is not empty even under the increasing returns to scale technologies.

Theorem 2.2. *Let $\mathcal{E}_C = (X_a, \prec_a, \omega_a, Y)$ be a coalition production economy which satisfies the continuity (CT), the convexity (CV), the monotonicity (MT), and the assumption of the consumer commodities (CC). Suppose that all consumers have strictly positive endowment vector (PE). The production set Y is assumed to be a closed distributive set which satisfies the assumption of the producer commodities (PC) and the boundedness condition,*

(BTP) (Bounded total production) the set $\{\boldsymbol{y} \in Y \mid \boldsymbol{y} \geq -\sum_{a=1}^{m} \omega_a\}$ is bounded.

Then, there exists a social equilibrium $(\boldsymbol{p}, (\boldsymbol{x}_a), \boldsymbol{y})$ for $\mathcal{E}_C$.

Proof. Let $\omega = \sum_{a=1}^{m} \omega_a$ and we define the set T such that:

$$T = \left\{ \sum_{i=1}^{k} \alpha_i \boldsymbol{y}_i \,\middle|\, \alpha_i \geq 0,\ \boldsymbol{y}_i \in Y, \text{ and } \boldsymbol{y}_i + \omega \in \Lambda,\ k = 1, 2, \ldots \right\},$$

which is the smallest convex cone with the vertex at zero containing $Y \cap (\Lambda - \omega)$. Consider the (private ownership) production economy $\mathcal{E}_T = (X_a, \prec_a, \omega_a, T)$. We will show that there exists a competitive equilibrium $(\boldsymbol{p}, (\boldsymbol{x}_a), \boldsymbol{y})$ for $\mathcal{E}_T$, and then it is a social equilibrium for $\mathcal{E}_C$.

In order to apply Theorem 1.2, it is suffice to show that the set T satisfies the free disposability (FD) and the NFP condition (see also the remark at the end of the proof of Theorem 1.2.) Since T is a closed convex cone with the vertex at zero, in order to prove (FD), it is sufficient to show that:

$$(-\delta^1 \cdots -\delta^\ell) \in T$$

for all sufficiently small $\delta^t > 0, t = 1 \cdots \ell$, and this is certainly true, since $\omega \gg \mathbf{0}$ and Y has the FD property.

In order to show (NFP) or $T \cap \mathbb{R}^\ell_+ = \{\mathbf{0}\}$, suppose that there exists $\boldsymbol{z} \in T \cap \mathbb{R}^\ell_+$ with $\boldsymbol{z} = (z^t) \neq \mathbf{0}$. It follows from the assumption (PC) that $z^t \leq 0$ for all $t = 1 \cdots k$. Hence, one has $\boldsymbol{z} = (0 \cdots 0, z^{t+1} \cdots z^\ell)$. Consider the vector $(-\omega^1 \cdots -\omega^k, z^{t+1} \cdots z^\ell)$. We claim that this latter vector must belong to Y, for if not, we can apply Theorem 2.1, and obtain a vector $\boldsymbol{p} \geq \mathbf{0}$ such that:

$$\boldsymbol{pz} = -\sum_{t=1}^{k} p^t \omega^t + \sum_{t=k+1}^{\ell} p^t z^t > 0,$$
$$\boldsymbol{py} \leq 0 \ \text{ for all } \boldsymbol{y} \in Y \quad \text{with } y^t \geq -\omega^t, \ \ t = 1 \cdots k.$$

But this implies that $\boldsymbol{py} \leq 0$ for all $\boldsymbol{y} \in T$ and therefore $\boldsymbol{pz} \leq 0$. This contradicts the previous inequality. Therefore:

$$(-\omega^1 \cdots -\omega^k, z^{t+1} \cdots z^\ell) \in Y.$$

Repeating the same argument with $\lambda \boldsymbol{z}$ instead of $\boldsymbol{z}$, one sees that:

$$(-\omega^1 \cdots -\omega^k, \lambda z^{t+1} \cdots \lambda z^\ell) \in Y, \quad \text{for all } \lambda,$$

contradicting the assumption (BTP). Therefore, there exists a competitive equilibrium $(\boldsymbol{p}, (\boldsymbol{x}_a), \boldsymbol{y})$ for the economy $\mathcal{E}_T$, which satisfies:

$$\boldsymbol{px}_a \leq \boldsymbol{p}\omega_a \quad \text{and} \quad \boldsymbol{x}_a \succsim_a \boldsymbol{x} \quad \text{whenever } \boldsymbol{px} \leq \boldsymbol{p}\omega_a, \ a = 1 \cdots m,$$
$$\boldsymbol{py} = 0 \quad \text{and} \quad \boldsymbol{pz} \leq 0 \quad \text{for all } \boldsymbol{z} \in T,$$

and we have:

$$\sum_{a \in A} \boldsymbol{x}_a \leq \sum_{a \in A} \omega_a + \boldsymbol{y}.$$

From the profit maximization over T, we see that:

$$\boldsymbol{pz} \leq 0 \text{ for all } \boldsymbol{z} \in Y \quad \text{such that } z^t \geq -\sum_{a=1}^{m} \omega_a^t \quad \text{for } t = 1 \cdots k.$$

In order to show that $(\boldsymbol{p}, (\boldsymbol{x}_a), \boldsymbol{y})$ is a social equilibrium for $\mathcal{E}_C$, we need only to demonstrate that $\boldsymbol{y} \in Y$. First note that $\boldsymbol{y} + \omega \geq \boldsymbol{0}$, since $\boldsymbol{x}_a \geq \boldsymbol{0}$ for all $a = 1 \cdots m$. We now claim that if $y^t > -\omega^t$, then $p^t = 0$. To see this, suppose that $y^1 > -\omega^1$ and $p^1 > 0$. Then for some a, $x_a^1 > 0$, since the feasibility condition holds with the equality, $\sum_a \boldsymbol{x}_a = \omega + \boldsymbol{y}$ by the monotonicity (MT) of preferences. We define $\hat{\boldsymbol{x}}_a = (0 \cdots 0, x_a^{k+1} + \delta \cdots x_a^{\ell} + \delta)$ for some $\delta > 0$ with $\delta \sum_{t=k+1}^{\ell} p^t \leq p^1 x_a^1$. Then, we have $\boldsymbol{p}\hat{\boldsymbol{x}}_a \leq \boldsymbol{p}\omega_a$ and by the conditions (MT) and (CC), we see that $\boldsymbol{x}_a \prec_a \hat{\boldsymbol{x}}_a$, a contradiction. This means that if $y^t > -\omega^t$ for $1 \leq t \leq k$, we can reduce y^t without disturbing the preferences, maintaining $\boldsymbol{py} = 0$ and $\boldsymbol{y} \in T$, since T has the FD.

So, assume that $y^t = -\omega^t$ for $1 \leq t \leq k$. Now, suppose that $\boldsymbol{y} \notin Y$. Applying Theorem 2.1, we obtain a non-zero vector $\pi \geq \boldsymbol{0}$ such that $\pi \boldsymbol{y} > 0$ and $\pi \boldsymbol{z} \leq 0$ for all $\boldsymbol{z} \in T$. Since $\boldsymbol{y} \in T$, this is a contradiction which establishes $\boldsymbol{y} \in Y$. □

4.3. PRODUCTION ECONOMIES WITH EXTERNAL INCREASING RETURNS

Let L be an ordered vector space (Appendix H) which is to be understood as the commodity space. In this section, L is the ℓ-dimensional space $\mathbb{R}^\ell$. In Section 5.5 of Chap. 5, we will discuss an economy in which the commodity space L is an infinite-dimensional space. Let $L_+ = \{\boldsymbol{x} \in L \mid \boldsymbol{x} \geq \boldsymbol{0}\}$ be the non-negative orthant of L. Consider the function:

$$F: L_+ \times L_+ \to L_+, \;\; (\boldsymbol{z}, \boldsymbol{k}) \mapsto \boldsymbol{y} = F(\boldsymbol{z}, \boldsymbol{k}),$$

where $\boldsymbol{z} \in L_+$ stands for input vector, $\boldsymbol{y} \in L_+$ output, and $\boldsymbol{k} \in L_+$ is a parameter. The function F is called a technology function, which is distinguished from the (standard) production function.

Definition 3.1. The technology function $F(\boldsymbol{z}, \boldsymbol{k})$ is said to exhibit external increasing returns to scale (or social increasing returns to scale) if $F(\boldsymbol{z}, \boldsymbol{k})$ is homogeneous of degree 1 in $\boldsymbol{z}$,

$$F(\lambda \boldsymbol{z}, \boldsymbol{k}) = \lambda F(\boldsymbol{z}, \boldsymbol{k}) \;\; \text{for all } \lambda \geq 0 \quad \text{and all} \quad (\boldsymbol{z}, \boldsymbol{k}) \in L_+ \times L_+,$$

and it is of monotonically increasing in $\boldsymbol{k}$,

$$\boldsymbol{k} \leq \boldsymbol{k}' \text{ implies } F(\mathbf{z},\boldsymbol{k}) \leq F(\mathbf{z},\boldsymbol{k}') \quad \text{for all } \mathbf{z} \in L_+.$$

The firms will be assumed to take the parameter $\boldsymbol{k}$ as given when it maximizes the profit; on the other hand, $\boldsymbol{k}$ will be determined endogenously at the aggregate input level, $\boldsymbol{k} = \sum \mathbf{z}$ in equilibrium, where the summation is over all firms in the industry, or sometimes in the economy as a whole, depending on the range of externalities. When the function F is of external returns to scale, we see that it exhibits the "increasing returns" socially, for:

$$\lambda F(\mathbf{z},\boldsymbol{k}) = F(\lambda \mathbf{z},\boldsymbol{k}) \leq F(\lambda \mathbf{z}, \lambda \boldsymbol{k}) \quad \text{for all } \lambda \geq 1.$$

In the following, the technology function will be sometimes given by the (usual) production function with a parameter,

$$f\colon \mathbb{R}_+ \times \mathbb{R}_+ \to \mathbb{R}_+, \quad (z,k) \mapsto y = f(z,k).$$

We say that the production function f is also of the the external increasing returns if $f(z,k)$ is homogeneous of degree 1 in z, and it is monotonically increasing in k.

Example 3.1 (Chipman, 1970). There exist ℓ industries, each producing a single homogeneous commodity from a single factor, or the labor z^0. Every firm in the industry $t(=1 \cdots \ell)$ has the same production function:

$$y^t = k_t^{\epsilon_t - 1} z^t, \quad \epsilon_t > 0$$

which is of the external increasing returns to scale. If there exist ν_t firms in the industry t, assuming that they operate under the same level of input, $k_t = \nu_t z^t$ will hold in equilibrium. Then the firm's objective production function is:

$$y^t = \kappa_t (z^t)^{\epsilon_t}, \quad \kappa_t = (\nu_t)^{\epsilon_t - 1}, \; t = 1 \cdots \ell.$$

Chipman assumed that there existed m consumers with the same utility function:

$$u(x^1 \cdots x^\ell) = \sum_{t=1}^{\ell} \beta_t \log x^t.$$

Note that the utility function does not depend on the commodity zero, the labor. Hence, each consumer supplies the labor inelastically up to the amount which he/she initially holds. Assume that all consumer has one unit of labor and zero amount of other commodities; hence, the initial endowment vector of all consumers is $(1, 0, \ldots 0) \in \mathbb{R}^{\ell+1}$.

The competitive equilibrium is a triple of vectors $(\boldsymbol{p},\boldsymbol{x},\boldsymbol{z}) = ((p^t),(x^t),(z^t)) \in \mathbb{R}^\ell_+ \times \mathbb{R}^\ell_+ \times \mathbb{R}^\ell_+$ which satisfies that:

$$\boldsymbol{px} = w \quad \text{and} \quad \beta^t = \lambda p^t x^t,$$
$$\boldsymbol{py} = \boldsymbol{z}, \quad \text{where } \boldsymbol{y} = (y^t) = (k_t^{\epsilon_t - 1} z^t),$$
$$\sum_{t=1}^{\ell} v_t z^t = m, \quad k_t = v_t z^t,$$

where λ is the multiplier, and from the above equations, we have $w\lambda = \sum_{t=1}^{\ell} \beta^t$. In the next section, we will discuss the welfare property of the competitive equilibrium for a (generalized) Chipman model.

Example 3.2 (Romer, 1986). For simplicity, we present a discrete time version of the model of Romer (1986), which is an optimal growth model of the infinite time horizon. At each time period $t+1 \geq 1$, the output commodity y^{t+1} is produced from the input commodity z^t which is available at the previous period $t \geq 0$ through the external increasing returns to scale technology,

$$y^{t+1} = f(z^t, k^t), \quad t = 0, 1, \ldots$$

and in equilibrium, $k^t = z^t$ for all t should hold. The representative consumer's preference is given by a time-separable utility function:

$$U(\boldsymbol{x}) = \sum_{t=0}^{\infty} \delta^t u(x^t) \quad \text{for } \boldsymbol{x} = (x^t),$$

where $0 \leq \delta \leq 1$ is a discount factor. At the initial period $t = 0$, he/she has the initial endowment $\omega^0 > 0$. The competitive equilibrium is a triple of vectors $(\boldsymbol{p},\boldsymbol{x},\boldsymbol{z})$ which is characterized as:

$$\boldsymbol{px} \leq p^0\omega^0 \quad \text{and} \quad U(\boldsymbol{x}) \geq U(\boldsymbol{x}') \text{ whenever } \boldsymbol{px}' \leq p^0\omega^0,$$
$$p^{t+1} f(z^t, z^t) - p^t z^t = 0, \quad t = 0, 1, \ldots$$
$$\boldsymbol{x} + \boldsymbol{z} \leq (\omega^0, y^1, y^2 \ldots), \quad y^{t+1} = f(z^t, z^t), \ t = 0, 1 \ldots.$$

In Chap. 5, we will prove the existence of the competitive equilibrium for a (generalized) Romer model.

In the following, suppose that there exist m consumers in the economy $a = 1 \cdots m$. As usual, the consumer a has the preference-consumption set pair $(X_a, \prec_a)$ and the initial endowment vector $\omega_a \in \mathbb{R}^\ell$. The consumption set X_a is assumed to be a closed and convex subset of $\mathbb{R}^\ell$ which is bounded from below. We also assume that there exist n firms in the economy indexed by $b = 1 \cdots n$. The production technology of the firm b is described by the external increasing returns to scale technology function:

$$F_b \colon \mathbb{R}^\ell_+ \times \mathbb{R}^\ell_+ \to \mathbb{R}^\ell_+, \quad (\boldsymbol{z}_b, \boldsymbol{k}_b) \mapsto \boldsymbol{y}_b = F_b(\boldsymbol{z}_b, \boldsymbol{k}_b), \quad b = 1 \cdots n.$$

Then, the production set of the firm b, $Y(\boldsymbol{k}_b)$ is defined by:

$$Y_b(\boldsymbol{k}_b) = \{\boldsymbol{y}_b - \boldsymbol{z}_b \in \mathbb{R}^\ell \mid \boldsymbol{y}_b, \boldsymbol{z}_b \geq \boldsymbol{0},\ \boldsymbol{y}_b \leq F_b(\boldsymbol{z}_b, \boldsymbol{k}_b)\}, \quad b = 1 \cdots n,$$

which depends on the parameter $\boldsymbol{k}_b \in \mathbb{R}^\ell_+$.

Note that for each $\boldsymbol{k}_b$, the set $Y_b(\boldsymbol{k}_b)$ is a convex cone with the vertex at zero, and satisfies for each $\boldsymbol{k}_b \in \mathbb{R}^\ell_+$, the following:

(NP) (No production) $\boldsymbol{0} \in Y_b(\boldsymbol{k}_b)$, and
(FD) (Free disposability) $\mathbb{R}^\ell_- \subset Y_b(\boldsymbol{k}_b)$.

Obviously, the correspondence:

$$Y_b \colon \mathbb{R}^\ell_+ \to \mathbb{R}^\ell_+, \quad \boldsymbol{k}_b \mapsto Y(\boldsymbol{k}_b)$$

is continuous if and only if the function $F_b \colon \mathbb{R}^\ell_+ \times \mathbb{R}^\ell_+ \to \mathbb{R}^\ell_+$ is continuous.

We call the $(3m + n)$-tuple $\mathcal{E}_{\boldsymbol{k}} = (X_a, \prec_a, \omega_a, F_b)$ an economy with the external increasing returns or simply an economy.

The equilibrium concept for the economy $\mathcal{E}_{\boldsymbol{k}} = (X_a, \prec_a, \omega_a, F_b)$ is the competitive equilibrium with the externalities.

Definition 3.2. An $(m+2n+1)$-tuple $(\boldsymbol{p}, (\boldsymbol{x}_a), (\boldsymbol{y}_b, \boldsymbol{z}_b))$ is said to consist of a competitive equilibrium if and only if

(E-1) $\boldsymbol{p}\boldsymbol{x}_a \leq \boldsymbol{p}\omega_a$ and $\boldsymbol{x}_a \succsim_a \boldsymbol{x}'$ whenever $\boldsymbol{p}\boldsymbol{x}' \leq \boldsymbol{p}\omega_a$, $a = 1 \cdots m$,
(E-2) $\boldsymbol{p}\boldsymbol{y} \leq \boldsymbol{p}\boldsymbol{y}_b = 0$ for all $\boldsymbol{y} \in Y_b\big(\sum_{c=1}^{n} z_c\big)$, $b = 1 \cdots n$, and
(E-3) $\sum_{a=1}^{m} \boldsymbol{x}_a \leq \sum_{b=1}^{n} (\boldsymbol{y}_b - \boldsymbol{z}_b) + \sum_{a=1}^{m} \omega_a$.

Note that by the assumption of the homogeneity of F_b in the first variable, the profit of each firm in the competitive equilibrium is equal to zero. The next theorem plays a significant role in Chap. 5.

Theorem 3.1. *Suppose that an economy $\mathcal{E}_{\boldsymbol{k}} = (X_a, \prec_a, \omega_a, F_b)_{a \in A, b \in B}$ satisfies the continuity (CT), the convexity (CV), local non-satiation (LNS), and the minimum income condition (MI) for every $a = 1 \cdots m$, and for every $b = 1 \cdots n$, the function:*

$$F_b \colon \mathbb{R}^\ell_+ \times \mathbb{R}^\ell_+ \to \mathbb{R}^\ell_+, \quad (\boldsymbol{z}, \boldsymbol{k}) \mapsto \boldsymbol{y} = F_b(\boldsymbol{z}, \boldsymbol{k}),$$

is continuous. Finally, we assume that the set of feasible allocations:

$$\mathcal{F} = \left\{ ((\boldsymbol{x}_a), (\boldsymbol{y}_b, \boldsymbol{z}_b)) \in \prod_{a=1}^{m} X_a \times \mathbb{R}_+^{2\ell} \,\middle|\, \sum_{a=1}^{m} \boldsymbol{x}_a \right.$$

$$\left. \leq \sum_{b=1}^{n} (\boldsymbol{y}_b - \boldsymbol{z}_b) + \sum_{a=1}^{m} \omega_a, \ \boldsymbol{y}_b \leq F_b\left(\boldsymbol{z}_b, \sum_{c=1}^{n} \boldsymbol{z}_c\right) \right\}$$

is bounded. Then, there exists a competitive equilibrium $(\boldsymbol{p}^*, (\boldsymbol{x}_a^*), (\boldsymbol{y}_b^*, \boldsymbol{z}_b^*))$ *for* $\mathcal{E}$.

Proof. We will construct an $(m+n+1)$-person game Γ with the strategy space $\mathbb{R}^{2\ell}$, and apply Theorem 7.1 of Chap. 2.

Since the set $\mathcal{F}$ is bounded, we can take a compact and convex set $K \subset \mathbb{R}^{\ell}$ such that if $((\boldsymbol{x}_a), (\boldsymbol{y}_b, \boldsymbol{z}_b)) \in \mathcal{F}$, then $\boldsymbol{x}_a, \boldsymbol{y}_b, \boldsymbol{z}_b \in K$ for all $a \in A$ and all $b \in B$. We define:

$$\hat{X}_a = \{X_a \cap K\} \times \{\boldsymbol{0}\} \subset \mathbb{R}^{\ell} \times \mathbb{R}^{\ell}, \quad a = 1 \cdots m,$$

$$\hat{\Delta} = \Delta \times \{\boldsymbol{0}\}, \quad \text{where } \Delta = \left\{ (p^t) \in \mathbb{R}_+^{\ell} \,\middle|\, \sum_{t=1}^{\ell} p^t = 1 \right\}.$$

The first m players are described as follows. The player $a(=1 \cdots m)$ has the chouse set $\hat{X}_a$, the constraint correspondence $\mathcal{A}_a$: $\prod_{c=1}^{m} \hat{X}_c \times K^{2n} \times \hat{\Delta} \to \hat{X}_a$ which is defined by:

$$\mathcal{A}_a((\boldsymbol{x}_a, \boldsymbol{0}), (\boldsymbol{y}_b, \boldsymbol{z}_b), (\boldsymbol{p}, \boldsymbol{0})) = \{(\boldsymbol{x}, \boldsymbol{0}) \in \hat{X}_a \mid \boldsymbol{p}\boldsymbol{x} \leq \boldsymbol{p}\omega_a\},$$

and a has the preference correspondence P_a: $\prod_{c=1}^{m} \hat{X}_c \times K^{2n} \times \hat{\Delta} \to \hat{X}_a$ which is defined by:

$$P_a((\boldsymbol{x}_a, \boldsymbol{0}), (\boldsymbol{y}_b, \boldsymbol{z}_b), (\boldsymbol{p}, \boldsymbol{0})) = \{(\boldsymbol{x}, \boldsymbol{0}) \in \hat{X}_a \mid \boldsymbol{x}_a \prec_a \boldsymbol{x}\}.$$

By Proposition 3.2 of Chap. 2, the correspondence $\mathcal{A}_a$ is a nonempty, compact and convex-valued, and continuous correspondence. By the assumption LNS, the set $P_a((\boldsymbol{x}_a, \boldsymbol{0}), (\boldsymbol{y}_b, \boldsymbol{z}_b), (\boldsymbol{p}, \boldsymbol{0}))$ is nonempty for all $((\boldsymbol{x}_a, \boldsymbol{0}), (\boldsymbol{y}_b, \boldsymbol{z}_b), (\boldsymbol{p}, \boldsymbol{0})) \in \prod_{c=1}^{m} \hat{X}_c \times K^{2n} \times \hat{\Delta}$. In the proof of Theorem 5.2 of Chap. 2, we showed that the set $P_a((\boldsymbol{x}_a, \boldsymbol{0}), (\boldsymbol{y}_b, \boldsymbol{z}_b), (\boldsymbol{p}, \boldsymbol{0}))$ is convex, and since the preference $\prec_a$ is irreflexive, we have $\boldsymbol{x}_a \notin P_a((\boldsymbol{x}_a, \boldsymbol{0}), (\boldsymbol{y}_b, \boldsymbol{z}_b), (\boldsymbol{p}, \boldsymbol{0}))$. The open graph of P_a comes from the continuity (CT) of $\prec_a$.

The player $b(=1\cdots n)$ is described as follows. The choice set of the player b is $K\times K = K^2$, and he/she has the constraint correspondence $\mathcal{A}_b$: $\prod_{a=1}^m \hat{X}_a \times K^{2n} \times \hat{\Delta} \to K^2$ which is defined by:

$$\mathcal{A}_b((\boldsymbol{x}_a,\mathbf{0}),(\boldsymbol{y}_b,\boldsymbol{z}_b),(\boldsymbol{p},\mathbf{0})) = \left\{(\boldsymbol{y},\boldsymbol{z}) \in K^2 \,\middle|\, \boldsymbol{y} \leq F_b\left(\boldsymbol{z},\sum_{c=1}^n \boldsymbol{z}_c\right)\right\},$$

and the preference correspondence P_b: $\prod_{a=1}^m \hat{X}_a \times K^{2n} \times \hat{\Delta} \to K^2$, which is defined by:

$$P_b((\boldsymbol{x}_a,\mathbf{0}),(\boldsymbol{y}_b,\boldsymbol{z}_b),(\boldsymbol{p},\mathbf{0})) = \{(\boldsymbol{y},\boldsymbol{z}) \in K^2 \mid \boldsymbol{p}(\boldsymbol{y}-\boldsymbol{z}) > \boldsymbol{p}(\boldsymbol{y}_b-\boldsymbol{z}_b)\}.$$

The correspondence $\mathcal{A}_b$ is continuous. Indeed, let:

$$((\boldsymbol{x}_a(k),\mathbf{0}),(\boldsymbol{y}_b(k),\boldsymbol{z}_b(k)),(\boldsymbol{p}(k),\mathbf{0})) \in \prod_{a=1}^m \hat{X}_a \times K^{2n} \times \hat{\Delta}, \quad k = 0,1,\ldots$$

be a sequence which converges to $((\boldsymbol{x}_a,\mathbf{0}),(\boldsymbol{y}_b,\boldsymbol{z}_b),(\boldsymbol{p},\mathbf{0}))$, and let $(\boldsymbol{y}(k),\boldsymbol{z}(k))$ be a sequence in K^2 such that $\boldsymbol{y}(k) \leq F_b(\boldsymbol{z}(k),\sum_{c=1}^n \boldsymbol{z}_c(k))$ for all k, and $(\boldsymbol{y}(k),\boldsymbol{z}(k)) \to (\boldsymbol{y},\boldsymbol{z})$. Since, K is closed and the function F_b is continuous, one has $(\boldsymbol{y},\boldsymbol{z}) \in K^2$ and $\boldsymbol{y} \leq F_b(\boldsymbol{z},\sum_{c=1}^n \boldsymbol{z}_c)$. This shows that $\mathcal{A}_b$ is closed. Since K is compact, it is upper hemi-continuous.

Next, for a sequence $((\boldsymbol{x}_a(k),\mathbf{0}),(\boldsymbol{y}_b(k),\boldsymbol{z}_b(k)),(\boldsymbol{p}(k),\mathbf{0}))$ in $\prod_{a=1}^m \hat{X}_a \times K^{2n} \times \hat{\Delta}$ which converges to $((\boldsymbol{x}_a,\mathbf{0}),(\boldsymbol{y}_b,\boldsymbol{z}_b),(\boldsymbol{p},\mathbf{0})) \in \prod_{a=1}^m \hat{X}_a \times K^{2n} \times \hat{\Delta}$, let $(\boldsymbol{y},\boldsymbol{z}) \in K^2$ be a point with $\boldsymbol{y} \leq F_b(\boldsymbol{z},\sum_{c=1}^n \boldsymbol{z}_c)$. We define a sequence $(\boldsymbol{y}(k),\boldsymbol{z}(k)) \in K^2$ in the following way. For each k, set $\boldsymbol{z}(k) = \boldsymbol{z}$. For $t(=1\cdots \ell)$ such that $y^t = F^t(\boldsymbol{z},\sum_{c=1}^n \boldsymbol{z}_c)$, set $y^t(k) = F^t(\boldsymbol{z}(k),\sum_{c=1}^n \boldsymbol{z}_c(k))$, $k = 0,1\ldots$. For t such that $y^t < F^t(\boldsymbol{z},\sum_{c=1}^n \boldsymbol{z}_c)$, set $y^t(k) = y^t$, $k = 0,1\ldots$. Then, $(\boldsymbol{y}(k),\boldsymbol{z}(k)) \to (\boldsymbol{y},\boldsymbol{z})$ and:

$$\boldsymbol{y}(k) \leq F_b\left(\boldsymbol{z}(k),\sum_{c=1}^n \boldsymbol{z}_c(k)\right)$$

for all k sufficiently large. Therefore, $\mathcal{A}_b$ is lower hemi-continuous, hence continuous.

It follows from the homogeneity of degree zero in the first variable that $\mathcal{A}_b$ is nonempty, since $\mathbf{0} \in \mathcal{A}_b((\boldsymbol{x}_a(k),\mathbf{0}),(\boldsymbol{y}_b(k),\boldsymbol{z}_b(k)),(\boldsymbol{p}(k),\mathbf{0}))$, and convex-valued (it is a convex cone). Obviously, P_b has an open graph, convex-valued, and $(\boldsymbol{y}_b,\boldsymbol{z}_b) \notin P_b((\boldsymbol{x}_a,\mathbf{0}),(\boldsymbol{y}_b,\boldsymbol{z}_b),(\boldsymbol{p},\mathbf{0}))$.

The last player, called as the market player, has the choice set $\hat{\Delta}$ and a constraint correspondence:

$$\mathcal{A}((\boldsymbol{x}_a,\mathbf{0}),(\boldsymbol{y}_b,\boldsymbol{z}_b),(\boldsymbol{p},\mathbf{0})) = \hat{\Delta},$$

which is a natural projection, and the preference correspondence P: $\prod_{a=1}^{m} \hat{X}_a \times K^{2n} \times \hat{\Delta} \to \hat{\Delta}$ which is defined by:

$$P((\boldsymbol{x}_a,0),(\boldsymbol{y}_b,\boldsymbol{z}_b),(\boldsymbol{p},0))$$
$$= \left\{ (q,0)) \in \hat{\Delta} \,\middle|\, q\left(\sum_{a=1}^{m} \boldsymbol{x}_a - \sum_{b=1}^{n} (\boldsymbol{y}_b - \boldsymbol{z}_b) - \sum_{a=1}^{m} \omega_a\right) \right.$$
$$\left. > \boldsymbol{p}\left(\sum_{a=1}^{m} \boldsymbol{x}_a - \sum_{b=1}^{n} (\boldsymbol{y}_b - \boldsymbol{z}_b) - \sum_{a=1}^{m} \omega_a\right)\right\}.$$

It is clear that the correspondence $\mathcal{A}$ is nonempty and convex-valued, continuous correspondence and the correspondence P has an open graph, and it is nonempty, convex-valued, and $(\boldsymbol{p},0) \notin P((\boldsymbol{x}_a,0),(\boldsymbol{y}_b,\boldsymbol{z}_b),(\boldsymbol{p},0))$. We can apply Theorem 7.1 of Chap. 2 and obtain a Nash equilibrium $((\boldsymbol{x}_a,0),(\boldsymbol{y}_b,\boldsymbol{z}_b),(\boldsymbol{p},0))$ of the game Γ. We will show that it is a competitive equilibrium of the economy $\mathcal{E}_{\boldsymbol{k}}$. It is immediate that the conditions (E-1) and (E-2) are met. Then, by the condition (E-1), we have $\boldsymbol{p}\boldsymbol{x}_a \leq \boldsymbol{p}\omega_a$, $a = 1 \cdots m$. By the condition (E-2), $\boldsymbol{p}(\boldsymbol{y}_b - \boldsymbol{z}_b) = 0$, $b = 1 \cdots n$. Hence, it follows that:

$$\boldsymbol{p}\left(\sum_{a=1}^{m} \boldsymbol{x}_a - \sum_{b=1}^{n} (\boldsymbol{y}_b - \boldsymbol{z}_b) - \sum_{a=1}^{m} \omega_a\right) \leq 0 \quad \text{for all } \boldsymbol{p} \in \Delta.$$

Since the market player maximizes the value of the excess demand, taking $\boldsymbol{p} = \boldsymbol{e}_k = (\delta_k^t)$, where $\delta_k^t = 1$ for $t = k$ and $\delta_k^t = 0$ for $t \neq k$, we obtain:

$$\sum_{a=1}^{m} \boldsymbol{x}_a \leq \sum_{b=1}^{n} (\boldsymbol{y}_b - \boldsymbol{z}_b) + \sum_{a=1}^{m} \omega_a.$$

Therefore the condition (E-3) is met. Hence $(\boldsymbol{p},(\boldsymbol{x}_a),(\boldsymbol{y}_b,\boldsymbol{z}_b))$ is a competitive equilibrium for the economy $(\hat{X}_a, \omega_a, \prec_a |_K, F_b|_K)$.

Take compact and convex sets $K_n \subset \mathbb{R}^\ell$ such that $K \subset K_0 \cdots \subset K_n \subset \ldots$ and $\cup_{n=0}^{\infty} K_n = \mathbb{R}^\ell$. Repeating the above discussion for the game Γ_n which is obtained from Γ by replacing K by K_n, we get a sequence of the Nash equilibria or the competitive equilibria $(\boldsymbol{p}(n),(\boldsymbol{x}_a(n)),(\boldsymbol{y}_b(n),\boldsymbol{z}_b(n)))$, $n \in \mathbb{N}$. As in the same way of the proof of Theorem 1.2, we can show that the limit $(\boldsymbol{p}^*,(\boldsymbol{x}_a^*),(\boldsymbol{y}_b^*,\boldsymbol{z}_b^*))$ is a competitive equilibrium of the economy $\mathcal{E}_{\boldsymbol{k}}$. □

4.4. PARETO OPTIMALITY AND TAX POLICIES

In this section, we will discuss the Pareto optimality of the competitive equilibria in the (generalized) Chipman economy with the external increasing returns. The commodity space is assumed to be $\mathbb{R}^\ell = \mathbb{R}^{n+1}$. Let $(x^0, x^1 \cdots x^n) \in \mathbb{R}^{n+1}$. There exist n industries indexed by $b = 1 \cdots n$. The industry b produces the commodity b, and the commodity zero is called the labor or factor, and it is used as an input and not producible.

In the industry b, there exist finitely many identical firms, each of which has the increasing returns to scale production function:

$$f_b: \mathbb{R}_+ \times \mathbb{R}_+ \to \mathbb{R}_+, \quad (z,k) \mapsto y = f(z,k).$$

Let $z_{b\nu}$ be the input level of the ν-th firm in the industry b. The aggregate input level of the industry is then $z^b = \sum_\nu z^b_\nu$. Since the firms are identical, the aggregate output level can be written as $y^b = \sum_\nu f_b(z^b_\nu, z^b) = f_b\left(\sum_\nu z^b_\nu, z^b\right) = f_b(z^b, z^b)$. Hence, we do not have to specify the individual firm input level z^b_ν. For notational simplicity, we denote $g_b(z^b) = f_b(z^b, z^b)$.

There exist m consumers, such that $a = 1 \cdots m$. The consumer a has the utility function on the consumption set $X_a = \{0\} \times \mathbb{R}^n_+ \approx \mathbb{R}^n_+$,

$$u_a: \mathbb{R}^n_+ \to \mathbb{R}, \quad \boldsymbol{x} \mapsto u_a(\boldsymbol{x}), \ a = 1 \cdots m.$$

Note that we assume that the consumer's utility does not depend on the production factor (labor), the commodity zero. Therefore, the consumers supply the amount of the labor, which they initially own inelastically. We assume that the consumer a owns initially the labor only, hence the endowment vector of the consumer is of the form $\omega_a = (\omega^0_a, 0 \cdots 0) \in \mathbb{R}^{n+1}$, $\omega^0_a > 0$. Further, we assume that the utility function is monotone,

(UMT) (Monotonicity) if $\boldsymbol{x}, \boldsymbol{z} \in X_a$ and $\boldsymbol{x} < \boldsymbol{z}$, then $u_a(\boldsymbol{x}) < u_a(\boldsymbol{z})$.

Although this economy does not satisfy the minimum income condition (MI), we can demonstrate that there exists a competitive equilibrium using the Negishi method of proof. Since it exploits the welfare theorem, it gives us a lot of information on the optimality of equilibria, and it is a basis of our discussion of tax policies implementing the social optima.

Let $\mathcal{F}$ be the set of feasible allocations:

$$\mathcal{F} = \left\{ ((\boldsymbol{x}_a), \boldsymbol{z}) = ((x^b_a), (z^b)) \in \mathbb{R}^{mn}_+ \times \mathbb{R}^n_+ \,\middle|\, \sum_{a=1}^m \boldsymbol{x}_a \leq \boldsymbol{y} = (y^b), \right.$$
$$\left. y^b \leq f_b\left(z^b, \sum_{c=1}^n z^c\right), \quad 0 \leq z^b \leq \sum_{a=1}^m \omega^0_a, \ b = 1 \cdots n \right\}.$$

Since $0 \leq \boldsymbol{x}_a \leq \sum_{a=1}^m \boldsymbol{x}_a \leq f_b\left(\sum_{c=1}^n \omega^0_c, \sum_{c=1}^n \omega^0_c\right)$, the set $\mathcal{F}$ is bounded. Let K be a compact and convex subset of $\mathbb{R}^{mn} \times \mathbb{R}^n$ such that $\mathcal{F} \subset$ interior K.

Consider the following constrained social optimization problem $P(k)$.

$$P(k):\ \text{Given } k = (k_b) \in \mathbb{R}^n_+ \text{ and } \alpha = (\alpha^a) \in \mathbb{R}^m_+ \text{ with } \sum_{a=1}^{m} \alpha^a = 1,$$

$$\text{maximize } \sum_{a=1}^{m} \alpha^a u_a(\boldsymbol{x}_a) \text{ subject to}$$

$$\sum_{a=1}^{m} x_a^b \leq f_b(z^b, k^b), \quad b = 1 \cdots m, \sum_{b=1}^{n} z^b \leq \sum_{a=1}^{m} \omega_a^0, \quad ((\boldsymbol{x}_a), \boldsymbol{z}) \in K.$$

Since the set K is compact and convex, the problem $P(k)$ has a solution when the utility function $u_a(\cdot)$ is continuous and concave. By the Kuhn–Tucker theorem (Theorem A5) of Appendix A, the solution of the problem $P(k)$ is a saddle point of the Lagrangian[1]

$$\mathcal{L}_{\alpha,\boldsymbol{k}}((\boldsymbol{x}_a), z, \boldsymbol{p}, w) = \sum_{a=1}^{m} \alpha^a u_a(\boldsymbol{x}_a) + \sum_{b=1}^{n} p^b \left(\sum_{b=1}^{n} f_b(z^b, k_b) - \sum_{a=1}^{m} x_a^b \right)$$
$$+ w \left(\sum_{a=1}^{m} \omega_a^0 - \sum_{b=1}^{n} z^b \right),$$

where $\boldsymbol{p} = (p^b) \in \mathbb{R}^n_+$ and $w \in \mathbb{R}_+$ are the Lagrangian multipliers.

Let $((\hat{\boldsymbol{x}}_a), \hat{\boldsymbol{z}}, \hat{\boldsymbol{p}}, \hat{w})$ be the saddle point. We can take a constant $b > 0$ which satisfies:

$$b > \sup \left\{ \sum_{a=1}^{m} | w\omega_a^0 - \boldsymbol{p}\boldsymbol{x}_a | \ \middle|\ (\boldsymbol{p}, w) \in \Delta, ((\boldsymbol{x}_a), \boldsymbol{z}) \in \mathcal{F} \right\}.$$

For given $((\alpha^a), (\boldsymbol{x}_a), \boldsymbol{p}, w)$, we define $\hat{\alpha} = (\hat{\alpha}^a)$ by:

$$\hat{\alpha}^a = \frac{\max\{0, \alpha^a + (1/b)(w\omega_a^0 - \boldsymbol{p}\boldsymbol{x}_a)\}}{\sum_{a=1}^{m} \max\{0, \alpha^a + (1/b)(w\omega_a^0 - \boldsymbol{p}\boldsymbol{x}_a)\}}, \quad a = 1 \cdots m.$$

Let $Z = \{\boldsymbol{z} \in \mathbb{R}^n_+ \mid ((\boldsymbol{x}_a), \boldsymbol{z}) \in K \text{ for some } (\boldsymbol{x}_a) \in \mathbb{R}^{mn}\}$. Obviously, Z is a compact and convex subset of $\mathbb{R}^n$. Then, we can show that the competitive equilibrium is a fixed

[1] In order to guarantee Slater's condition, it is suffice to assume that $f_b(z, 0) > 0$ for $z > 0$. But this point is not so important for the subsequent discussions on optimality.

point of the mapping Φ of $\Delta \times Z \times K \times \Delta$ to itself defined by:

$$\Phi: (\alpha, k, (\boldsymbol{x}_a), \boldsymbol{z}, \boldsymbol{p}, w) \mapsto (\hat{\alpha}, \hat{\boldsymbol{z}}, (\hat{\boldsymbol{x}}_a), \hat{\boldsymbol{z}}, \hat{\boldsymbol{p}}, \hat{w}).$$

By the Kakutani's fixed-point theorem (Theorem D1) of Appendix D, there exists a fixed point:

$$(\hat{\alpha}, \hat{\boldsymbol{z}}, (\hat{\boldsymbol{x}}_a), \hat{\boldsymbol{z}}, \hat{\boldsymbol{p}}, \hat{w}) \in \Phi(\hat{\alpha}, \hat{\boldsymbol{z}}, (\hat{\boldsymbol{x}}_a), \hat{\boldsymbol{z}}, \hat{\boldsymbol{p}}, \hat{w}).$$

We will show that the fixed point is a competitive equilibrium. Since the fixed point is a saddle point of the Lagrangian, we have:

$$\mathcal{L}_{\hat{\alpha}\hat{\boldsymbol{z}}}((\boldsymbol{x}_a), \boldsymbol{z}, \hat{\boldsymbol{p}}, \hat{w}) \leq \mathcal{L}_{\hat{\alpha}\hat{\boldsymbol{z}}}((\hat{\boldsymbol{x}}_a), \hat{\boldsymbol{z}}, \hat{\boldsymbol{p}}, \hat{w}) \leq \mathcal{L}_{\hat{\alpha}\hat{\boldsymbol{z}}}((\hat{\boldsymbol{x}}_a), \hat{\boldsymbol{z}}, \boldsymbol{p}, w)$$

for every $((\boldsymbol{x}_a), \boldsymbol{z}) \in K$, and every $\boldsymbol{p} \geq \boldsymbol{0}$ and $w \geq 0$.

It follows from the monotonicity of the utility functions (UTM) that $\boldsymbol{p} \gg \boldsymbol{0}$ and $w > 0$. For if not, we have a contradiction to the first inequality of the saddle point property, which implies that the maximality of $\mathcal{L}_{\hat{\alpha}\hat{\boldsymbol{z}}}$ with respect to $((\boldsymbol{x}_a), \boldsymbol{z})$. Setting $\boldsymbol{x}_a = \hat{\boldsymbol{x}}_a, a = 1 \cdots m$, and $z_c = \hat{z}_c, c \neq b$, it follows from the first inequality and from the homogeneity of the production function with respect to the first variable,

$$\hat{p}^b f_b\left(\hat{z}^b, \sum_{c=1}^{n} \hat{z}^c\right) - \hat{w}\hat{z}^b = 0, \quad b = 1 \cdots n.$$

Therefore, the profit maximization of the firms in the industry $b = 1 \cdots n$ is established. Next, the monotonicity of the utility function implies that $\hat{\alpha}^a > 0$, $a = 1 \cdots m$. For if not, then $\hat{\alpha}^a = 0$ for some a. By the definition of $\hat{\alpha}$, for such an a we have $0 = \max\{0, (1/b)(\hat{w}\omega_a^0 - \hat{\boldsymbol{p}}\hat{\boldsymbol{x}}_a)\}$, hence, $0 < \hat{w}\omega_a^0 \leq \hat{\boldsymbol{p}}\hat{\boldsymbol{x}}_a$. Setting $\boldsymbol{x}_a = \boldsymbol{0}$ and $\boldsymbol{x}_b = \hat{\boldsymbol{x}}_a + \hat{\boldsymbol{x}}_b$ for a consumer b such that $\hat{\alpha}^b > 0$, we have a contradiction to the first inequality of the saddle point property. Therefore, it follows that:

$$\hat{\alpha}^a = \frac{\hat{\alpha}^a + (1/b)(\hat{w}\omega_a^0 - \hat{\boldsymbol{p}}\hat{\boldsymbol{x}}_a)}{\sum_{c=1}^{m}\{\hat{\alpha}^c + (1/b)(\hat{w}\omega_c^0 - \hat{\boldsymbol{p}}\hat{\boldsymbol{x}}_c)\}}, \quad a = 1 \cdots m,$$

and from this, we obtain:

$$\hat{w}\omega_a^0 - \hat{\boldsymbol{p}}\hat{\boldsymbol{x}}_a = \hat{\alpha}^a \sum_{c=1}^{m} (\hat{w}\omega_c^0 - \hat{\boldsymbol{p}}\hat{\boldsymbol{x}}_c), \quad a = 1 \cdots m.$$

On the other hand, by the second inequality which minimizes the Lagrangian with respect to the multipliers (prices),

$$0 = \sum_{a=1}^{m} \left(\hat{w}\omega_a^0 - \hat{\boldsymbol{p}}\hat{\boldsymbol{x}}_a\right) + \sum_{b=1}^{n} \left(\hat{p}^b f_b(\hat{z}^b, \hat{z}^b) - \hat{w}\hat{z}^b\right)$$

$$= \sum_{a=1}^{m} \left(\hat{w}\omega_a^0 - \hat{\boldsymbol{p}}\hat{\boldsymbol{x}}_a\right),$$

hence one obtains:

$$\hat{\boldsymbol{p}}\hat{\boldsymbol{x}}_a = \hat{w}\omega_a^0, \quad a = 1 \cdots m.$$

Setting $\boldsymbol{x}_a = \boldsymbol{x}$ and $\boldsymbol{x}_b = \hat{\boldsymbol{x}}_b$ for $b \neq a$, we have from the first inequality that:

$$\hat{\alpha}^a u_a(\boldsymbol{x}) - \boldsymbol{p}\boldsymbol{x} \leq \hat{\alpha}^a u_a(\hat{\boldsymbol{x}}) - \hat{\boldsymbol{p}}\hat{\boldsymbol{x}}_a = \hat{\alpha}^a u_a(\hat{\boldsymbol{x}}_a) - \hat{w}\omega_a^0,$$

or we have:

$$\hat{w}\omega_a^0 - \hat{\boldsymbol{p}}\boldsymbol{x} \leq \hat{\alpha}^a (u_a(\hat{\boldsymbol{x}}_a) - u_a(\boldsymbol{x})).$$

Hence, $\hat{\boldsymbol{p}}\boldsymbol{x} \leq \hat{w}\omega_a^0$ implies that $u_a(\boldsymbol{x}) \leq u_a(\hat{\boldsymbol{x}}_a)$, $a = 1 \cdots m$. Therefore, the utility maximization for each consumer is established.

Finally, the market conditions are obvious from the constraint conditions of the problem $P(k)$; indeed they hold with the exact equality by virtue of $\hat{\boldsymbol{p}} \gg \boldsymbol{0}$ and $\hat{w} > 0$.

We have obtained the next theorem.

Theorem 4.1. *Suppose that the utility functions are continuous, concave, and monotone, and each consumer has a strictly positive amount of labor $\omega_a^0 > 0$ as the initial endowment.*[2] *Then, there exists a competitive equilibrium for the generalized Chipman model.*

In the following paragraphs, we discuss the Pareto optimality of the competitive equilibria. For definiteness, we assume that the utility function is of the Cobb–Douglas form:

$$u_a((x_a^b)) = \sum_{b=1}^{n} \beta_a^b \log x_a^b, \quad \sum_{b=1}^{n} \beta_a^b = 1, \quad \beta_a^b \geq 0, \quad a = 1 \cdots m, \quad b = 1 \cdots n,$$

---

[2]See also the previous foot-note for Slater's condition.

and we calculate the equilibrium explicitly. The first-order conditions for the saddle point property of $\mathcal{L}_{\hat{\alpha},\hat{z}}$ are given by:

$$\hat{\alpha}^a \frac{\beta_a^b}{\hat{x}_a^b} - \hat{p}^b = 0, \quad a = 1 \cdots m, \;\; b = 1 \cdots n,$$

$$\hat{p}^b f_b(\hat{z}^b, \hat{z}^b) - \hat{w}\hat{z}^b = 0, \quad b = 1 \cdots n.$$

Summing the first mn equations over b and using $\hat{\boldsymbol{p}}\hat{\boldsymbol{x}}_a = \hat{w}\omega_a^0$, we obtain:

$$\hat{\alpha}^a = \hat{w}\omega_a^0, \quad a = 1 \cdots m.$$

Summing the same equations over a with the help of the above m identities,

$$\hat{w}\sum_{a=1}^m \beta_a^b \omega_a^0 = \sum_{a=1}^m \hat{\alpha}^a \beta_a^b = \sum_{a=1}^m \hat{p}^b \hat{x}_a^b$$
$$= \hat{p}^b f_b(\hat{z}^b, \hat{z}^b) = \hat{w}\hat{z}^b,$$

hence, we have:

$$\hat{z}^b = \sum_{a=1}^m \beta_a^b \omega_a^0, \quad b = 1 \cdots n.$$

The equilibrium (relative) prices are then determined by:

$$\frac{\hat{p}^b}{\hat{w}} = \frac{\hat{z}^b}{f_b(\hat{z}^b, \hat{z}^b)}, \quad b = 1 \cdots n,$$

and the equilibrium consumptions are:

$$\hat{x}_a^b = \left(\frac{\hat{w}}{\hat{p}^b}\right)\beta_a^b \omega_a^0 = \left(\frac{\beta_a^b \omega_a^0}{\sum_{a=1}^m \beta_a^b \omega_a^0}\right) f_b(\hat{z}^b, \hat{z}^b), \quad a = 1 \cdots m, \;\; b = 1 \cdots n.$$

We now turn to the Pareto-optimal allocations. Let $g_b(z^b) = f_b(z^b, z^b)$ and consider the social optimization problem:

P: Given $\alpha = (\alpha^a) \in \mathbb{R}_+^m$,

$$\text{maximize} \quad \sum_{a=1}^m \alpha^a \sum_{b=1}^n \beta_a^b \log x_a^b \text{ subject to:}$$

$$\sum_{a=1}^m x_a^b \leq g_b(z^b), \;\; b = 1 \cdots m, \quad \sum_{b=1}^n z^b \leq \sum_{a=1}^m \omega_a^0, \quad ((\boldsymbol{x}_a), \boldsymbol{z}) \in K.$$

Note that the problem P does not depend on k any more. We want to compare the solution $((\tilde{\boldsymbol{x}}_a), (\tilde{z}^b))$ of P with the equilibrium allocation $((\hat{\boldsymbol{x}}_a), (\hat{z}^b))$. The problem is what a social weight $\alpha = (\alpha^a)$, should we use for the social welfare maximization problem P? Fortunately, we have already obtained the answer, namely that we should use $\alpha^a = \hat{\alpha}^a = \hat{w}\omega_a^0$, $a = 1\cdots m$. Then, the first-order conditions for the problem P are written as:

$$\hat{\alpha}^a \frac{\beta_a^b}{\tilde{x}_a^b} - \lambda^b = 0, \quad a = 1\cdots m, \ b = 1\cdots n,$$

$$\lambda^b g_b'(\tilde{z}^b) - \mu = 0, \quad b = 1\cdots n,$$

where λ^b and μ are the Lagrangian multipliers.

Summing the first mn equations over a, we obtain:

$$\lambda^b = \frac{\sum_{a=1}^m \hat{\alpha}^a \beta_a^b}{\sum_{a=1}^m \tilde{x}_a^b} = \frac{\sum_{a=1}^m \hat{\alpha}^a \beta_a^b}{g_b(\tilde{z}^b)}, \quad b = 1\cdots n.$$

Substituting these equalities into the second group of n equations of the first-order conditions, we have:

$$\begin{aligned}\mu &= \left(\sum_{a=1}^m \hat{\alpha}^a \beta_a^b\right)(g_b'(\tilde{z}^b)/g_b(\tilde{z}^b)) \\ &= \left(\hat{w}\sum_{a=1}^m \beta_a^b \omega_a^0\right)(g_b'(\tilde{z}^b)/g_b(\tilde{z}^b)), \quad b = 1\cdots n.\end{aligned}$$

Multiplying these equations by $\tilde{z}^b$ and summing over b with the help of $\sum_{b=1}^n \tilde{z}_b = \sum_{a=1}^m \omega_a^0$, we obtain:

$$\begin{aligned}\mu &= \hat{w}\sum_{b=1}^n \left(\frac{\sum_{a=1}^m \beta_a^b \omega_a^0}{\sum_{a=1}^m \omega_a^0}\right)\left(\frac{g_b'(\tilde{z}^b)\tilde{z}^b}{g_b(\tilde{z}^b)}\right) \\ &= \hat{w}\sum_{b=1}^n \left(\frac{\sum_{a=1}^m \beta_a^b \omega_a^0}{\sum_{a=1}^m \omega_a^0}\right)\epsilon_b(\tilde{z}^b),\end{aligned}$$

where $\epsilon_b(\tilde{z}^b) = g_b'(\tilde{z}^b)\tilde{z}^b/g_b(\tilde{z}^b)$ is the elasticity of the production for the firm b. From this and the previous n equations, we obtain:

$$\tilde{z}^b = \left(\frac{\epsilon_b(\tilde{z}^b)}{\sum_{b=1}^n \left(\sum_{a=1}^m \beta_a^b \omega_a^0 / \sum_{a=1}^m \omega_a^0\right)\epsilon_b(\tilde{z}^b)}\right)\sum_{a=1}^m \beta_a^b \omega_a^0, \quad b = 1\cdots n,$$

and

$$\tilde{x}_a^b = \frac{\hat{\alpha}^a \beta_a^b}{\lambda^b} = \left(\frac{\beta_a^b \omega_a^0}{\sum_{a=1}^m \beta_a^b \omega_a^0}\right) g_b(\tilde{z}^b), \quad a = 1\cdots m, \ b = 1\cdots n.$$

In view of $\hat{\mathbf{x}}_a, \hat{\mathbf{z}}, \tilde{\mathbf{x}}_a$, and $\tilde{\mathbf{z}}$, we see that:

$$\tilde{x}_a^b \geq \hat{x}_a^b \quad \text{if and only if } \tilde{z}^b \geq \hat{z}^b, \quad a = 1 \cdots m, \quad b = 1 \ldots n.$$

This is equivalent to:

$$\epsilon_b(\tilde{z}^b) \geq \sum_{b=1}^{n} \chi_b \epsilon_b(\tilde{z}^b), \quad b = 1 \cdots n,$$

where $\chi_b = \sum_{a=1}^m \beta_a^b \omega_a^0 / \sum_{a=1}^m \omega_a^0, b = 1 \cdots n.$

Therefore, we have obtained the next fundamental result on the optimality of the competitive equilibria.

Theorem 4.2. *Under the above assumptions, optimal output of the b-th producer is greater than, equal to, or less than the competitive output according as the elasticity of the industry (at optimum) is greater than, equal to, or less than the weighted average of elasticities of all industries. If all industries' elasticities are equal to the weighted average, the competitive equilibrium is Pareto optimal.*

In the following paragraphs, we assume that the objective production function is of the constant elasticity form:

$$g_b(z) = \kappa_b z^{\epsilon_b}, \quad b = 1 \cdots n$$

and we will discuss the tax policies to implement the optimal competitive equilibrium. We recall the consumers' and the producers' optimal conditions,

$$p^b x_a^b = \hat{w} \beta_a^b \omega_a^0, \quad a = 1 \cdots m,$$

where we set $\alpha^a = \hat{\alpha}^a = \hat{w} \omega_a^0$, and we have:

$$q^b g_b(z^b) = \hat{w} z^b, \quad b = 1 \cdots n.$$

First, we consider the case of per-unit exercise taxes,

$$p^b - q^b = t^b, \quad b = 1 \cdots n.$$

Then, we have:

$$t^b g_b(z^b) = p^b \sum_{a=1}^{m} x_a^b - q^b g(z^b)$$

$$= \hat{w} \left(\sum_{a=1}^{m} \beta_a^b \omega_a^0 - z^b \right), \quad b = 1 \cdots n.$$

Hence, $\sum_{b=1}^{n} t^b g_b(\hat{z}^b) = 0$, or the total subsidies paid is equal to the total taxes collected. Setting $z^b = \tilde{z}^b$, the optimal tax rates are calculated as:

$$\tilde{t}^b = \frac{\hat{w}}{\kappa_b}\left(1 - \frac{\epsilon_b}{\sum_{b=1}^{n} \chi_b \epsilon_b}\right)^{-\epsilon_b}\left(\sum_{a=1}^{m} \beta_a^b \omega_a^0\right)^{1-\epsilon_b}, \quad b = 1 \cdots n.$$

$\tilde{t}^b \geq 0$ if and only if $\sum_{b=1}^{n} \chi_b \epsilon_b \geq \epsilon_b$. When $\sum_{b=1}^{n} \chi_b \epsilon_b = 1$, this reduces to Marshall's rule followed by Pigou (1912) which asserts that one should levy taxes on the industry, which exhibits the decreasing returns to scale ($\epsilon_b < 1$) and should subsidize the increasing returns to scale industries.

Next, we consider the case of *ad valorem* taxes,

$$p^b - q^b = \tau^b p^b, \quad b = 1 \cdots n.$$

In this case, the optimal tax rates are calculated as:

$$\tilde{\tau}^b = 1 - \frac{q^b g_b(\tilde{z}^b)}{p^b \sum_{a=1}^{m} \tilde{x}_a^b}$$

$$= 1 - \frac{\tilde{z}^b}{\sum_{a=1}^{m} \beta_a^b \omega_a^0} = 1 - \frac{\epsilon_b}{\sum_{b=1}^{n} \chi_b \epsilon_b}, \quad b = 1 \cdots n.$$

Consider the (objective) marginal cost:

$$\delta_b = \frac{d}{dy^b}\hat{w} z^b = \frac{\hat{w}}{g_b'(z^b)}, \quad b = 1 \cdots n,$$

and the market value of the objective marginal product:

$$\rho_b = p^b g_b'(z^b), \quad b = 1 \cdots n.$$

Then, it follows that:

$$\frac{p^b}{\delta_b} = \frac{\rho_b}{\hat{w}} = \frac{p^b g_b'(z^b) z^b}{\hat{w} z^b} = \epsilon_b \frac{p^b g_b(z^b)}{\hat{w} z^b}$$

$$= \epsilon_b \frac{p^b \sum_{a=1}^{m} x_a^b}{\hat{w} z^b} = \epsilon_b \frac{\sum_{a=1}^{m} \beta_a^b \omega_a^0}{z^b}, \quad b = 1 \cdots m.$$

Hence, in the competitive equilibrium, setting $z^b = \hat{z}^b = \sum_{a=1}^{m} \beta_a^b \omega_a^0$, we obtain:

$$\frac{\hat{p}^b}{\delta_b} = \frac{\rho_b}{\hat{w}} = \epsilon_b, \quad b = 1 \cdots n,$$

whereas, the optimal competitive equilibrium $z^b = \tilde{z}^b = (\epsilon_b / \sum_{b=1}^n \chi_b \epsilon_b) \sum_{a=1}^m \beta_a^b \omega_a^0$ requires:

$$\frac{\tilde{p}^b}{\delta_b} = \frac{\tilde{\rho}_b}{\hat{w}} = \sum_{b=1}^{n} \chi_b \epsilon_b, \quad b = 1 \cdots n.$$

This is the proportionality rule developed by Pigou and Kahn; for all commodities, price must be proportional to the marginal costs, and the wage rate to the value of the marginal product, the factor of proportionality being the average of the production elasticities, $\sum_{b=1}^n \chi_b \epsilon_b$. Pigou wrote in his *The Economics of Welfare* 4th edition, p. 225:

> "When it was urged above that in certain industries a wrong amount of resources is being invested because the value of the marginal social net product there differs from the value of the marginal private net product, it was tacitly assumed that in the main body of industries these two values are equal .... If in all industries that value of marginal social and marginal private net product differed to exactly the same extent, the *optimum* distribution of resources would always be attained and there would be ... no case for financial interference."

If the average value $\sum_{b=1}^n \chi_b \epsilon_b$ is equal to one, we have the special case in which price should be set equal to the marginal costs, and the wage rate to the marginal product. This is nothing but the marginal cost pricing (MCP) rule.

4.5. NOTES

Section 4.1: The contents of this section are standard. The most basic reference is Debreu (1959), Chaps. 5 and 6. See also Nikaido (1968), Chap. 5 and Arrow and Hahn (1971), Chaps. 3 and 5, and McKenzie (2002), Chaps. 5 and 6.

Section 4.2: All of the expositions are proposed by Scarf (1986). His analysis is definitive in the sense that he proved a converse of Theorem 2.2.

Theorem. *Suppose that Y be a closed production set satisfying the conditions* (NP), (FD), (PC), (BTP). *Assume that there exists a non-empty core for every economy in which the consumers' preferences satisfy* (MT), (CV), (CT), *and* (CC), *and in which the initial endowments of all consumers are strictly positive* (PE). *Then, Y is a distributive set.*

Dehez and Dreze (1988b) proposed a related concept of "output-distributive" production sets. In their terminology, the Scarf's definition is called "input distributive" sets.

Section 4.3: As stated in Section 1.5, the concept of the external increasing returns was originally proposed by Edgeworth. It was formulated in the modern theoretical framework by Chipman (1970) and Romer (1986). Extending their results to the case of several consumers, the general equilibrium analysis was started by Suzuki (1992). The expositions in this section are based on the works of Suzuki (1996).

Section 4.4: The expositions of this section are essentially proposed by Chipman (1970). In *Principles*, p. 389, Marshall wrote:

> "... By similar reasoning, it may be shown that a tax on a commodity which obeys the law of increasing return is more injurious to the consumer than if levied on one which obeys the law of constant return. For it lessens the demand and therefore the output. It thus probably increases the expenses of manufacture somewhat: sends up the price by more than the amount of the tax; and finally diminishes consumers' surplus by much more than the total payments which it brings in to the exchequer. On the other hand, a bounty on such a commodity causes so great a fall in its price to the consumer, that the consequent increase of consumers' surplus may exceed the total payments made by the State to the producers; and certainly will do so in case the law of increasing return acts at all sharply."

In the light of the analysis of this section, the theoretical background of the so called "Pigou tax" originally proposed by Marshall cited above seems to become clear.

ECONOMIES WITH INFINITELY MANY COMMODITIES

Chapter **5**

5.1. MARKETS WITH INFINITELY MANY COMMODITIES

In this chapter, we will discuss the markets in which infinitely many commodities are traded. As discussed in Section 2.1 of Chap. 2, a commodity is distinguished from each other by its physical character, the location, and the time at, which it is available to traders. If any of these properties is to be indexed by infinitely many parameters, the commodity space will be naturally infinite-dimensional.

Let us start with the available time period of commodities. As Debreu stated (see Section 1.6 of Chap. 1), "there are conceptual difficulties in postulating a predetermined instant beyond which all economic activity either ceases or is outside the scope of analysis." Therefore, a natural candidate for the commodity space in order to handle this situation is a sequence space for discrete time model or a function space when the model is of continuous time. In the following, we will discuss the case of discrete time models, hence we will work with the spaces of sequences. Then, the commodity vector is represented by a sequence:

$$\boldsymbol{x} = (x^0, x^1 \cdots x^t \cdots).$$

The t-th coordinate x^t of $\boldsymbol{x}$ stands for the x^t amount of the goods which is available to the market at the date (period) t. All contracts are performed "now", or at the initial date zero. Hence, for the price vector $\boldsymbol{p}$ which is also given by a sequence:

$$\boldsymbol{p} = (p^0, p^1 \cdots p^t \cdots),$$

the t-th coordinate p^t of $\boldsymbol{p}$ means that p^t amount of money for one unit of good, which will be available at the period t. Note that the money is paid "now", not at the date t.

In order to describe this kind of market structure, the most natural space seems to be $\mathbb{R}^\infty$; the space of all sequences endowed with the topology of pointwise convergence, or the product topology. It seems natural that the commodity space contains all streams of commodities, since we do not have any particular reason to exclude some commodity stream from the analysis *a priori*. However, the space $\mathbb{R}^\infty$ is technically not manageable very much. This space is so large that the dual space of $\mathbb{R}^\infty$ is the space of vectors, which have all but finite number of zero coordinates. This means that if we require that all commodity bundles $\boldsymbol{x}$ should have finite value $\boldsymbol{px} < \infty$, we have to restrict the price vectors $\boldsymbol{p} = (p^t)$ within the class of vectors with only finite number of p^t are nonzero. If we allow price vectors with infinitely many non-zero coordinates, the market value of some of the commodity vectors will be infinite, $\boldsymbol{px} = \pm\infty$. In such a model, obviously, we would be in trouble technically to handle the budget constraints of the consumers and/or the profit maximization of the firms.

Bewley (1970) proposed the space ℓ^∞ which is the space of all bounded sequences as the commodity space. The space ℓ^∞ is a Banach space with the supremum norm. The dual space of ℓ^∞ is the space *ba*, the set of finitely additive set functions on $\mathbb{N}$ with the total variation norm (Appendix H). The space *ba* is not a sequence space. From the economic point of view, however, it is more natural that the price vector is an element of the subspace *ca* of *ba*, which is the space of countably additive set functions on $\mathbb{N}$. The space *ca* is identified with the space ℓ^1, which is also a Banach space with the norm:

$$\|\boldsymbol{p}\| = \sum_{t=0}^{\infty} |p^t|, \quad \boldsymbol{p} = (p^t) \in \ell^1,$$

where we set $p^t = p(\{t\})$. For the mathematical properties of these spaces and related results, see Appendix H. In Section 5.2, we will define and prove the existence of the competitive equilibrium for an exchange economy with the commodity space ℓ^∞.

Although the space ℓ^∞ is mathematically effective, it is not adequate for our purpose of establishing the competitive equilibrium for the production economy with the external increasing returns. Since the external increasing returns make the economy to grow without bound as Young had already pointed out in 1928, the commodity space of the bounded streams is too small. Therefore, in Section 5.5, we will extend the space to include the sequences growing without bound. Let $\rho = (\rho^t)$ be a vector of the maximal growth path of the economy, which will be defined precisely in Section 5.5. Then, we can define the space ℓ^∞_ρ which is weighted by ρ as:

$$\ell^\infty_\rho = \left\{ \boldsymbol{x} = (x^t) \,\middle|\, \sup_{t \geq 0} \mid x^t/\rho^t \mid < \infty \right\}.$$

Then, we will see that the feasible allocations are contained in the unit ball of ℓ^∞_ρ, which is a nice convex and compact set (in the weak* topology, see Appendix H). The dual space of the space ℓ^∞_ρ is the weighted *ba* space of finitely additive set functions on $\mathbb{N}$, denoted by ba_ρ. Mathematically speaking, the weighted spaces ℓ^∞_ρ

and ba_ρ are isometrically isomorphic to the spaces ℓ^∞ and ba, respectively. Hence, all mathematical techniques used in Section 5.2 can also be applied in Section 5.5, and indeed we will do so.

Another case in which the commodity space is to be infinite-dimensional vector space is the model of commodity differentiation. In this case, the commodities are described by specifying their characteristics, which are points in an *a priori* given compact metric space K. In other words, a commodity bundle is represented by a distribution (finite measure) on K. For example, let $K = [0, A] \times [0, P]$, where $A > 0$ and $P > 0$ are sufficiently large numbers, and the commodities under consideration could be all foods indexed by their content of Vitamin A and proteins. Then, one unit of carrot would be represented by the Dirac measure:

$$\delta_c(B) = \begin{cases} 1 & \text{if } c \in B, \\ 0 & \text{otherwise,} \end{cases}$$

for every Borel set $B \subset K$. Here, the point $c \in K$ representing the carrot will have a large first coordinate, and similarly, steak is represented by δ_s, where $s \in K$ has a large second coordinate.

More sophisticated (and delicious?) "food" such as an expensive French dinner would be represented by a more complicated measure such as $\sum_{t_i \in K} x_i \delta_{t_i}$, where each $t_i \in K$ represents a particular food material used in the French cuisine. Since every Borel measure $\boldsymbol{x} \in ca(K)$ is a weak* limit of linear combination of the Dirac measures (see Appendix H), or $\sum_{t_i \in K} x_i \delta_{t_i} \to \boldsymbol{x}$, it would be natural to postulate that the "food", in general, is defined by a Borel measure $\boldsymbol{x}$ on the set of characteristics K.

This example clearly shows that the reason why the measures are more appropriate than the (measurable) functions on K in order to describe the commodity differentiation situations. The choice problem here is not typically how much of each (perhaps individually insignificant for the consumer's preferences) commodity characteristics to buy, but which commodity bundle to buy.

Let $ca(K, \mathcal{B}(K))$ be the set of countably additive set functions (or measure) on K. Then, the dual space of $ca(K, \mathcal{B}(K))$ is $C(K)$, the set of continuous (real-valued) functions on K. Therefore, in the models with the commodity space $ca(K, \mathcal{B}(K))$, the price vector will be a continuous function on K, and for a commodity vector $\boldsymbol{x} \in ca(K, \mathcal{B}(K))$ and a price vector $\boldsymbol{p} \in C(K)$, the market value of $\boldsymbol{x}$ evaluated by $\boldsymbol{p}$ is given by the inner product:

$$\boldsymbol{px} = \int_K \boldsymbol{p}(t) d\boldsymbol{x}.$$

We will discuss the model of commodity differentiation for an exchange economy in Section 5.3.

The proof of the existence of the competitive equilibria for these models with infinite-dimensional commodity spaces will be performed by reducing them to the proof for the classical (finite) economies discussed in Chap. 2. At least two alternative methods of proof are available.

The first one is to consider the projection of the original economies $\mathcal{E}^\infty$ to an ℓ-dimensional subeconomies $\mathcal{E}^\ell$, $\ell = 1, 2 \ldots$. That is, the commodity space L_ℓ of the economy $\mathcal{E}^\ell$ is the linear subspace of the commodity space L of $\mathcal{E}^\infty$, which is spanned by (at least) ℓ vectors $\boldsymbol{x}_1 \cdots \boldsymbol{x}_\ell$ in L. The preferences of $\mathcal{E}^\ell$ are obtained by the restrictions of those in $\mathcal{E}^\infty$ to L_ℓ. The endowment vectors $\omega_{a,\ell}$ will be defined appropriately such that $\omega_{a,\ell} \to \omega_a$ as $\ell \to \infty$, where ω_a is the endowment vector of $\mathcal{E}^\infty$, $a = 1 \cdots m$. Intuitively, we obtain a sequence of finite-dimensional economies $\mathcal{E}^\ell$, which converges to the infinite-dimensional economy $\mathcal{E}^\infty$. By the results of the previous chapters, for every ℓ, there exists a competitive equilibrium $(\boldsymbol{p}_\ell, (\boldsymbol{x}_{a,\ell}))$ of the economy $\mathcal{E}^\ell$. Under suitable conditions, there exists a limit point $(\boldsymbol{p}, (\boldsymbol{x}_a))$ of the sequence $(\boldsymbol{p}_\ell, (\boldsymbol{x}_{a,\ell}))$ such that $(\boldsymbol{p}, (\boldsymbol{x}_a))$ is a competitive equilibrium of the original economy $\mathcal{E}^\infty$. We will apply this method in Sections 5.2, 5.3, and 5.5.

The second proof is to invoke the Negishi type proof which exploits the fundamental theorems of welfare economics. This method will do its jobs on the utility possibility frontier of the m consumers by constructing the fixed-point mapping on the $m-1$-dimensional simplex. Therefore, this proof will work well for the economies with a finite number of consumers, even if the commodity space is infinite-dimensional. In Section 5.4, we will prove the existence of competitive equilibria for an exchange economy, by applying the Negishi type proof. In this section, we will assume that the commodity space is to be a general Banach lattice, which is a Banach space with the lattice structure (see Appendix H). The spaces ℓ^∞ and $ca(K)$ are examples of the Banach lattice. We will see that the lattice structure of the space is relevant to the existence of the competitive equilibrium. This is interesting, since one would not know it when he/she works only within the finite-dimensional models.

Section 5.5 is a main body of this entire book. We will prove the existence of the competitive equilibria for an infinite time horizon economy with the external increasing returns.

A mathematical difficulty arising from the infinite-dimensional commodity spaces is that the inner product $\boldsymbol{px}$ may not be jointly continuous with respect to the topology under consideration. That is to say, even if $\boldsymbol{p}_n \to \boldsymbol{p}$ and $\boldsymbol{x}_n \to \boldsymbol{x}$, they do not imply that $\boldsymbol{p}_n\boldsymbol{x}_n \to \boldsymbol{px}$. This problem is particularly serious when one deals with the externalities in the production.

Fortunately, as we will see, the structure, which is specific to the external increasing returns, is sufficient to rescue the proof. For example, the producers earn zero profit because of the homogeneity of their production function, and the periodic structure of the whole production process, $y^{t+1} = f_b(z^t, k^t)$, $t = 0, 1 \ldots$ is "just enough" to rescue the proof.

Finally, in Section 5.6, we will discuss the differentiability of the consumer demand on the infinite-dimensional commodity spaces. We will present a result proposed by Araujo (1987), which states that if a demand function is differentiable, then the commodity space must have the Hilbert space structure. Therefore, if the commodity space is assumed to be ℓ^p, $1 \leq p \leq +\infty$, and the demand function is smooth, then p is equal to 2. This result shows that mathematical properties of the demand function will impose a strong condition on the underlying commodity space, and it is also interesting, since one would not recognize such a phenomenon when one works only within the finite-dimensional setup.

5.2. EXCHANGE ECONOMIES WITH INFINITE TIME HORIZON

As explained in Section 5.1, the commodity space L is assumed to be ℓ^∞, the space of all bounded sequences:

$$\ell^\infty = \left\{ \boldsymbol{x} = (x^t) \,\middle|\, \sup_{t \geq 0} | x^t | < +\infty \right\}.$$

Since we discuss the exchange economies in this section, the assumption that all consumption streams are restricted to be bounded seems to be natural and appropriate.

The price vector π is to be a continuous linear functional on ℓ^∞, which is a bounded finitely additive set function on $\mathbb{N}$,

$$\pi \colon 2^{\mathbb{N}} \to \mathbb{R}, \quad \pi(A \cup B) = \pi(A) + \pi(B)$$

for all $A, B \subset \mathbb{N}$ such that $A \cap B = \emptyset$.

The set of all bounded finitely additive set functions is denoted by *ba*, which is a Banach space with the total variation norm:

$$\|\pi\| = \sup \left\{ \sum_{i=1}^{n} |\pi(E_i)| \mid E_i \subset \mathbb{N} \text{ and } E_i \cap E_j = \emptyset (i \neq j),\, n \geq 1 \right\}.$$

From the economic point of view, however, it is more natural that the price vector is countably additive. In this case, we set:

$$\pi(\{t\}) = p^t, \quad t \in \mathbb{N}$$

and the price vector π is represented by the infinite number of coordinates as:

$$(p^0, p^1 \cdots p^t \cdots).$$

Denoting $\boldsymbol{p} = (p^t)$, the variation norm for the countably additive set function $\pi = \boldsymbol{p} = (p^t)$ is reduced to the ℓ^1 norm:

$$\|\boldsymbol{p}\| = \sum_{t=0}^{\infty} |p^t|.$$

The set of all countably additive set functions is a subset of *ba* and denoted by ℓ^1, which is a Banach space with the ℓ^1 norm.

To sum up:

The commodity space of the infinite time horizon economy is given by the space of all bounded sequences ℓ^∞. The price vector of this economy is an element of ℓ^1, the space of all countably additive set functions, or summable sequences.

As usual, there exist m consumers in the economy, indexed by $a = 1 \cdots m$. Throughout this chapter and hence in this section, the consumption set X_a is assumed to be the non-negative orthant (positive cone) of the commodity space, or:

$$X_a = \ell^\infty_+ = \{\boldsymbol{x} = (x^t) \in \ell^\infty \mid x^t \geq 0, t \in \mathbb{N}\}, \quad a = 1 \cdots m.$$

Each consumer owns a strictly positive vector $\omega_a \gg \mathbf{0}$ as the initial endowment vector. In fact, we assume a stronger assumption for the initial endowments.

(PE_∞) (Positive endowment on ℓ^∞) there exists a positive number $\gamma_a > 0$, such that:

$$\omega_a^t \geq \gamma_a \quad \text{for all } t \in \mathbb{N}, \ a = 1 \cdots m.$$

The consumer a has the preference relation $\prec_a \subset X_a \times X_a$, which is an irreflexive and transitive binary relation on X_a. The continuity assumption on the infinite-dimensional commodity space is more intricate than that on the finite-dimensional space, since there are several topologies on the space. Then, we assume that the preferences are continuous with respect to the Mackey topology.

(CT_∞) (Mackey continuity) the preference relation $\prec_a \subset X_a \times X_a$ is open relative to $X_a \times X_a$ in the $\tau(\ell^\infty, \ell^1)$-topology.

Note that the weaker the topology, the stronger is the assumption itself. As we will see, the relevant topology in the proof of the existence theorem is the weak* topology $\sigma(\ell^\infty, \ell^1)$, which is weaker than the Mackey topology. The reason why the Mackey continuity is enough is based on the fact that both the topologies coincide on a bounded set of ℓ^∞. In fact on a bounded set of ℓ^∞, the Mackey topology coincides with the product topology (see Appendix H). We also assume that:

(CV) (Convexity) for every $\boldsymbol{x} \in X_a$, the set $\{\boldsymbol{z} \in X_a \mid \boldsymbol{z} \succsim_a \boldsymbol{x}\}$ is convex, $a = 1 \cdots m$.

As usual, we obtain a non-negative equilibrium price vector thanks to:

(MT) (Monotonicity) for every $\boldsymbol{x},\boldsymbol{z} \in X_a$, $\boldsymbol{x} \leq \boldsymbol{z}$ and $\boldsymbol{z} \neq \boldsymbol{x}$ imply $\boldsymbol{x} \prec_a \boldsymbol{z}$, $a = 1 \cdots m$.

The $2m$-tuple $(\prec_a, \omega_a)_{a=1}^m$ is called an infinite time horizon economy, or simply an economy and denoted by $\mathcal{E}^\infty$. An m-tuple of consumption vectors $(\boldsymbol{x}_1 \cdots \boldsymbol{x}_m)$ such that $\boldsymbol{x}_a \in X_a = \ell_+^\infty$ is called an allocation. As usual, the allocation $(\boldsymbol{x}_a)_{a=1}^m$ is said to be feasible if it satisfies that:

$$\sum_{a=1}^m \boldsymbol{x}_a \leq \sum_{a=1}^m \omega_a.$$

The definition of the competitive equilibrium is now clear enough.

Definition 2.1. A pair $(\boldsymbol{p}, (\boldsymbol{x}_a))$ of a price vector $\boldsymbol{p} \in \ell_+^1$ and an allocation $(\boldsymbol{x}_a)_{a\in A} \in \prod_{a=1}^m X_a$ is said to consist of a competitive equilibrium if and only if

(E-1) $\boldsymbol{p}\boldsymbol{x}_a \leq \boldsymbol{p}\omega_a$ and $\boldsymbol{x}_a \succsim_a \boldsymbol{y}$ for all $\boldsymbol{y} \in X_a$ such that $\boldsymbol{p}\boldsymbol{y} \leq \boldsymbol{p}\omega_a$, $a = 1 \cdots m$,
(E-2) $\sum_{a=1}^m \boldsymbol{x}_a \leq \sum_{a=1}^m \omega_a$.

Let $(\boldsymbol{x}_a)_{a=1}^m$ be a feasible allocation. The concept of the Pareto optimality is defined for $\mathcal{E}^\infty$ as exactly in the same way as that in the case of the classical exchange economies (see Section 2.5 of Chap. 2). Then, we have:

Proposition 2.1. *Let $(\boldsymbol{p}, \boldsymbol{x}_1 \cdots \boldsymbol{x}_m) \in \ell_+^1 \times \prod_{a\in A} X_a$ be a competitive equilibrium of an economy $\mathcal{E}^\infty$ which satisfies the local nonsatiation:*

(LNS_∞) *(Local nonsatiation on ℓ^∞) for every $\boldsymbol{x} \in X_a$ and every neighborhood U of $\boldsymbol{x}$ in the $\sigma(\ell^\infty, \ell^1)$-topology, there exists a vector $\boldsymbol{z} \in U$ such that $\boldsymbol{x} \prec_a \boldsymbol{z}$, $a = 1 \cdots m$, and the negative transitivity (NTR) of preferences for all $a \in A$. Then, the allocation $(\boldsymbol{x}_1 \cdots \boldsymbol{x}_m)$ is Pareto optimal.*

Note that in the definition of the local nonsatiation (LNS), the weak* neighborhoods are enough. The proof of Proposition 2.1 is the same as that of Proposition 5.1 of Chap. 2.

The fundamental existence theorem for the economy $\mathcal{E}^\infty$ now reads as follows.

Theorem 2.1. *Suppose that an economy $\mathcal{E}^\infty = (\prec_a, \omega_a)$ satisfies the continuity (CT_∞), the convexity (CV), the monotonicity (MT), and the (strong) positive endowment assumption (PE_∞) for every $a = 1 \cdots m$. Then, there exists a competitive equilibrium $(\boldsymbol{p}, \xi_1 \cdots \xi_m)$ for $\mathcal{E}^\infty$.*

Proof. We define the set of sequences whose coordinates are zero after T,

$$K^T = \{\boldsymbol{x} = (x^t) \mid 0 = x^{T+1} = x^{T+2} = \cdots\} \subset L, \quad T = 0, 1, \ldots.$$

The set K^T is naturally identified with $\mathbb{R}^{T+1}$. Let:

$$X_a^T = X_a \cap K^T \approx \mathbb{R}_+^{T+1}, \quad a = 1 \cdots m,$$

and we have a preference relation $\prec_a^T$ on X_a^T which is the restriction of $\prec_a$ on the truncated consumption set $X_a^T, a = 1 \cdots m$. Let $\omega_a(T) \in K^T \approx \mathbb{R}^{T+1}$ be the truncated initial endowment vector which is defined by:

$$\omega_a^t(T) = \begin{cases} \omega_a^t & \text{for } 0 \leq t \leq T, \\ 0 & \text{for } t \geq T+1. \end{cases}$$

Then, we have obtained the T-period economy $\mathcal{E}^T = \left((X_a^T, \prec_a^T), \omega_a(T)\right)_{a=1}^m$. The next lemma is easy.

Lemma 2.1. *Under the assumptions of Theorem 2.1, there exists a competitive equilibrium* $(\boldsymbol{p}(T), (\boldsymbol{x}_a(T))_{a=1}^m)$ *for the T-period economy* $\mathcal{E}^T, T = 0, 1 \ldots$.

Proof. It is clear that the truncated T-period preference relation $\prec_a^T$ satisfies the continuity (CT) (in the usual topology on $\mathbb{R}^{T+1}$), the monotonicity (MT) and the convexity (CV). The strong positive endowment assumption (PE_∞) implies that $\omega_a(T) \gg \mathbf{0}$ for every $a = 1 \cdots m$. Therefore, the minimum income condition:

$$\boldsymbol{p}\omega_a(T) > 0 = inf \boldsymbol{p}\mathbb{R}_+^{T+1} \quad \text{for every } \boldsymbol{p} \geq \mathbf{0} \text{ with } \boldsymbol{p} \neq \mathbf{0}$$

holds for all $a = 1 \cdots m$.

Hence, it follows from Theorem 5.1 of Chap. 2 that there exists a competitive equilibrium $(\boldsymbol{p}(T), (\boldsymbol{x}_a(T))_{a=1}^m)$ for $\mathcal{E}^T$. □

Let $(\boldsymbol{p}(T), (\boldsymbol{x}_a(T))_{a=1}^m)$ be an equilibrium for $\mathcal{E}^T$. Note that the monotonicity assumption (MT) implies that $\boldsymbol{p}(T) \geq \mathbf{0}$ for all $T \in \mathbb{N}$. Of course $\boldsymbol{p}(T) \neq \mathbf{0}$. Hence, we can normalize the equilibrium price vector $\boldsymbol{p}(T) = (p^t(T))$ as $\|\boldsymbol{p}(T)\| = \sum_{t=0}^T p^t(T) = 1$. Let $\pi(T) = (\boldsymbol{p}(T), 0, 0, \ldots)$. Then, $\pi(T) \in \ell^1 \subset ba$ and $\pi(T) \geq \mathbf{0}$ for all $T \in \mathbb{N}$. Hence, $\|\pi(T)\| = \sum_{t=1}^T p^t(T) = 1$, where the norm stands for ℓ^1-norm on the sequence space.

Similarly, we define $\xi_a(T) = (\boldsymbol{x}_a(T), 0, 0 \ldots), a = 1 \cdots m$. Since $(\boldsymbol{x}_a(T))_{a=1}^m$ is a feasible allocation, we have:

$$\mathbf{0} \leq \boldsymbol{x}_a(T) \leq \sum_{c=1}^m \omega_c(T), \quad a = 1 \cdots m,$$

so that:

$$\|\xi_a(T)\| \leq \left\| \sum_{c=1}^m \omega_c(T) \right\| \leq \left\| \sum_{c=1}^m \omega_c \right\|, \quad a = 1 \ldots m,$$

where the norm is the sup-norm on the space ℓ^∞.

Then by Alaoglu's theorem (Theorem H3 of Appendix H), norm-bounded sets are weakly compact in the weak* topologies, or the weak* closure of bounded sets are compact in the weak* topologies. Hence by Proposition B5 of Appendix B, there exist cluster points $\pi \in ba$ and $\boldsymbol{x}_a \in \ell^\infty$ of the sequences $\{\pi(T)\}$ and $\{\boldsymbol{x}_a(T)\}$, respectively. Since the positive orthant of ℓ^∞ is weak* closed, it follows that $\xi_a \in X_a = \ell^\infty_+$, $a = 1 \cdots m$. Similarly, we have $\pi \geq \mathbf{0}$. Let $\mathbf{1} = (1, 1, \dots) \in \ell^\infty$. Since $\pi(T)\mathbf{1} = 1$ for all $T \in \mathbb{N}$, it follows that $\pi\mathbf{1} = 1$ in the limit. Hence, $\pi \neq \mathbf{0}$.

We have almost reached the goal.

Lemma 2.2. *The vector $\pi \in ba$ and the allocation $(\xi_a)_{a=1}^m$ satisfies the following conditions.*

(BA-1) $\xi_a \succsim_a \mathbf{z}$ for all $\mathbf{z} \in X_a$ such that $\pi\mathbf{z} \leq \pi\omega_a$,
(E-2) $\sum_{a=1}^m \xi_a \leq \sum_{a=1}^m \omega_a$.

Proof. Suppose that the condition (BA-1) does not hold. Then for some a, there exists $\mathbf{z} \in X_a$ such that $\xi_a \prec_a \mathbf{z}$ and $\pi\mathbf{z} \leq \pi\omega_a$. Since $\omega_a^t \geq \beta_a > 0$ for all t and $\pi\mathbf{1} = 1$, we have $\pi\omega_a > 0$. Hence, we can take $\alpha \in (0, 1)$ such that $\xi_a \prec_a \alpha\mathbf{z}$ and $\pi\alpha\mathbf{z} < \pi\omega_a$. Let $\mathbf{z}(T) = (z^0, z^1 \cdots z^T, 0, 0 \dots)$ be the projection of $\mathbf{z}$ to $\mathbb{R}^{T+1}$. Given $\boldsymbol{q} = (q^t) \in \ell^1$, one has $\boldsymbol{q}\alpha\mathbf{z}(T) = \sum_{t=0}^T \alpha q^t z^t \to \sum_{t=0}^\infty \alpha q^t z^t = \boldsymbol{q}\alpha\mathbf{z}$ as $T \to +\infty$. Hence, $\alpha\mathbf{z}(T) \to \alpha\mathbf{z}$ in the $\sigma(\ell^\infty, \ell^1)$-topology by Proposition H2 of Appendix H. Since $\pi \geq \mathbf{0}$ and $\alpha\mathbf{z}(T) \leq \alpha\mathbf{z}$ for all T, $\pi\alpha\mathbf{z}(T) \leq \pi\alpha\mathbf{z} < \pi\omega_a$ for all T. Fix a $T_0 \in \mathbb{N}$ such that $\xi_a \prec_a \alpha\mathbf{z}(T_0)$. Let $Q \subset ba$ and $U \subset \ell^\infty$ be neighborhoods in the weak* topologies of π and ξ_a, respectively, which satisfy $\pi'\alpha\mathbf{z}(T_0) < \pi'\omega_a$ and $\boldsymbol{x} \prec_a \alpha\mathbf{z}(T_0)$ for all $\pi' \in Q$ and for all $\boldsymbol{x} \in U$. Since π and ξ_a are cluster points of the sequences $(\pi(T))$ and $(\xi_a(T))$ respectively, there exists $S \geq T_0$ such that $\pi(S) \in Q$ and $\xi_a(S) \in U$. On the other hand, $\pi(S)\alpha\mathbf{z}(T_0) < \pi(S)\omega_a = \sum_{t=0}^S p^t(S)\omega_a^t = \pi(S)\omega_a(S)$ and $\xi_a(S) \prec_a \alpha\mathbf{z}(T_0)$. This contradicts the fact that $\xi_a(S)$ or $\boldsymbol{x}_a(S)$ is an equilibrium consumption bundle of the consumer a in the economy $\mathcal{E}^S$. This establishes the condition (BA-1).

Since $\sum_{a=1}^m \xi_a(T) \leq \sum_{a=1}^m \omega_a(T)$ for all T, it follows that $\sum_{a=1}^m \xi_a \leq \sum_{a=1}^m \omega_a$ in the limit. Hence, the condition (E-2) is proved. □

The existence of an equilibrium price vector in ℓ^1 can be established by the next lemma.

Lemma 2.3. *There exists a vector $\boldsymbol{p} \in \ell^1_+$ such that $(\boldsymbol{p}, (\xi_a))_{a=1}^m$ satisfies the following conditions.*

(E-1) $\boldsymbol{p}\xi_a \leq \boldsymbol{p}\omega_a$ and $\xi_a \succsim_a \mathbf{z}$ for all $\mathbf{z} \in X_a$ such that $\boldsymbol{p}\mathbf{z} \leq \boldsymbol{p}\omega_a$,
(E-2) $\sum_{a=1}^m \xi_a \leq \sum_{a=1}^m \omega_a$.

Proof. We have already established the condition (E-2) in Lemma 2.2. Since $\pi \geq \mathbf{0}$, it follows from Yosida–Hewitt theorem (Theorem H4 of Appendix H) that we can write:

$$\pi = \pi_c + \pi_p,$$

where $\pi_c \geq \mathbf{0}$ and $\pi_p \geq \mathbf{0}$ are the countably additive part and the purely finitely additive part of π, respectively.

For given $a = 1 \cdots m$, take $\mathbf{z} = (z^t) \in X_a = \ell^\infty_+$ with $\xi_a \prec_a \mathbf{z}$ and let $\mathbf{z}(T) = (z^0, z^1 \cdots z^T, 0, 0 \cdots)$ be the projection of $\mathbf{z}$ to $\mathbb{R}^{T+1}$. Then, as in the proof of Lemma 2.2, we can show that $\xi_a \prec_a \mathbf{z}(T)$ for T sufficiently large. By Lemma 2.2, we have $\pi\mathbf{z}(T) > \pi\omega_a$. Since π_p is purely finitely additive, $\pi_p(\{0, 1, \ldots, T\}) = 0$. It follows from this and $\pi_c \geq \mathbf{0}$ that:

$$\pi\mathbf{z}(T) = (\pi_c + \pi_p)\mathbf{z}(T) = \pi_c\mathbf{z}(T) \leq \pi_c\mathbf{z},$$

since $\mathbf{z}(T) \leq \mathbf{z}$. On the other hand, $\pi_p \geq \mathbf{0}$ and $\omega_a \geq \mathbf{0}$ imply that $\pi\omega_a = (\pi_c + \pi_p)\omega_a \geq \pi_c\omega_a$, and consequently, we have:

$$\pi_c\mathbf{z} > \pi_c\omega_a.$$

Denoting $\pi_c(\{t\}) = p^t$, we get a vector $\boldsymbol{p} = (p^t)$ in ℓ^1, which satisfies:

$$(*) \ \ \xi_a \prec_a \mathbf{z} \text{ implies } \boldsymbol{p}\mathbf{z} > \boldsymbol{p}\omega_a, \quad a = 1 \cdots m.$$

By the monotonicity (MT) of preferences, one can take $\mathbf{z} \in X_a = \ell^\infty$, which is arbitrarily close to ξ_a such that $\xi_a \prec_a \mathbf{z}$. Thus, (*) implies that:

$$\boldsymbol{p}\xi_a \geq \boldsymbol{p}\omega_a, \quad a = 1 \cdots m.$$

On the other hand, since $\sum_{a=1}^m (\xi_a - \omega_a) \leq \mathbf{0}$ by Lemma 2.2, we have:

$$0 \leq \sum_{a=1}^m \boldsymbol{p}(\xi_a - \omega_a) \leq 0.$$

Consequently, we have $\boldsymbol{p}(\xi_a - \omega_a) = 0$ for all a, hence:

$$\boldsymbol{p}\xi_a = \boldsymbol{p}\omega_a, \quad a = 1 \cdots m.$$

This and (*) prove the condition (E-1). □

Recall that a feasible allocation $(\boldsymbol{x}_a)_{a=1}^m$ is said to be an equilibrium relative to the price vector $\boldsymbol{p}$ if and only if for all $\mathbf{z} \in X_a$,

$$\boldsymbol{x}_a \prec_a \mathbf{z} \text{ implies that } \boldsymbol{p}\boldsymbol{x}_a < \boldsymbol{p}\mathbf{z}, \quad a = 1 \cdots m.$$

See Definition 5.2 of Chap. 2. The second fundamental theorem of welfare economics, which is a converse of Proposition 2.1, still holds as it did for the classical exchange economies (Theorem 5.2 of Chap. 2). The key mathematical theorem used there was the separating hyperplane theorem (Theorem A1 of Appendix A), and it is extended to the infinite-dimensional vector spaces, which is known as the Hahn–Banach theorem (Theorem E3 of Appendix E) with a cost that the separated convex set has to have an interior point.

A mathematical advantage of the space ℓ^∞ is that this interiority condition is easily obtained. For instance, the positive orthant of ℓ^∞ has an interior point in the norm topology, namely $\mathbf{1} = (1, 1 \cdots)$. This is in contrast with other Banach spaces such as ℓ^p for $p \neq \infty$ or $ca(K)$. In these spaces, we have to impose an additional assumption on the preferences (and production sets) to prove the existence theorem and/or the second welfare theorem (see the subsequent sections).

Theorem 2.2. *Let $\mathcal{E}^\infty$ be an infinite time horizon exchange economy. For every consumer $a \in A$, suppose that the preference relation satisfies the negative transitivity (NTR), the continuity (CT_∞), the convexity (CV), and the monotonicity (MT). Then, every Pareto-optimal allocation is a quasi-equilibrium relative to some price vector $\boldsymbol{p} \in \ell^1_+$.*

Proof. Let $(\boldsymbol{x}_a)_{a=1}^m$ be a Pareto-optimal allocation. Then, since it is a feasible allocation, the condition (E-2) is met. For each consumer a, we define the strictly preferred set $P_a(\boldsymbol{x})$ by:

$$P_a(\boldsymbol{x}) = \{\mathbf{z} \in X_a \mid \boldsymbol{x} \prec_a \mathbf{z}\}.$$

The set $P_a(\boldsymbol{x})$ is nonempty for all $\boldsymbol{x} \in X_a$, since $\boldsymbol{x}+\mathbf{1} \in P_a(\boldsymbol{x})$ by the assumption (MT), where $\mathbf{1} = (1, 1 \cdots)$. As in the proof of Theorem 5.2 of Chap. 2, we can show that the set $P_a(\boldsymbol{x})$ is convex for all $\boldsymbol{x} \in X_a$. Moreover, the set $P_a(\boldsymbol{x})$ has an interior point $\boldsymbol{x}+\mathbf{1}$. For the ball $B(\boldsymbol{x}+\mathbf{1}, 1/2)$ centered at $\boldsymbol{x}+\mathbf{1}$ with the radius $1/2$ is contained in $P_a(\boldsymbol{x})$.

Let $Q = \sum_{a=1}^m P_a(\boldsymbol{x}_a)$. Then, Q is convex as the sum of the convex sets and interior $Q \neq \emptyset$. Since the allocation $(\boldsymbol{x}_a)_{a=1}^m$ is Pareto optimal, $\sum_{a=1}^m \omega_a \notin Q$. Hence, we can apply the Hahn–Banach theorem (Theorem E3 of Appendix E), we get a vector $\pi \in ba$ with $\pi \neq \mathbf{0}$ and,

$$\pi \sum_{a=1}^m \omega_a \leq \pi \sum_{a=1}^m \mathbf{z}_a$$

for every $\mathbf{z}_a \in P_a(\boldsymbol{x}_a)$, $a = 1 \cdots m$. By the monotonicity (MT), it follows that $\pi \geq \mathbf{0}$. Fix an a. For $b \neq a$, we can take $\mathbf{z}_b$ arbitrarily close to $\boldsymbol{x}_b$ in the norm topology. Hence:

$$\pi \sum_{a=1}^m \omega_a \leq \pi \sum_{b \neq a} \boldsymbol{x}_b + \pi \mathbf{z}_a \leq \pi \sum_{a=1}^m \omega_a - \pi \boldsymbol{x}_a + \pi \mathbf{z}_a,$$

and from this, we obtain:

$$\pi \boldsymbol{x}_a \leq \pi \mathbf{z}_a \quad \text{for every } \mathbf{z}_a \in P_a(\boldsymbol{x}_a), \ a = 1 \cdots m.$$

Since $\pi \geq \mathbf{0}$, it follows from the Yosida–Hewitt theorem (Theorem H4 of Appendix H) that we can write $\pi = \pi_c + \pi_p$, where $\pi_c \geq \mathbf{0}$ and $\pi_p \geq \mathbf{0}$ are the countably additive part and the purely finitely additive part of π, respectively. Then, for given $\mathbf{z}_a \in P_a(\mathbf{x}_a)$, we have $\pi_c \mathbf{x}_a \leq (\pi_c + \pi_p)\mathbf{x}_a \leq \pi \mathbf{z}_a$. Suppose that $\pi_c \mathbf{x}_a > \pi_c \mathbf{z}$ for some $\mathbf{z} \in P_a(\mathbf{x}_a)$. Let $\mathbf{z}(T) = (z^0 \cdots z^T, 0 \cdots)$ be the projection of the vector $\mathbf{z}$ to $\mathbb{R}^{T+1}$. Then $\mathbf{z}(T) \to \mathbf{z}$ in the Mackey topology by Proposition H5 of Appendix H. Since π_p is purely finitely additive, $\pi_p \mathbf{z}(T) = 0$ for all T. Hence, for T sufficiently large, we have $\mathbf{z}(T) \in P_a(\mathbf{x}_a)$ and $\pi \mathbf{x}_a \geq \pi_c \mathbf{x}_a > \pi \mathbf{z}(T) = \pi_c \mathbf{z}(T)$, a contradiction. Denoting $\mathbf{p} = (p^t) = (\pi(\{t\}))$, we have obtained:

$$\mathbf{p}\mathbf{x}_a \leq \mathbf{p}\mathbf{z} \quad \text{whenever } \mathbf{x}_a \prec_a \mathbf{z}, \; a = 1 \cdots m.$$

This shows that the optimal allocation $(\mathbf{x}_a)$ is a quasi-equilibrium relative to $\mathbf{p}$. In fact, it is an equilibrium relative to $\mathbf{p}$. Since we assumed the monotonicity (MT), it follows that $\mathbf{p} \gg \mathbf{0}$, or $p^t > 0$ for all t. Since $X_a = \ell^\infty_+$, one has $\mathbf{p}\mathbf{x}_a = \inf \mathbf{p}X_a = 0$ only when $\mathbf{x}_a = \mathbf{0}$. Therefore, we have:

$$\mathbf{p}\mathbf{x}_a < \mathbf{p}\mathbf{z} \quad \text{whenever } \mathbf{x}_a \prec_a \mathbf{z}, \; a = 1 \cdots m.$$

This proves the theorem. □

Finally, we give a result on the continuity of the preferences on ℓ^∞. In applications of the infinite time horizon economies, the preferences of the consumers are sometimes given by the time-separable form,

$$u(\mathbf{x}) = \sum_{t=0}^{\infty} v(x^t, t) \quad \text{for } \mathbf{x} = (x^t) \in X = \ell^\infty_+.$$

As a special case, the utility function of the discounted sum of a stationary one-period utility:

$$u(\mathbf{x}) = \sum_{t=0}^{\infty} \beta^t v(x^t) \quad \text{for } \mathbf{x} = (x^t) \in X = \ell^\infty_+$$

is often discussed. The next theorem implies that the time-separable utility functions are continuous in the Mackey topology under very general assumptions.

Theorem 2.3. *Suppose that the one-period utility function,*

$$v : \mathbb{R}_+ \times \mathbb{N} \to \mathbb{R}, \quad (x, t) \mapsto v(x, t)$$

satisfies that (a) for each $t \in \mathbb{N}$, $v(\cdot, t)$ is a continuous, non-decreasing, concave function with $v(0, t) = 0$, and (b) for each $x \in \mathbb{R}_+$, $\sum_{t=0}^{\infty} v(x, t) < \infty$. Then, the utility function:

$$u(\mathbf{x}) = \sum_{t=0}^{\infty} v(x^t, t) \quad \text{for } \mathbf{x} = (x^t) \in X = \ell^\infty_+$$

is continuous with respect to $\tau(\ell^\infty, \ell^1)$-topology.

Proof. First, we show that if a net $(\boldsymbol{x}_\alpha)$ of non-negative sequences, or $\boldsymbol{x}_\alpha \geq \mathbf{0}$ for all α converges to $\boldsymbol{x}$ in the $\tau(\ell^\infty, \ell^1)$-topology, then $|\boldsymbol{x}_\alpha - \boldsymbol{x}| \to \mathbf{0}$ in the $\sigma(\ell^\infty, \ell^1)$-topology, where setting $\boldsymbol{x}_+ = (x^t_+)$ and $\boldsymbol{x}_- = (x^t_-)$ as $x^t_+ = \max\{0, x^t\}$ and $x^t_- = \max\{0, -x^t\}$, respectively, or $|\boldsymbol{x}| = \boldsymbol{x}_+ + \boldsymbol{x}_-$. In order to see this, first note that a net $\{\boldsymbol{z}_\alpha\}$ of non-negative vectors which converges to a vector $\boldsymbol{z}_* \geq \mathbf{0}$ in the $\tau(\ell^\infty, \ell^1)$-topology is bounded in the ℓ^∞ norm. To see this, for each α, take $\boldsymbol{p}_\alpha \in S_+ \equiv \{\boldsymbol{p} \in \ell^1 \mid \boldsymbol{p} \geq \mathbf{0}, \|\boldsymbol{p}\| = \boldsymbol{p}\mathbf{1} = 1\}$ such that $\|\boldsymbol{z}_\alpha\| = \sup\{|\boldsymbol{p}\boldsymbol{z}_\alpha| \mid \boldsymbol{p} \in \ell^1, \|\boldsymbol{p}\| = 1\} = \boldsymbol{p}_\alpha \boldsymbol{z}_\alpha$, since $\boldsymbol{z}_\alpha \geq \mathbf{0}$ and the set S_+ is $\sigma(\ell^\infty, \ell^1)$-compact. Let $\boldsymbol{p}_*$ be a limit point of the net $\{\boldsymbol{p}_\alpha\} \subset S_+$. Since the evaluation map $\boldsymbol{p}\boldsymbol{z}$ is $\sigma(\ell^1, \ell^\infty) \times \tau(\ell^\infty, \ell^1)$ jointly continuous by Proposition H6 of Appendix H, it follows that $\|\boldsymbol{z}_\alpha\| = \boldsymbol{p}_\alpha \boldsymbol{z}_\alpha \to \boldsymbol{p}_* \boldsymbol{z}_*$. Hence, the set $\{\boldsymbol{z}_\alpha\}$ is bounded. Let $\boldsymbol{x}_\alpha \to \boldsymbol{x}$ in the $\tau(\ell^\infty, \ell^1)$-topology with $\boldsymbol{x}, \boldsymbol{x}_\alpha \geq \mathbf{0}$ for all α. Then, $\|(\boldsymbol{x}_\alpha - \boldsymbol{x})_+\| \leq \|\boldsymbol{x}_\alpha - \boldsymbol{x}\| \leq \|\boldsymbol{x}_\alpha\| + \|\boldsymbol{x}\|$, hence the net $(\boldsymbol{x}_\alpha - \boldsymbol{x})_+$ is norm-bounded and converges to $\mathbf{0}$ pointwise. Hence, by Proposition H5 of Appendix H, $(\boldsymbol{x}_\alpha - \boldsymbol{x})_+ \to \mathbf{0}$ in the $\sigma(\ell^\infty, \ell^1)$-topology. Therefore, $(\boldsymbol{x}_\alpha - \boldsymbol{x})_-$ and $|\boldsymbol{x}_\alpha - \boldsymbol{x}|$ do so as well.

Suppose that $\{\boldsymbol{x}_\alpha\}$ is a net of non-negative vectors in ℓ^∞ converging to $\boldsymbol{x}$ in the Mackey topology $\tau(\ell^\infty, \ell^1)$. Let $\tilde{\boldsymbol{x}}_\alpha = (\tilde{x}^t_\alpha)$ and $\hat{\boldsymbol{x}}_\alpha = (\hat{x}^t_\alpha)$ be defined by:

$$\tilde{x}^t_\alpha = \max\{0, x^t - |x^t_\alpha - x^t|\},$$
$$\hat{x}^t_\alpha = \max\{0, x^t + |x^t_\alpha - x^t|\},$$

respectively. Then, $\tilde{\boldsymbol{x}}_\alpha \leq \boldsymbol{x} \leq \hat{\boldsymbol{x}}_\alpha$ and by what was proved in the previous paragraph, both $\tilde{\boldsymbol{x}}_\alpha$ and $\hat{\boldsymbol{x}}_\alpha$ converge to $\boldsymbol{x}$ in the $\sigma(\ell^\infty, \ell^1)$-topology. Since $v(x, t)$ is nondecreasing by the assumption, we have:

$$|v(x^t, t) - v(x^t_\alpha, t)| \leq v(\hat{x}^t_\alpha, t) - v(x^t, t) + v(x^t, t) - v(\tilde{x}^t_\alpha, t).$$

It is therefore sufficient to prove:

Case 1. $\boldsymbol{x}_\alpha \leq \boldsymbol{x}$ for all α and $\sum_{t=0}^{\infty} (v(x^t, t) - v(x^t_\alpha, t)) \to 0$.
Case 2. $\boldsymbol{x}_\alpha \geq \boldsymbol{x}$ for all α and $\sum_{t=0}^{\infty} (v(x^t_\alpha, t) - v(x^t, t)) \to 0$.

We first show that Case 1 and suppose that $\boldsymbol{x}_\alpha \leq \boldsymbol{x}$ for all α. First of all, for all $\boldsymbol{z} = (z^t) \in \ell^\infty$, the utility function $u(\boldsymbol{z})$ is summable and well defined, since $v(z^t, t) \leq v(\|\boldsymbol{z}\|, t)$.

Let $\epsilon > 0$ and for each $\beta \geq 0$, let $P(\beta) = \{t \in \mathbb{N} \mid x^t \geq \beta\}$. When $\beta_1 \geq \beta_2 \geq \cdots \to 0$, we have $P(\beta_1) \subset P(\beta_2) \subset \cdots \subset P(0) = \mathbb{N}$. Moreover, we have $\sum_{t \in \mathbb{N}} v(\beta, t) \to 0$ as $\beta \to 0$. Therefore, there exists $\beta > 0$ such that $\sum_{t \notin P(\beta)} v(x^t, t) < \epsilon$. Since $v(x^t, t)$ is concave,

$$\frac{v(x^t, t) - v(x^t_\alpha, t)}{x^t - x^t_\alpha} \leq \frac{v(x^t, t)}{x^t} \leq \frac{v(x^t, t)}{\beta}$$

for all $t \in P(\beta)$, hence:

$$\sum_{t \in \mathbb{N}} (v(x^t, t) - v(x^t_\alpha, t)) \le (1/\beta) \sum_{t \in P(\beta)} v(x^t, t)(x^t - x^t_\alpha) + \sum_{t \notin P(\beta)} (v(x^t, t) - v(x^t_\alpha, t))$$

$$\le (1/\beta) \sum_{t \in \mathbb{N}} v(x^t, t)(x^t - x^t_\alpha) + \sum_{t \notin P(\beta)} v(x^t, t).$$

Since $\boldsymbol{x}_\alpha \to \boldsymbol{x}$ in the $\sigma(\ell^\infty, \ell^1)$-topology, we have:

$$\sum_{t \in \mathbb{N}} v(x^t, t)(x^t - x^t_\alpha) < \beta\epsilon$$

for α large enough. Hence, $u(\boldsymbol{x}) - u(\boldsymbol{x}_\alpha) \le \epsilon + \epsilon$ for α large enough and Case 1 is proved.

Now suppose that $\boldsymbol{x} \le \boldsymbol{x}_\alpha$ for all α. Let $\epsilon > 0$. Then, we can choose $\beta > 0$ such that:

$$\sum_{t=0}^{\infty} v(\beta, t) < \epsilon,$$

$$\sum_{t=0}^{\infty} \left(v(x^t + \beta, t) - v(x^t, t)\right) < \epsilon.$$

If $\beta \le x^t$, then:

$$v(x_\alpha, t) - v(x^t, t) \le \frac{1}{\beta} v(x^t, t)(x^t_\alpha - x^t).$$

If $x^t < \beta \le x^t_\alpha$, then we have:

$$v(x^t_\alpha, t) - v(x^t, t) \le v(x^t_\alpha, t) - v(x^t, t) + (v(\beta, t) - v(x^t, t))$$

$$\le (1/\beta) v(\beta, t)(x^t_\alpha - x^t) + v(\beta, t)$$

by the concavity and the monotonicity of $v(\cdot, t)$.

Finally, if $x^t_\alpha < \beta$, it follows that:

$$v(x^t_\alpha, t) - v(x^t, t) \le v(\beta, t).$$

Therefore, in all cases,

$$v(x^t_\alpha, t) - v(x^t, t) \le (1/\beta) v(x^t, t)(x^t_\alpha - x^t) + (1/\beta) v(\beta, t)(x^t_\alpha - x^t) + v(\beta, t).$$

Since $\boldsymbol{x}_\alpha \to \boldsymbol{x}$ in the $\sigma(\ell^\infty, \ell^1)$-topology, we can take α_0 such that for all $\alpha \ge \alpha_0$,

$$\sum_{t=0}^{\infty} v(x^t, t)(x^t_\alpha - x^t) < \beta\epsilon, \quad \sum_{t=0}^{\infty} v(\beta, t)(x^t_\alpha - x^t) < \beta\epsilon.$$

Therefore, if $\alpha \geq \alpha_0$, one obtains:

$$u(x_\alpha) - u(x) < (1/\beta)\beta\epsilon + (1/\beta)\beta\epsilon + \epsilon = 3\epsilon.$$

This verifies Case 2, and the proof of Theorem 2.3 is established. □

5.3. EXCHANGE ECONOMIES WITH DIFFERENTIATED COMMODITIES

Let (K, d) be a compact metric space, or K is a compact space and d is a non-negative function on $K \times K$ such that $d(s,t) = 0$ only if $s = t$, $d(s,t) = d(t,s)$ for all $s, t \in K$, and $d(s,t) \leq d(s,u) + d(u,t)$ for all $s, t, u \in K$ (see Appendix C).

The intended economic meaning of the set K is that it should be understood as the set of commodity characteristics, namely that an element $t \in K$ is interpreted as a complete description of all the economically relevant characteristics of commodities.

Example 3.1 (Hotelling, 1929). $K = [0,1]$, the unit interval. Every $t \in K$ represents a location.

Example 3.2 (Lancaster, 1971; Rosen, 1974). $K = \{t = (t^i) \in \mathbb{R}^\ell_+ \mid \|t\|^2 = \sum_{i=1}^{\ell} (t^i)^2 = 1\}$. Each $t \in K$ represents a characteristic profile which is identified with a unit of the commodity containing t^i units of the characteristics $i(= 1 \cdots \ell)$.

Example 3.3 (Jones, 1984). $K = [0,1] \times [0,1]$. Every $t = (v, p) \in K$ represents a food which is characterized by its contents of vitamins and proteins. See also Section 5.1.

In this section, the commodity vector or the commodity bundle is defined as a Borel-signed measure on K, or more precisely, denoting the collection of all Borel-measurable subsets of K as $\mathcal{B}(K)$, the commodity vector $\boldsymbol{x}$ is a bounded countably additive set function on the measurable space $(K, \mathcal{B}(K))$. Then, the economic interpretation of $\boldsymbol{x}$ is that for every $B \in \mathcal{B}(K)$, $\boldsymbol{x}(B)$ is the total amount of the commodity with its characteristics in B. Let $ca(K, \mathcal{B}(K))$ be the set of all Borel-signed measures on K. For $\boldsymbol{x} \in ca(K, \mathcal{B}(K))$, $\boldsymbol{x} \geq \mathbf{0}$ means that $\boldsymbol{x}(B) \geq 0$ for all $B \in \mathcal{B}(K)$. $\boldsymbol{x} > \mathbf{0}$ means that $\boldsymbol{x} \geq \mathbf{0}$ and $\boldsymbol{x} \neq \mathbf{0}$. Let $ca_+(K, \mathcal{B}(K)) = \{\boldsymbol{x} \in ca(K, \mathcal{B}(K)) \mid \boldsymbol{x} \geq \mathbf{0}\}$ be the positive orthant of $ca(K, \mathcal{B}(K))$. $ca_+(K, \mathcal{B}(K))$ is nothing but the set of all Borel measures on $(K, \mathcal{B}(K))$, or $ca_+(K, \mathcal{B}(K)) = \mathcal{M}(K, \mathcal{B}(K))$. The support of $\boldsymbol{x} \in \mathcal{M}(K, \mathcal{B}(K))$ is the smallest closed subset of K with a full $\boldsymbol{x}$-measure and denoted by support $(\boldsymbol{x})$.

In this market, the natural candidate of a price vector is to be a continuous function $\boldsymbol{p}(t)$ on K. Let $C(K)$ be the set of all continuous functions on K. The space $C(K)$ is a Banach space with the norm $\|\boldsymbol{p}\| = \sup\{|p(t)|| \ t \in K\}$. See Appendix H.

The market value of a commodity vector $\boldsymbol{x} \in ca(K, \mathcal{B}(K))$ evaluated by the price system $\boldsymbol{p} \in C(K)$ is defined by the inner product:

$$\boldsymbol{px} = \int_K \boldsymbol{p}(t) d\boldsymbol{x}.$$

By this inner product, we can show that the dual space of $C(K)$ is equal to $ca(K, \mathcal{B}(K))$ (Proposition H10 of Appendix H). Then by the definition of the norm (Appendix H), the norm of $\boldsymbol{x} \in \mathcal{M}(K, \mathcal{B}(K))$ is given by $\|\boldsymbol{x}\| = \sup\{|\boldsymbol{px}| \mid \boldsymbol{p} \in C(K),\ \|\boldsymbol{p}\| = 1\}$. It is well known that this norm coincides with the variation norm:

$$\|\boldsymbol{x}\| = \sup\left\{\sum_{i=1}^{n} |\boldsymbol{x}(E_i)| \mid E_1 \cdots E_n \text{ are mutually disjoint, arbitrary finite subsets of } \mathbb{N}\right\}.$$

Note that for $\boldsymbol{x} \geq \boldsymbol{0}$, $\|\boldsymbol{x}\| = \boldsymbol{1x} = \boldsymbol{x}(K)$, where $\boldsymbol{1}(t) = 1$ for all $t \in K$.

We can consider an alternative topology on $ca(K, \mathcal{B}(K))$, namely the weak* topology which is weaker than the norm topology. It is the topology of pointwise convergence on $C(K)$. Hence, a net $\{\boldsymbol{x}_\alpha\}$ in $ca(K, \mathcal{B}(K))$ converges to $\boldsymbol{x}$ in the weak* topology if and only if

$$\int_K \boldsymbol{p}(t) d\boldsymbol{x}_\alpha \to \int_K \boldsymbol{p}(t) d\boldsymbol{x} \quad \text{for every } \boldsymbol{p} \in C(K).$$

It is well known that norm-bounded subsets of $ca(K, \mathcal{B}(K))$ are compact and metrizable in the weak* topology (Proposition H11 of Appendix H). Throughout this section, we will use the weak* topology unless otherwise specified.

A mathematical advantage of the weak* topology is that the set of finite linear combinations of the Dirac measures is dense in $ca(K, \mathcal{B}(K))$ (Proposition H12 of Appendix H). A mathematical disadvantage of the space $ca(K, \mathcal{B}(K))$ is that the norm interior of the positive orthant $ca_+(K, \mathcal{B}(K))$ is empty.

There exist m consumers in the economy indexed by $a = 1 \cdots m$, or $a \in A = \{1 \cdots m\}$. The consumption set X_a of the consumer a is assumed to be the positive (non-negative) orthant of the commodity space $ca(K, \mathcal{B}(K))$,

$$X_a = ca_+(K, \mathcal{B}(K)) = \mathcal{M}(K, \mathcal{B}(K)), \quad a = 1 \cdots m.$$

The preference relation $\prec_a \subset X_a \times X_a$ is irreflexive and transitive binary relation on X_a. We assume that for all $a \in A$,

($\mathrm{CT}_{\text{weak*}}$) (Continuity in the weak* topology) the set $\prec = \{(\boldsymbol{x}, \boldsymbol{y}) \in X_a \times X_a \mid \boldsymbol{x} \prec_a \boldsymbol{y}\}$ is open in $X_a \times X_a$ in the weak* topology,

(CV) (Convexity) for every $\boldsymbol{x} \in X_a$ the set $\{\boldsymbol{y} \in X_a \mid \boldsymbol{y} \succsim_a \boldsymbol{x}\}$ is convex, and

(MT) (Monotonicity) if $\boldsymbol{x}, \boldsymbol{y} \in X_a$ and $\boldsymbol{x} < \boldsymbol{y}$, then $\boldsymbol{x} \prec_a \boldsymbol{y}$.

In Section 5.2, we saw that the non-negative orthant of the space ℓ^∞ has an interior point $\boldsymbol{1} = (1, 1, \dots)$ in the norm topology and the vector $\boldsymbol{1}$ played a crucial role in the proofs of the existence theorem and the second welfare theorem. However, since the norm interior of $ca_{+}(K, \mathcal{B}(K))$ is empty, we need an additional assumption in order to prove the existence of a competitive equilibrium on the space $ca(K, \mathcal{B}(K))$.

(BRS) (Bounded marginal rate of substitution) for all sequences $t_n, s_n \in K$, $a_n, b_n > 0$, and $\boldsymbol{x}_n \in \mathcal{M}(K, \mathcal{B}(K))$ such that $\lim_n t_n = \lim_n s_n$, $\lim_n \boldsymbol{x}_n = \boldsymbol{x}$, and $\lim_n a_n/b_n > 1$, there exists an N such that $\boldsymbol{x}_N + a_N\delta_{t_N} \succsim_a \boldsymbol{x}_N + b_N\delta_{t_N}$, where the limit of $\{\boldsymbol{x}_n\}$ is taken with respect to the weak* topology, and δ_t is the Dirac measure defined by:

$$\delta_t(B) = \begin{cases} 1 & \text{if } t \in B, \\ 0 & \text{otherwise.} \end{cases}$$

The assumption (BRS) implies that if t and s are close enough in K, the consumer a is willing to accept any trade of t and s in which the "terms" are strongly greater than one. Later in Proposition 3.1, we will show that if the preference relation is represented by a smooth utility function, and its second derivatives are uniformly bounded, then the assumption (BRS) is satisfied.

The consumer a has the initial endowment vector $\omega_a \in \mathcal{M}(K, \mathcal{B}(K))$. Since we consider an exchange economy, all characteristics should be available initially in the market. Hence, we assume:

(AE) (Adequate endowment) $\omega_a \geq \boldsymbol{0}$ for all $a = 1 \cdots m$, and

$$\text{support}\left(\sum_{a=1}^{m} \omega_a\right) = K.$$

A $2m$-tuple of the preferences and the initial endowment vectors $(\prec_a, \omega_a)_{a=1}^m$ is called an economy with differentiated commodities and denoted by $\mathcal{E}_{\mathcal{M}}$.

The definition of the competitive equilibrium for $\mathcal{E}_{\mathcal{M}}$ should be clear.

Definition 3.1. A pair $(\boldsymbol{p}, (\boldsymbol{x}_a))$ of a price vector $\boldsymbol{p} \in C(K)$ and an allocation $(\boldsymbol{x}_a)_{a=1}^m \in \prod_{a=1}^m X_a$ is said to consist of a competitive equilibrium if and only if the following conditions hold,

(E-1) $\boldsymbol{p}\boldsymbol{x}_a \leq \boldsymbol{p}\omega_a$ and $\boldsymbol{x}_a \succsim_a \boldsymbol{y}$ for all $\boldsymbol{z} \in X_a$ such that $\boldsymbol{p}\boldsymbol{z} \leq \boldsymbol{p}\omega_a$, $a = 1 \cdots m$,
(E-2) $\sum_{a=1}^m \boldsymbol{x}_a \leq \sum_{a=1}^m \omega_a$.

The equilibrium existence theorem for the economy $\mathcal{E}_{\mathcal{M}}$ now reads:

Theorem 3.1. *Suppose that an economy $\mathcal{E}_{\mathcal{M}} = (\prec_a, \omega_a)$ satisfies the negative transitivity (NTR), the continuity (CT_{weak*}), the convexity (CV), the monotonicity (MT), and the bounded marginal rate of substitution (BRS). Furthermore, the adequate endowment assumption (AE) for every $a = 1 \cdots m$ is satisfied. Then, there exists a competitive equilibrium $(\boldsymbol{p}, \boldsymbol{x}_1 \cdots \boldsymbol{x}_m)$ for $\mathcal{E}_{\mathcal{M}}$.*

Proof. Since a compact metric space is separable (Proposition C3 of Appendix C), there exists a countable dense subset $\{t_1, t_2 \cdots\}$ of K. Let $K^\ell = \{t_1 \cdots t_\ell\}$ and $\text{LS}(K^\ell)$ be the linear subspace spanned by $\delta_{t_1} \cdots \delta_{t_\ell}$. We now define the ℓ-dimensional subeconomy $\mathcal{E}^\ell$ as follows.

First, let $X_a^\ell = X_a \cap LS(K^\ell)$ and $\prec_a^\ell = \prec_a \cap (X_a^\ell \times X_a^\ell)$. For each ℓ, take disjoint measurable sets $M_k^\ell, k = 1 \cdots \ell$ with $\cup_{k=1}^\ell M_k^\ell = K, t_k \in M_k^\ell$ for all k, ℓ, and $B(t_k, r_k) \subset M_k^\ell$ for some $r_k > 0$, and finally $\sup\{\text{diam}\, M_k^\ell\} \to 0$ as $\ell \to \infty$, where $\text{diam}\, M_k^\ell = \sup\{d(\boldsymbol{x}, \boldsymbol{y}) \mid \boldsymbol{x}, \boldsymbol{y} \in M_k^\ell\}$.

We set $\omega_a^\ell = \sum_{k=1}^\ell \omega_a(M_k^\ell)\delta_{t_k}$ Then, it is easy to show that $\omega_a^\ell \in LS(K^\ell)$ and $\omega_a^\ell \gg \mathbf{0}$ for all ℓ, and $\omega_a^\ell \to \omega_a$ in the weak* topology, $a = 1 \cdots m$. The $3m$-tuple $(X_a^\ell, \prec_a^\ell, \omega_a^\ell)$ is called the ℓ-dimensional subeconomy and it is denoted by $\mathcal{E}^\ell$. □

Lemma 3.1. *There exists a competitive equilibrium $(\boldsymbol{p}^\ell, (\boldsymbol{x}_a^\ell))$ for $\mathcal{E}^\ell$.*

Proof. The preference relation $\prec_a^\ell$ satisfies the continuity (CT), the monotonicity (MT) and the convexity (CV) on the ℓ-dimensional space $LS(K^\ell) \approx \mathbb{R}^\ell$, and the minimum income condition (MI) is satisfied, since $\omega_a^\ell \gg \mathbf{0}$. Then, by Theorem 5.1 of Chap. 2, there exists a competitive equilibrium $(\boldsymbol{p}^\ell, (\boldsymbol{x}_a^\ell))$ for $\mathcal{E}^\ell$. □

Without loss of generality, we can normalize the price vector as $\|\boldsymbol{p}\| = \max\{|p^\ell(t)| \mid t \in K\} = 1$. Recall that a sequence $(K^\ell, \boldsymbol{p}^\ell)$ is said to be equicontinuous if for all $\epsilon > 0$, there is a $\delta > 0$ such that for all ℓ and $t, s \in K^\ell$ with $d(t, s) < \delta$, it follows that $|p^\ell(t) - p^\ell(s)| < \epsilon$. See Appendix H. We have:

Lemma 3.2. *The sequence $(K^\ell, \boldsymbol{p}^\ell)$ is equicontinuous.*

Proof. Suppose not. Then, there exist sequences $t_{\ell_k}, s_{\ell_k} \in K_{\ell_k}$ such that $t_{\ell_k} \to t$, $s_{\ell_k} \to t$ and for some $r > 0$, $p^{\ell_k}(s_{\ell_k}) > p^{\ell_k}(t_{\ell_k}) + r$. Hence, $\lim p^{\ell_k}(s_{\ell_k})/p^{\ell_k}(t_{\ell_k}) \geq 1 + r'$ for some $r' > 0$. Henceforth, we drop the subscript k for ℓ and denote the sequences such as $t_\ell, s_\ell \in K^\ell$. By the monotonicity (MT) of the preferences, the feasibility condition for $(\boldsymbol{x}_a)$ holds with the equality. Hence, $\sum_{a=1}^m \boldsymbol{x}_a^\ell(\{s_\ell\}) = \sum_{a=1}^m \omega_a^\ell(\{s_\ell\}) > 0$, so that there exists a such that $\boldsymbol{x}_a^\ell(\{s_\ell\}) > 0$ for infinitely many ℓ's. We take $1 < \gamma < 1 + r'$ and define:

$$\boldsymbol{z}_a^\ell = \boldsymbol{x}_a^\ell - \boldsymbol{x}_a^\ell(\{s_\ell\})\delta_{s_\ell} + \gamma \boldsymbol{x}_a^\ell(\{s_\ell\})\delta_{t_\ell}.$$

Then, we have $\boldsymbol{p}^\ell \boldsymbol{z}_a^\ell < \boldsymbol{p}^\ell \boldsymbol{x}_a^\ell$ and $\boldsymbol{z}_a^\ell$ is in the a's budget set for infinitely many ℓ's. By the assumption (BRS), for such an ℓ, one has $\boldsymbol{z}_a^\ell \succsim_a^\ell \boldsymbol{x}_a^\ell$. By the monotonicity (MT), we can define $\tilde{\boldsymbol{z}}_a^\ell$ such that $\boldsymbol{p}\tilde{\boldsymbol{z}}_a^\ell < \boldsymbol{p}^\ell \boldsymbol{x}_a^\ell$ and $\boldsymbol{z}_a^\ell \prec_a \tilde{\boldsymbol{z}}_a^\ell$, hence by the negative transitivity (NTR), we have $\boldsymbol{x}_a^\ell \prec_a \tilde{\boldsymbol{z}}_a^\ell$, a contradiction. Therefore the proof is complete. □

By applying Proposition H14 of Appendix H, there is a price vector $\boldsymbol{p} \in C(K)$ with $(K^\ell, \boldsymbol{p}^\ell) \to (K, \boldsymbol{p})$. Recall that $(K^\ell, \boldsymbol{p}^\ell) \to (K, \boldsymbol{p})$ means that $K^\ell \to K$ in the closed convergence (Appendix F) and for all sequences $t_\ell \in K^\ell$ with $t_\ell \to t, \boldsymbol{p}^\ell(t_\ell) \to \boldsymbol{p}(t)$. Then by Proposition H13 of Appendix H, $\boldsymbol{p}^\ell \omega_a^\ell \to \boldsymbol{p}\omega_a$ and $\boldsymbol{p}^\ell \boldsymbol{x}_a^\ell \to \boldsymbol{p}\boldsymbol{x}_a$ for all a.

Since $\sum_{a=1}^{m} x_a^\ell \le \sum_{a=1}^{\ell} \omega_a^\ell$ for all ℓ, $\sum_{a=1}^{m} x_a \le \sum_{a=1}^{\ell} \omega_a$ in the limit. Hence, the condition (E-2) is met. We claim that for each a, x_a is maximal for $\prec_a$ in the budget set.

For some t, $p(t) = 1$. Then, by the assumption of adequate endowment (AE), $p\omega_a > 0$ for some a. If there is a vector $z \in X_a$ with $pz \le p\omega_a$ and $x_a \prec_a z$, then by the continuity ($CT_{\text{weak}*}$), we can take $z' \in X_a$ such that $pz' < p\omega_a$ and $x_a \prec_a z'$. Hence, for ℓ large enough, $p^\ell z' < p^\ell \omega_a^\ell$ and $x_a^\ell \prec_a^\ell z'$, a contradiction. By the monotonicity (MT), $p(t) > 0$ for all $t \in K$, hence $p\omega_a > 0$ for all a. Therefore,

$$px_a \le p\omega_a \quad \text{and} \quad \text{if } pz \le p\omega_a, \ \text{then } x_a \succsim_a z, \ a = 1 \cdots m.$$

This completes the proof. □

We now give a few remarks on the assumption (BRS). Suppose that the preference relation $\prec$ is represented by a utility function $u: \mathcal{M}(K, \mathcal{B}(K)) \to \mathbb{R}$ and consider the (Gateaux) derivative defined by:

$$D_t u(x) = \lim_{h \to 0} \frac{u(x + h\delta_t) - u(x)}{h}$$

if the limit exists, where it is understood that $h \to 0^+$ if $x(\{t\}) = 0$. $D_t u(x)$ is called the Gateaux derivative or the directional derivative of u at x in the direction of δ_t. See Appendix E. Let $R(x, t, h)$ be the remainder term after approximating $u(x + h\delta_t)$ by the first term in a Taylor series,

$$R(x, t, h) = u(x + h\delta_t) - (u(x) + hD_t u(x)).$$

If the second derivative exists and uniformly bounded, then $(1/h)R(x, h, t) \to 0$ as $h \to 0$. Now, consider the sequences as described in the assumption (BRS). If $D_t u(x)$ exists for all $x \in \mathcal{M}(K, \mathcal{B}(K))$ and $t \in K$,

$$u(x_\ell + a_\ell \delta_{s_\ell}) \approx u(x_\ell) + a_\ell D_{t_\ell} u(x_\ell),$$

$$u(x_\ell + b_\ell \delta_{t_\ell}) \approx u(x_\ell) + b_\ell D_{s_\ell} u(x_\ell).$$

Hence, $u(x_\ell + a_\ell \delta_{t_\ell}) \ge u(x_\ell + b_\ell \delta_{s_\ell})$ if $a_\ell / b_\ell \ge D_{s_\ell} u(x_\ell) / D_{t_\ell} u(x_\ell)$ as long as $D_t u(x)$ is well behaved, for example, $D_t u(x) > 0$ and $D_t u(x)$ is jointly continuous with respect to (t, x). This discussion would reveal the reason of the "bounded marginal rate of substitution". Summing up, we have obtained:

Proposition 3.1. *If the utility function $u(x)$ representing the preference $\prec$ is of class (Gâeaux) C^2 such that:*

(i) *$D_t u(x) > 0$ and $D_t u(x)$ is jointly continuous with respect to (t, x),*
(ii) *there exists $b > 0$ with $\|D_t^2 u(x)\| \le b$ for all $x \in \mathcal{M}_+(K, \mathcal{B}(K))$,*

then $\prec$ satisfies the assumption (BRS).

The first welfare theorem still holds for the economy $\mathcal{E}_{\mathcal{M}}$, however, the second welfare theorem is more intricate, since the norm interior of $\mathcal{M}(K, \mathcal{B}(K))$ is empty. We will discuss this more generally in Section 5.4.

5.4. EXCHANGE ECONOMIES ON A GENERAL BANACH SPACE

In this section, we will discuss a market with a mathematically general commodity space which contains the space $\ell^p(1 \leq p \leq +\infty)$ and $\mathcal{M}(K, \mathcal{B}(K))$ as special cases. By this mathematical generalization, the structure and the logic of the equilibrium existence theorems and the welfare theorems will be more transparent.

The commodity space which we consider in this section is assumed to be a Banach lattice. It is a Banach space L with an order relation $\leq$, denoted by $(L, \leq)$. The ordered linear space or the linear space with an order relation $\leq$ is called a vector lattice (Riesz space) if and only if for any $\boldsymbol{x}, \boldsymbol{y} \in L$, there exist elements of L, denoted by $\boldsymbol{x} \vee \boldsymbol{y}$ and $\boldsymbol{x} \wedge \boldsymbol{y}$ such that $\boldsymbol{x} \leq \boldsymbol{z}$ and $\boldsymbol{y} \leq \boldsymbol{z}$ ($\boldsymbol{z} \leq \boldsymbol{x}$ and $\boldsymbol{z} \leq \boldsymbol{y}$) implies $\boldsymbol{x} \vee \boldsymbol{y} \leq \boldsymbol{z}$ ($\boldsymbol{z} \leq \boldsymbol{x} \wedge \boldsymbol{y}$). Then, for any $\boldsymbol{x} \in L$, we can define the positive part $\boldsymbol{x}_+ = \boldsymbol{x} \vee \boldsymbol{0}$, and the negative part $\boldsymbol{x}_- = (-\boldsymbol{x}) \vee \boldsymbol{0}$, and the absolute value $|\boldsymbol{x}| = \boldsymbol{x}_+ + \boldsymbol{x}_-$. A vector lattice is called a Banach lattice if it is also a Banach space satisfying $|\boldsymbol{x}| \leq |\boldsymbol{y}|$ implies that $\|\boldsymbol{x}\| \leq \|\boldsymbol{y}\|$ for all $\boldsymbol{x}, \boldsymbol{y} \in L$. Note that the positive orthant $L_+ = \{\boldsymbol{x} \in L \mid \boldsymbol{0} \leq \boldsymbol{x}\}$ is closed in the norm topology. In this section, we will use the norm topology on L unless otherwise specified.

The spaces such as ℓ^p $(1 \leq p \leq +\infty)$ or $\mathcal{M}(K, \mathcal{B}(K)$ are all examples of the Banach lattices. A Banach space which is not a Banach lattice is $C^1[0,1]$, the space of all continuously differentiable functions (differentiable functions with the continuous first derivatives) on the unit interval $[0,1]$.

An element of the dual space L^*, or a continuous linear functional on L is called a price vector or a price functional.

There exist m consumers $a = 1 \cdots m$. The consumer a has the consumption set $X_a = L_+$, the preference relation $\prec_a$ which is an irreflexive, transitive and negatively transitive binary relation on $X_a \times X_a$, and the initial endowment vector $\omega_a \in L_+$. As usual, we assume that for all a, the preference $\prec_a$ satisfies:

- ($\mathrm{CT}_{\mathrm{norm}}$) (Continuity in the norm topology) $\prec_a \subset X_a \times X_a$ is open (in the norm topology) relative to $X_a \times X_a$,
- (CV) (Convexity) the set $\{\boldsymbol{z} \in X_a \mid \boldsymbol{z} \succsim_a \boldsymbol{x}\}$ is convex for all $\boldsymbol{x} \in X_a$, and
- (MT) (Monotonicity) $\boldsymbol{x} < \boldsymbol{z}$ (or $\boldsymbol{x} \leq \boldsymbol{z}$ and $\boldsymbol{x} \neq \boldsymbol{z}$) implies that $\boldsymbol{x} \prec_a \boldsymbol{z}$.

Note that the continuity assumption (CT) is weak in the sense that the relevant topology is the norm topology. For this point, we will give a few remarks with respect to the previous results of this chapter. The $2m$-tuple $\mathcal{E} = (\prec_a, \omega_a)$ is called an (exchange) economy.

The first problem in a Banach lattice has already appeared in Section 5.3 where the commodity space is $ca(K)$, although we did not point out the problem explicitly there. We just claimed that the bounded marginal rate of substitution assumption (BRS) is needed and proved the theorem using it. The next example proposed by Mas-Colell shows that in the space $ca(K)$, the non-zero supporting price vector, or the vector $\boldsymbol{p} \geq \mathbf{0}$ with $\boldsymbol{p} \neq \mathbf{0}$ such that $\boldsymbol{x} \succsim_a \boldsymbol{z}$ implies that $\boldsymbol{px} \geq \boldsymbol{pz}$ may cease to exist.

Example 4.1. Let $K = \mathbb{N} \cup \{\infty\}$ be the one point compactification of $\mathbb{N}$ (Theorem B2 of Appendix B). The commodity space is assumed to be $L = ca(K)$, the space of signed measures on K with the bounded variation norm (see Section 5.3). For $\boldsymbol{x} \in L$ and $t \in K$, let $x^t = x(\{t\})$ for $0 \leq t \leq \infty$.

For every $t \in K$, we define a function $u_t: \mathbb{R}_+ \to \mathbb{R}_+$ by

$$u_t(s) = \begin{cases} 2^t s & \text{for } s \leq 2^{-2t}, \\ 2^{-t} - 2^{-2t} + s & \text{otherwise.} \end{cases}$$

See Fig. 5.1.

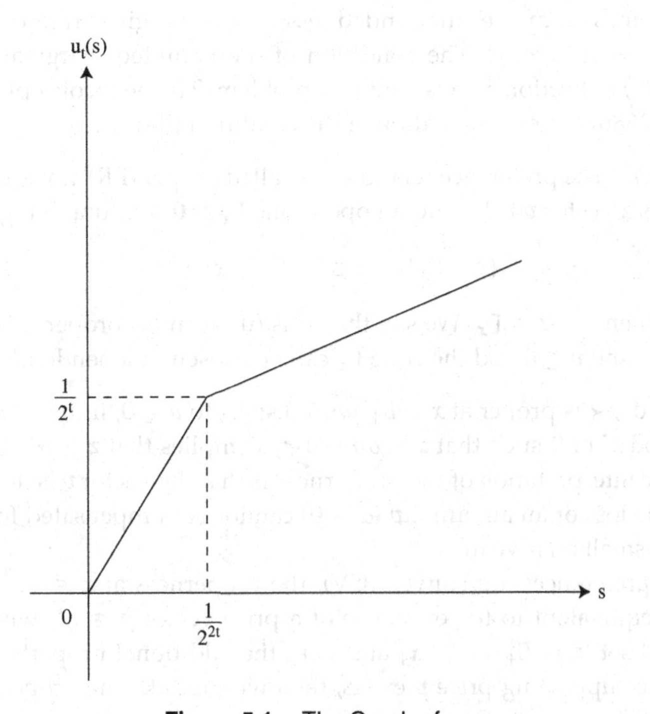

Figure 5.1. The Graph of u_t.

The preference relation $\prec$ on the consumption set L_+ is then defined by the utility function:

$$U(\boldsymbol{x}) = \sum_{t=0}^{\infty} u_t(x^t) \quad \text{for } \boldsymbol{x} = (x^t).$$

It is easy to see that $U(\boldsymbol{x})$ satisfies the monotonicity and the continuity ($\text{CT}_{\text{weak*}}$) in the weak* topology.

Now, let $\omega = (\omega^t) \in L_+$ be defined by $\omega^t = 2^{-2t}$ for $t < \infty$ and $\omega^\infty = 1$. Then, we can show that there does not exist non-zero continuous linear functional $\boldsymbol{p}$ such that $\boldsymbol{pz} \geq \boldsymbol{p}\omega$ whenever $\boldsymbol{z} \succsim \omega$. Indeed, let $\boldsymbol{p}$ be such a functional. For $\boldsymbol{x} \geq \boldsymbol{0}$, one has $\omega + \boldsymbol{x} \succsim \omega$. Hence, $\boldsymbol{px} \geq 0$, or $\boldsymbol{p}$ is a positive linear functional. For $t \in K$, denote $p^t = \boldsymbol{pe}_t$, where $\boldsymbol{e}_t = (e_t^s) = (\delta_t^s)$ and $\delta_t^s = 1$ for $t = s$, and $\delta_t^s = 0$, otherwise. From the first-order condition of the utility maximization, we have:

$$u_t'(\omega^t) = 2^t = \lambda p^t, \quad t = 0, 1, \ldots,$$

hence the only candidates of supporting price vectors are the scalar multiples of the vector $\boldsymbol{q} = (2^t)$. Since this is obviously unbounded, it is not continuous by Proposition H1 of Appendix H.

In Example 4.1, the nonexistence of a non-zero continuous supporting price obviously comes from the unboundedness of the marginal rate of substitution $u_t'(\omega^t)/u_0'(\omega^0) = 2^t$, $t \in \mathbb{N}$. The condition of the bounded marginal rate of substitution (BRS) of Section 5.3 rescued this problem. The next concept proposed by Mas-Colell (1986) is a generalization of the condition (BRS).

Definition 4.1. The preference relation $\prec$ is called proper (PR) at $\boldsymbol{x} \in L_+$ if and only if there exists a vector $\boldsymbol{v} \in L_+$ and an open cone $\Gamma_{\boldsymbol{x}}$ at $\boldsymbol{0} \in L$ containing $\boldsymbol{v}$ such that:

$$\{\boldsymbol{x} - \Gamma_{\boldsymbol{x}}\} \cap \{\boldsymbol{z} \in L_+ \mid \boldsymbol{z} \succsim \boldsymbol{x}\} = \emptyset,$$

or if $\boldsymbol{z} \succsim \boldsymbol{x}$, then $\boldsymbol{x} - \boldsymbol{z} \notin \Gamma_{\boldsymbol{x}}$. We say that $\prec$ is (uniformly) proper if it is proper at every $\boldsymbol{x} \in L_+$, and $\boldsymbol{v} \geq \boldsymbol{0}$ and the cone $\Gamma_{\boldsymbol{x}}$ can be chosen independently of $\boldsymbol{x}$.

In other words, $\prec$ is proper at $\boldsymbol{x} \in L_+$ with respect to $\boldsymbol{v} \geq \boldsymbol{0}$, if there exists an open neighborhood V of $\boldsymbol{0}$ such that $\boldsymbol{x} - \alpha\boldsymbol{v} + \boldsymbol{z} \succsim \boldsymbol{x}$ implies that $\boldsymbol{z} \notin \alpha V$ for all $\alpha > 0$. An economic interpretation of the properness is that the vector $\boldsymbol{v}$ is desirable in the sense that the loss of an amount $\alpha\boldsymbol{v}$ ($\alpha > 0$) cannot be compensated for any vector which is too small relative to $\alpha\boldsymbol{v}$.

When preferences are convex (CV), the properness at $\boldsymbol{x} \in L_+$ with respect to $\boldsymbol{v} \geq \boldsymbol{0}$ is equivalent to the existence of a price vector $\boldsymbol{p} \in L^*$, which supports the preferred set $\{\boldsymbol{z} \in L_+ \mid \boldsymbol{z} \succsim \boldsymbol{x}\}$ at $\boldsymbol{x}$ with the additional property that $\boldsymbol{pv} > 0$. Indeed, if the supporting price $\boldsymbol{p}$ exists, then we can take the properness cone as $\Gamma_{\boldsymbol{x}} = \{\boldsymbol{z} \in L \mid \boldsymbol{pz} > 0\}$. Conversely, if $\prec$ is proper at $\boldsymbol{x}$ with respect to $\boldsymbol{v}$, then

$\{z \in L_+ \mid z \succsim x\}$ and $x - \Gamma_x$ are disjoint convex sets and interior $(x - \Gamma_x) \neq \emptyset$, hence the Hahn–Banach separation theorem (Theorem E3 of Appendix E) provides a continuous linear functional $p \in L^*$ that separates them, or $p(x - y) \leq pz$ for each $y \in \Gamma_x$ and $z \succsim x$. Since Γ_x is an open cone at zero and contains v, it follows that $py \geq 0$ for each $y \in \Gamma_x$, hence $pv > 0$ and $pz \geq px$ for $z \succsim x$.

It can be easily shown that if the positive cone L_+ of the commodity space L has a non-empty interior such as the spaces ℓ^∞ or $L^\infty[0,1]$ with the sup-norm, then the monotonicity (MT) implies the properness (PR). In Section 5.2, we could have recognized that the monotonicity (MT), the convexity (CV), and the (Mackey) continuity (PE_∞) yield the properness (PR) at strictly positive vectors $x \gg \mathbf{0}$. This is also true for the space $L^\infty[0,1]$, and it is indeed a reason for the success of Bewley (1970) for his proof of the existence of competitive equilibria on $L^\infty[0,1]$.

A feasible allocation is an m-tuple of consumption vectors $(x_a) \in L_+^m$ such that $\sum_{a=1}^m x_a \leq \sum_{a=1}^m \omega_a$. Let $\mathcal{F}$ be the set of feasible allocations,

$$\mathcal{F} = \left\{ (x_a) \in L_+^m \;\middle|\; \sum_{a=1}^m x_a \leq \sum_{a=1}^m \omega_a \right\}.$$

A feasible allocation $(x_a) \in \mathcal{F}$ is said to be weakly Pareto optimal if there exists no other feasible allocation $(z_a) \in \mathcal{F}$ such that $x_a \prec_a z_a$ for all $a = 1 \cdots m$.

In this section, we are mainly concerned with a quasi-equilibrium rather than competitive equilibrium, since we will prove the existence theorem via the Negishi type proof, which utilizes the second fundamental theorem of welfare economics, and the former is more convenient for this method.

Definition 4.2. A pair $(p, (x_a)) \in L_+^* \times L_+^m$ of a price vector $p \in L_+^*$ and an allocation $(x_a)_{a=1\cdots m}$ is said to be a quasi-equilibrium if and only if

(QE-1) $px_a \leq p\omega_a$ and $pz \geq p\omega_a$ whenever $z \succsim_a x_a$ for all $a = 1 \cdots m$,
(E-2) $\sum_{a \in A} x_a \leq \sum_{a \in A} \omega_a$.

Under the assumption of the monotonicity (MT), the condition (Q-1) is reduced to the condition:

(E-1) $px_a \leq p\omega_a$ and $pz > p\omega_a$ whenever $x_a \prec_a z$ for all $a = 1 \cdots m$,

for the consumer satisfying that $p\omega_a > 0$. This can be easily shown as in corollary of Theorem 5.2 of Chap. 2. We can show that the first fundamental theorem of welfare economics still holds without any modification of Proposition 5.1 of Chap. 2. On the infinite-dimensional commodity space, however, the second welfare theorem will be technically more intricate. In order to prove this, we recall the concepts of the competitive equilibrium $(p, (x_a)) \in L_+^* \times L_+^m$ and the quasi-equilibrium $(x_a) \in L_+^m$ relative to the price vector $p \in L_+^*$, which are defined as Definitions 5.2 and 5.3 in Chap. 2, respectively.

Definition 4.3. An allocation $(\boldsymbol{x}_a)_{a=1\cdots m}$ is said to be a quasi-equilibrium relative to a price vector $\boldsymbol{p} \in L_+^*$ if and only if

(Q-1) $\boldsymbol{z} \succsim_a \boldsymbol{x}_a$ implies that $\boldsymbol{pz} \geq \boldsymbol{px}_a$ for all $a = 1 \cdots m$,
(E-2) $\sum_{a \in A} \boldsymbol{x}_a \leq \sum_{a \in A} \omega_a$.

As we saw, the properness condition (PR) makes sure the individual supportability. One may have expected that (weakly) optimal feasible allocations can be supported by some price vector, that is to say, the second welfare theorem holds by virtue of the properness. The next example proposed by Jones, however, shows that this conjecture is wrong.

Example 4.2. Let $L = L^\infty[0,1]$ be the set of essentially bounded functions on $[0,1]$ and $X = C^1[0,1]$, the set of all continuously differentiable functions on $[0,1]$ (a function is called continuously differentiable if it has the first derivative which is continuous), respectively. We consider the dual pair (L, X) and consider the $\sigma(L, X)$-topology on L. This topology is described by the net as $\boldsymbol{x}_\alpha \to \boldsymbol{x}$ if and only if $\int_0^1 \boldsymbol{x}_\alpha(t)\boldsymbol{p}(t)dt \to \int_0^1 \boldsymbol{x}(t)\boldsymbol{p}(t)dt$ for every continuously differentiable function $\boldsymbol{p}(t) \in C^1[0,1]$. See Appendix H. There exist two consumers in the economy, $a = 1, 2$. The consumer 1 has the utility function:

$$u_1(\boldsymbol{x}_1) = \int_0^1 (1-t)\boldsymbol{x}(t)dt,$$

and the consumer 2's utility function is given by:

$$u_2(\boldsymbol{x}_2) = \int_0^1 t\boldsymbol{x}(t)dt.$$

The utility functions are linear and continuous, therefore, they are proper. Let $\omega(t) = 1$ for all t. The feasible allocation $(\boldsymbol{x}_1, \boldsymbol{x}_2)$ which is defined by $\boldsymbol{x}_1(t) = 1$ if $t \leq 1/2$, $\boldsymbol{x}_1(t) = 0$ if $t > 1/2$ and $\boldsymbol{x}_2(t) = 0$ if $t \leq 1/2$, $\boldsymbol{x}_1(t) = 1$ if $t > 1/2$, is easily verified to be weakly optimal. However, it cannot be supported by any continuous linear functional on X, since the only candidate of the supporting price is given by $\boldsymbol{p}(t) = 1 - t$ if $t \leq 1/2$, $\boldsymbol{p}(t) = t$ if $t > 1/2$, which is obviously not differentiable function of t hence, it is not continuous as a linear functional on L.

The topology which we consider in the above example is quite artificial. Indeed, we give the topology on L by the dual pairing of $C^1[0,1]$, which is a Banach space but lacks the lattice structure. The key property for the second fundamental theorem on Banach spaces is the lattice structure of the spaces (Riesz decomposition property, Theorem H5 of Appendix H).

Theorem 4.1. *Let $\mathcal{E} = (\prec_a, \omega_a)$ be an exchange economy on a Banach lattice L. For every consumer $a = 1 \cdots m$, suppose that the preference relation satisfies the negative transitivity (NTR), the continuity ($\mathrm{CT}_{\mathrm{norm}}$), the convexity (CV), the monotonicity (MT), and* *the properness (PR). Then, every weakly Pareto-optimal allocation is a quasi-equilibrium relative to some price vector* $\boldsymbol{p} \in L_+^*$ *such that* $\boldsymbol{p}\sum_{a=1}^m \boldsymbol{v}_a = 1$, *where* $\boldsymbol{v}_a \in L_+$ *is a vector as given in Definition* 4.1.

Proof. Let an allocation $((\boldsymbol{x}_a))_{a=1}^m$ be weakly Pareto optimal and we set:

$$Z = \left\{ \sum_{a=1}^{m} (\boldsymbol{z}_a - \boldsymbol{x}_a) \middle| \boldsymbol{z}_a \succsim_a \boldsymbol{x}_a, \ a = 1 \cdots m \right\}.$$

Clearly, the set Z is convex. Let $V_a \subset L$ be a neighborhood of zero such that: $\boldsymbol{x} - \alpha\boldsymbol{v}_a + \boldsymbol{z} \succsim_a \boldsymbol{x}$ implies that: $\boldsymbol{z} \notin \alpha V_a$ for all $\alpha > 0$, where $\boldsymbol{v}_a$ is a vector of the properness (Definition 4.1). Let $V = \cap_{a=1}^m V_a$ and $\boldsymbol{v} = \sum_{a=1}^m \boldsymbol{v}_a$. Without loss of generality, we can assume that V is convex, $-V = V$, and it is solid, that is, $|\boldsymbol{u}| \leq |\boldsymbol{v}|$ and $\boldsymbol{v} \in V$ imply that $\boldsymbol{u} \in V$. Let Γ be the open convex cone spanned by $\boldsymbol{v} + V$. Then, we claim that $Z \cap (-\Gamma) = \emptyset$.

Suppose on the contrary that $Z \cap (-\Gamma) \neq \emptyset$. Then, there exists $\boldsymbol{z}_a \succsim_a \boldsymbol{x}_a$ such that, denoting $\boldsymbol{z} = \sum_{a=1}^m \boldsymbol{z}_a$ and $\boldsymbol{x} = \sum_{a=1}^m \boldsymbol{x}_a$, we have $\boldsymbol{z} - \boldsymbol{x} \in -\Gamma$, or $\boldsymbol{z} - (\boldsymbol{x} - \alpha\boldsymbol{v}) \in \alpha V$ for some $\alpha > 0$. Set $\boldsymbol{y} = \boldsymbol{x} - \alpha\boldsymbol{v}$. Since $\boldsymbol{z} \geq \boldsymbol{0}$, we have $\boldsymbol{y} - \boldsymbol{z} \leq \boldsymbol{y} \leq \boldsymbol{x}$, hence $(\boldsymbol{y} - \boldsymbol{z})_+ = (\boldsymbol{y} - \boldsymbol{z}) \vee \boldsymbol{0} \leq \boldsymbol{x}$. Since $\boldsymbol{y} - \boldsymbol{z} = (\boldsymbol{y} - \boldsymbol{z})_+ - (\boldsymbol{y} - \boldsymbol{z})_-$, it follows that $\boldsymbol{z} = (\boldsymbol{y} - \boldsymbol{x}) + \boldsymbol{x} - (\boldsymbol{y} - \boldsymbol{z})_+ + (\boldsymbol{y} - \boldsymbol{z})_- \geq -\alpha\boldsymbol{v} + (\boldsymbol{y} - \boldsymbol{z})_-$, or $(\boldsymbol{y} - \boldsymbol{z})_- \leq \boldsymbol{z} + \alpha\boldsymbol{v}$.

Since $\boldsymbol{z} + \alpha\boldsymbol{v} = \sum_{a=1}^m (\boldsymbol{z}_a + \alpha\boldsymbol{v}_a)$, the decomposition property of the vector lattice (Theorem H5 of Appendix H) implies that we can write $(\boldsymbol{y} - \boldsymbol{z})_- = \sum_{a=1}^m \boldsymbol{w}_a$, where $\boldsymbol{0} \leq \boldsymbol{w}_a \leq \boldsymbol{z}_a + \alpha\boldsymbol{v}_a$ for each a. We define $\boldsymbol{z}_a' = \boldsymbol{z}_a + \alpha\boldsymbol{v}_a - \boldsymbol{w}_a \geq \boldsymbol{0}$. Suppose that $\boldsymbol{z}_a \succsim_a \boldsymbol{z}_a'$ for some a. Since the preference is proper (PR), this implies that $\boldsymbol{w}_a \notin \alpha V$. On the other hand, $\boldsymbol{0} \leq \boldsymbol{w}_a \leq (\boldsymbol{y} - \boldsymbol{z})_- \leq |\boldsymbol{y} - \boldsymbol{z}|$ and $\boldsymbol{y} - \boldsymbol{z} \in \alpha V$, hence $\boldsymbol{w}_a \in \alpha V$, a contradiction. Thus, $Z \cap (-\Gamma) = \emptyset$, or $\boldsymbol{0} \notin Z + \Gamma$.

Since $Z + \Gamma$ is convex and open set, and $\boldsymbol{0} \notin Z + \Gamma$, we can apply the Hahn–Banach separation theorem (Theorem E3 of Appendix E) and there exists a nonzero, continuous linear functional $\boldsymbol{p}$ such that $\boldsymbol{p}\boldsymbol{y} > 0$ for all $\boldsymbol{y} \in Z + \Gamma$. Since Γ is cone and $\boldsymbol{0} \in Z$, this yields that $\boldsymbol{p}\boldsymbol{z} \geq 0$ for all $\boldsymbol{z} \in Z$ and $\boldsymbol{p}\boldsymbol{y} > 0$ for all $\boldsymbol{y} \in \Gamma$. In particular, $\boldsymbol{p}\boldsymbol{v} > 0$, so without loss of generality we put $\boldsymbol{p}\boldsymbol{v} = 1$. If $\boldsymbol{z} \in V$, then $\boldsymbol{v} - \boldsymbol{z} \in \Gamma$ and so $\boldsymbol{p}\boldsymbol{z} \leq \boldsymbol{p}\boldsymbol{v} = 1$. Since the same argument applies to $-\boldsymbol{z}$, one obtains $|\boldsymbol{p}\boldsymbol{z}| \leq 1$. Finally, let $\boldsymbol{z} \succsim_a \boldsymbol{x}_a$. Then $\boldsymbol{z} - \boldsymbol{x}_a \in Z$, hence $\boldsymbol{p}\boldsymbol{z} \geq \boldsymbol{p}\boldsymbol{x}_a$. This shows that $(\boldsymbol{p}, (\boldsymbol{x}_a))$ is a quasi-equilibrium relative to $\boldsymbol{p}$. The positivity of $\boldsymbol{p}$, $\boldsymbol{p} \geq \boldsymbol{0}$ follows from the monotonicity (MT) of $\prec_a$. □

Note that in the proof of Theorem 4.1, the properness assumption (PR) and the lattice structure of L are crucial. The next example put forth by Araujo shows that the utility possibility frontier need not be closed.

Example 4.3. The commodity space L is assumed to be ℓ^∞, the space of bounded sequences with the supremum norm (see Section 5.2). Suppose that the total endowment $\omega = (1, 1 \ldots) = \boldsymbol{1}$. There exist two consumers. Consumer 1 has the utility function:

$$u_1(\boldsymbol{x}) = \liminf_{t\to\infty} x^t \quad \text{for } \boldsymbol{x} = (x^t) \in \ell_+^\infty,$$

and the consumer 2's utility function is given by:

$$u_2(\boldsymbol{x}) = \sum_{t=0}^{\infty} \frac{x^t}{2^{t+1}} \quad \text{for } \boldsymbol{x} = (x^t) \in \ell_+^\infty.$$

It can be easily seen that the utility possibility frontier:

$$U(\mathcal{F}) = \{(u_1(\boldsymbol{x}_1), u_2(\boldsymbol{x}_2)) \mid \boldsymbol{x}_1 + \boldsymbol{x}_2 \le \omega, \boldsymbol{x}_1, \boldsymbol{x}_2 \ge 0\}$$

is of the form depicted in Fig. 5.2. In the figure, the point $(0,1)$ is included in $U(\mathcal{F})$. We will exclude these kind of situations by the assumption:

(CL) (Closedness of the utility possibility frontier) let $(\boldsymbol{x}_a^k) \in \mathcal{F}$ be a sequence of feasible allocations such that $j > k$ implies that $\boldsymbol{x}_a^j \succsim_a \boldsymbol{x}_a^k$ for all $a = 1 \cdots m$. Then, there exists a feasible allocation $(\boldsymbol{x}_a) \in \mathcal{F}$ such that $\boldsymbol{x}_a \succsim_a \boldsymbol{x}_a^k$ for all $k \in \mathbb{N}$ and all $a = 1 \cdots m$.

It can be easily seen that the assumption (CL) is equivalent to the closedness of $U(\mathcal{F})$. Which conditions on the individual preferences yield the closedness condition (CL)? Let the sequences $(\boldsymbol{x}^k)$ be as in the condition (CL). Then $Z_k = \bigcap_{b=1}^m \{\boldsymbol{x} = (\boldsymbol{x}_a) \in \mathcal{F} \mid \boldsymbol{x}_b \succsim_b \boldsymbol{x}^k\}$ constitutes a decreasing sequence of closed subsets of $\mathcal{F}$. We want to find

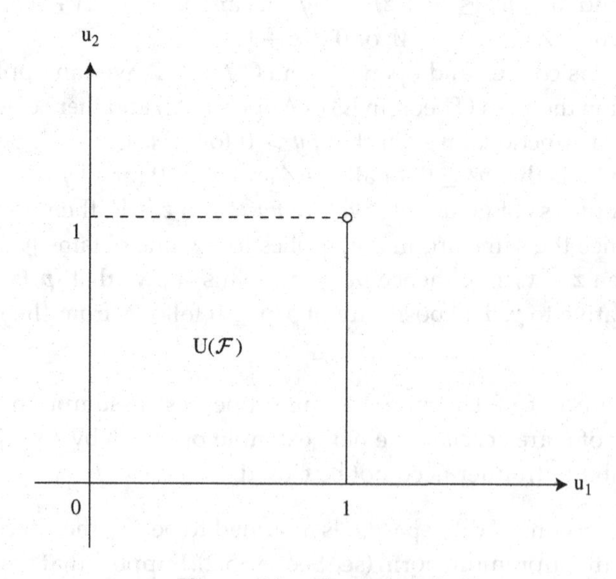

Figure 5.2. Example 4.3.

conditions that $\cap_{k=0}^{\infty} Z_k \neq \emptyset$. Obviously, this intersection is nonempty if the sets $\{\boldsymbol{x} \in L_+ \mid \boldsymbol{x} \succsim_a \boldsymbol{x}_a\}$ are closed for all a, and the feasible allocation set $\mathcal{F}$ is compact. When $L = ca(K)$, we assumed that the sets $\{\boldsymbol{x} \in L_+ \mid \boldsymbol{x} \succsim_a \boldsymbol{x}_a\}$ are closed in the weak* topology, and ensured the compactness of $\mathcal{F}$ in the weak* topology by Alaoglu's theorem (Theorem H3 of Appendix H). When $L = \ell^\infty$, we assumed that the sets $\{\boldsymbol{x} \in L_+ \mid \boldsymbol{x} \succsim_a \boldsymbol{x}_a\}$ are closed in the Mackey topology, but the Mackey topology is equal to the weak* topology on the convex sets by Theorem H2 of Appendix H. This is exactly the way by which we obtained the closedness condition (CL) in Sections 5.2 and 5.3.

By these preparations, we can eventually prove the existence theorem on a general Banach space. In the following theorem, we will apply that the preferences are represented by utility functions. (This assumption is, in fact, unnecessary. See Section 5.7.)

Theorem 4.2. *Let L be a Banach lattice and suppose that for each a, the preference of the economy $\mathcal{E} = (\prec_a, \omega_a)$ represented by a utility function $u_a: L_+ \to \mathbb{R}$ which is continuous in the norm topology, quasi-concave (UCV), monotone (MT) and (uniformly) proper (PR). We assume that the total endowment $\omega = \sum_{a=1}^m \omega_a$ is desirable in the sense that $u_a(\alpha\omega) > u_a(\mathbf{0})$ for all $\alpha > 0$. Furthermore, the closedness condition (CL) holds. Then, there exists a quasi-equilibrium $(\boldsymbol{p}, (\boldsymbol{x}_a))$ for $\mathcal{E}$ such that $\boldsymbol{p}\sum_{a=1}^m \boldsymbol{v}_a = 1$, where $\boldsymbol{v}_a$ is a vector of the properness condition (PR), $a = 1 \cdots m$.*

Proof. Consider the order interval on L, $[\mathbf{0}, \omega] = \{\boldsymbol{z} \in L \mid \mathbf{0} \leq \boldsymbol{z} \leq \omega\}$. Without loss of generality, we can assume that $u_a(\mathbf{0}) = 0$ and $u_a(\omega) = 1$, $a = 1 \cdots m$. We define a map $U: [\mathbf{0}, \omega]^m \to [0, 1]^m$ by $U((\boldsymbol{x}_a)) = (u_a(\boldsymbol{x}_a))$, and $f: \Delta \to \mathbb{R}$ by $f(\boldsymbol{v}) = \sup\{\alpha \in \mathbb{R} \mid \alpha\boldsymbol{v} \in U(\mathcal{F})\}$, where as usual, $\Delta = \{\boldsymbol{v} = (v^a) \in \mathbb{R}_+^\ell \mid \sum_{a=1}^m v^a = 1\}$ is the $m-1$-simplex. Then, we have:

Lemma 4.1. *$f: \Delta \to [0, m]$ is a well-defined continuous function with $f(\boldsymbol{v}) > 0$ for all $\boldsymbol{v} \in \Delta$.*

Proof. Since $\omega \geq \mathbf{0}$ is desirable, it follows that $0 = u_a(\mathbf{0}) < u_a(\alpha\omega)$ for all $\alpha > 0$, $a = 1 \cdots m$. Therefore, we have $f(\boldsymbol{v}) > 0$ for all $\boldsymbol{v} \in \Delta$.

Let $\boldsymbol{v}_n \to \boldsymbol{v} \in \Delta$ and $\alpha\boldsymbol{v} = U(\boldsymbol{x})$ for some $\boldsymbol{x} \in \mathcal{F}$ and $\alpha > 0$. Take $0 < \beta < \alpha$. We can assume without loss of generality that $\beta v_n^a < \alpha v_n^a$ for all n and for all a with $v^a > 0$. For a with $v^a = 0$, let $u_a(\gamma_n^a\omega) = \beta v_n^a$. Then, $\gamma_n^a \to 0$ for a with $v^a = 0$, hence, $\gamma_n = \sum_{v^a=0} \gamma_n^a \to 0$. We define an allocation $\boldsymbol{y}^n = (\boldsymbol{y}_a^n)$ by $\boldsymbol{y}_a^n = \gamma_n^a\omega$ if $v^a = 0$, and $\boldsymbol{y}_a^n = (1 - \gamma_n)\boldsymbol{x}_a$ otherwise. Then,

$$\sum_{a=1}^m \boldsymbol{y}_a^n = \sum_{v^a=0} \boldsymbol{y}_a^n + \sum_{v^a>0} \boldsymbol{y}_a^n$$

$$= \gamma_n\omega + (1 - \gamma_n)\sum_{v^a>0} \boldsymbol{x}_a \leq \gamma_n\omega + (1 - \gamma_n)\sum_{a=1}^m \boldsymbol{x}_a \leq \omega,$$

or $(\boldsymbol{y}_a^n) \in \mathcal{F}$. For n sufficiently large, we have $u_a(\boldsymbol{y}_a^n) \geq \beta v_n^a$ for $a = 1 \cdots m$. Take $\mu_a \in [0,1]$ such that $u_a(\mu_a \boldsymbol{y}_a^n) = \beta v_n^a$ for $a = 1 \cdots m$. Then $\sum_{a=1}^m \mu_a \boldsymbol{y}_a^n \leq \sum_{a=1}^m \boldsymbol{y}_a^n \leq \omega$, or we get a feasible allocation whose utility is precisely $\beta \boldsymbol{v}_n$. Since $\beta < \alpha$ is arbitrary, we conclude that $f(\boldsymbol{v}) \leq \liminf_{n\to\infty} f(\boldsymbol{v}_n)$. Similarly, we can show that $\limsup_{n\to\infty} f(\boldsymbol{v}_n) \leq f(\boldsymbol{v})$. □

By the closedness assumption (CL), for any $\boldsymbol{v} \in \Delta$, we can find a feasible allocation $\boldsymbol{x}(\boldsymbol{v}) = (\boldsymbol{x}_a(\boldsymbol{v})) \in \mathcal{F}$ such that $U(\boldsymbol{x}(\boldsymbol{v})) = f(\boldsymbol{v})\boldsymbol{v}$. By the monotonicity (MT), it follows that $\sum_{a=1}^m \boldsymbol{x}_a(\boldsymbol{v}) = \omega$. Then by Theorem 4.1, we can define:

$$P(\boldsymbol{v}) = \{\boldsymbol{p} \in L_+^* \mid \boldsymbol{p}\boldsymbol{x}_a(\boldsymbol{v}) \leq \boldsymbol{p}\boldsymbol{x} \text{ whenever } \boldsymbol{x} \succsim_a \boldsymbol{x}(\boldsymbol{v}), a = 1 \cdots m,$$
$$\boldsymbol{p}\boldsymbol{v} = 1, |\boldsymbol{p}\boldsymbol{z}| \leq 1 \text{ for all } \boldsymbol{z} \in V\}.$$

Theorem 4.1 implies that $P(\boldsymbol{v}) \neq \emptyset$ for all $\boldsymbol{v}$. Then, we define:

$$\sigma(\boldsymbol{v}) = \{(\boldsymbol{p}(\omega_1 - \boldsymbol{x}_1(\boldsymbol{v})) \cdots \boldsymbol{p}(\omega_m - \boldsymbol{x}_m(\boldsymbol{v}))) \in \mathbb{R}^m \mid \boldsymbol{p} \in P(\boldsymbol{v})\}.$$

If $\boldsymbol{s} = (s^a) \in \sigma(\boldsymbol{v})$, then $\sum_{a=1}^m s^a = 0$, and $\boldsymbol{0} \in \sigma(\boldsymbol{v})$ if and only if $(\boldsymbol{p}, \boldsymbol{x}(\boldsymbol{v}))$ is a quasi-equilibrium, where $\boldsymbol{p} \in P(\boldsymbol{v})$.

Lemma 4.2. *The correspondence $\sigma \colon \Delta \to \mathbb{R}^m$ is upper hemi-continuous.*

Proof. Let $\boldsymbol{v}_n \to \boldsymbol{v} \in \Delta$, $\boldsymbol{s}_n \in \sigma(\boldsymbol{v}_n)$ for all $n \in \mathbb{N}$ and $\boldsymbol{s}_n \to \boldsymbol{s}$. We want to show that $\boldsymbol{s} \in \sigma(\boldsymbol{v})$. Let $\boldsymbol{p}_n \in P(\boldsymbol{v}_n)$ be such that $\boldsymbol{s}_n = (s_n^a) = (\boldsymbol{p}_n(\omega_a - \boldsymbol{x}_a(\boldsymbol{v}_n)))$. Since $|\boldsymbol{p}_n\boldsymbol{z}| \leq 1$ for all $\boldsymbol{z} \in V$, we can assume that the sequence $(\boldsymbol{p}_n)$ has a weak* limit $\boldsymbol{p}$ by Alaoglu's theorem (Theorem H3 of Appendix H), or $\boldsymbol{p}_n\boldsymbol{x} \to \boldsymbol{p}\boldsymbol{x}$ for all $\boldsymbol{x} \in L$. Since $\boldsymbol{p}_n\boldsymbol{v} = 1$ for all n, it follows that $\boldsymbol{p}\boldsymbol{v} = 1$. Let $\boldsymbol{z} \succsim_a \boldsymbol{x}_a(\boldsymbol{v})$. Then, by the monotonicity (MT), there exists $\boldsymbol{z}'$ arbitrary close to $\boldsymbol{z}$ and $\boldsymbol{z} \prec_a \boldsymbol{z}'$. Hence, $\boldsymbol{x}_a(\boldsymbol{v}_n) \prec_a \boldsymbol{z}'$ for all n large enough. Therefore, $\boldsymbol{p}_n\boldsymbol{z}' \geq \boldsymbol{p}_n\boldsymbol{x}_a(\boldsymbol{v}) = \boldsymbol{p}_n\omega_a - s_n^a$ for all n sufficiently large. Taking the limit, we have $\boldsymbol{p}\boldsymbol{z}' \geq \boldsymbol{p}\omega_a - s^a = \boldsymbol{p}\boldsymbol{x}_a(\boldsymbol{v})$. Since $\boldsymbol{z}'$ is arbitrarily close to $\boldsymbol{z}$, we conclude that:

$$\boldsymbol{p}\boldsymbol{z} \geq \boldsymbol{p}\omega_a - s^a (= \boldsymbol{p}\boldsymbol{x}_a(\boldsymbol{v})) \quad \text{whenever } \boldsymbol{z} \succsim_a \boldsymbol{x}(\boldsymbol{v}), \ a = 1 \cdots m.$$

Indeed, if $\boldsymbol{p}\boldsymbol{x}_a(\boldsymbol{v}) = 0$, then $\boldsymbol{p}\boldsymbol{z} \geq 0$ holds trivially. Suppose that $\boldsymbol{z} \succsim_a \boldsymbol{x}_a(\boldsymbol{v})$ and $\boldsymbol{p}\boldsymbol{z} < \boldsymbol{p}\boldsymbol{x}_a(\boldsymbol{v})$. Then, by the monotonicity (MT), we can take $\boldsymbol{z}'$ close enough to $\boldsymbol{z}$ such that $\boldsymbol{z} \prec_a \boldsymbol{z}'$ and $\boldsymbol{p}\boldsymbol{z}' < \boldsymbol{p}\boldsymbol{x}_a(\boldsymbol{v})$, a contradiction.

Setting $\boldsymbol{z} = \boldsymbol{x}_a(\boldsymbol{v})$, we have $\boldsymbol{p}\boldsymbol{x}_a(\boldsymbol{v}) \geq \boldsymbol{p}\omega_a - s^a$ for all a. Since $\sum_{a=1}^m \boldsymbol{x}_a(\boldsymbol{v}) = \sum_{a=1}^m \omega_a$, it follows that $\boldsymbol{p}\boldsymbol{x}_a(\boldsymbol{v}) = \boldsymbol{p}\omega_a - s^a$. This proves $\boldsymbol{s} \in \sigma(\boldsymbol{v})$. □

The last step of the proof is now standard. For each $\boldsymbol{v}$, there exists $\epsilon > 0$ such that $\epsilon(\omega_a - \boldsymbol{x}_a(\boldsymbol{v})) \in V$, hence $|\boldsymbol{p}(\omega_a - \boldsymbol{x}_a(\boldsymbol{v}))| \leq 1/\epsilon$ for all $\boldsymbol{p} \in P(\boldsymbol{v})$. Since $\sigma(\boldsymbol{v})$ is upper hemi-continuous, it is compact-valued. Hence, by Proposition D3 of Appendix D,

$\sigma(\Delta)$ is a compact subset of $\mathbb{R}^m$, or we can define:

$$T = \left\{ \boldsymbol{s} = (s^a) \,\middle|\, \sum_{a=1}^{m} s^a = 0, \sum_{a=1}^{m} |s^a| \leq \bar{v} \right\}$$

for some $\bar{v} > 0$ such that $\sigma(\Delta) \subset T$. We define a function $\boldsymbol{v}: \Delta \times T \to \Delta$, $\boldsymbol{v}(\boldsymbol{v},\boldsymbol{s}) = (v^a(\boldsymbol{v},\boldsymbol{s}))$ by:

$$v^a(\boldsymbol{v},\boldsymbol{s}) = \frac{\max\{0, v^a + s^a\}}{\sum_{a=1}^{m} \max\{0, v^a + s^a\}}, \quad a = 1 \cdots m.$$

Since $\sum_{a=1}^{m} \max\{0, v^a + s^a\} \geq \sum_{a=1}^{m} (v^a + s^a) = 1 > 0$, the function $\boldsymbol{v}(\boldsymbol{v},\boldsymbol{s})$ is continuous. Finally, we define a correspondence:

$$\Phi: \Delta \times T \to \Delta \times T, \; \Phi(\boldsymbol{v},\boldsymbol{s}) = (\boldsymbol{v}(\boldsymbol{v},\boldsymbol{s}), \sigma(\boldsymbol{v})).$$

Since $\sigma(\boldsymbol{v})$ is convex for each $\boldsymbol{v}$, $\Phi(\boldsymbol{v},\boldsymbol{s})$ is upper hemi-continuous, compact, and convex-valued correspondence from a compact and convex set to itself. Therefore, by Kakutani's fixed-point theorem (Theorem D1 of Appendix D), there exists a fixed point $(\boldsymbol{v}^*,\boldsymbol{s}^*) \in \Phi(\boldsymbol{v}^*,\boldsymbol{s}^*)$, or $\boldsymbol{v}^* = \boldsymbol{v}(\boldsymbol{v}^*,\boldsymbol{s}^*)$ and $\boldsymbol{s}^* \in \sigma(\boldsymbol{v}^*)$. For a such that $v^{a*} = 0$, we have $s^{a*} \geq 0$. For $u_a(\boldsymbol{x}_a(\boldsymbol{v}^*)) = 0$, or $\boldsymbol{0} \succsim_a \boldsymbol{x}_a(\boldsymbol{v}^*)$ which implies that $0 \geq \boldsymbol{p}\boldsymbol{x}_a(\boldsymbol{v}^*) \geq 0$ for every $\boldsymbol{p} \in P(\boldsymbol{v}^*)$. Therefore, we have $\max\{0, v^{a*} + s^{a*}\} = v^{a*} + s^{a*}$ for all $a = 1 \cdots m$, hence,

$$v^{a*} = v^a(\boldsymbol{v}^*,\boldsymbol{s}^*) = \frac{v^{a*} + s^{a*}}{\sum_{a=1}^{m} (v^{a*} + s^{a*})} = v^{a*} + s^{a*}, \quad a = 1 \cdots m,$$

or $s^{a*} = 0$, $a = 1 \cdots m$. This proves that $\boldsymbol{0} \in \sigma(\boldsymbol{v}^*)$. Take $\boldsymbol{p}^* \in P(\boldsymbol{v}^*)$. Then, $(\boldsymbol{p}^*, (\boldsymbol{x}_a(\boldsymbol{v}^*))$ is a quasi-equilibrium. □

5.5. INFINITE TIME HORIZON ECONOMIES WITH EXTERNAL INCREASING RETURNS

As discussed in Section 4.2 of Chap. 4, we assume that there exist n firms indexed by $b(= 1 \cdots n)$, and the firm b has the external increasing returns to scale production function,

$$f_b: \mathbb{R}_+ \times \mathbb{R}_+ \to \mathbb{R}_+, \; (z^t, k^t) \mapsto y^{t+1} = f_b(z^t, k^t), \quad t = 0, 1, \ldots .$$

As the time coordinates suggest, the firms uses the input commodity at the date t and produces the output commodity at the next date $t+1$.

There are m consumers in the economy who are, as usual, indexed by $a(=1 \cdots m)$. Let $\omega = \sum_{a=1}^{m} \omega_a = (\omega^0, \omega^1 \ldots)$ be the total endowment vector, which is assumed to be strictly positive or belong to $\mathbb{R}^{\infty}_{++} = \{\boldsymbol{x} = (x^t) \in \mathbb{R}^{\ell} \mid x^t > 0, t = 0, 1, \dots\}$. We define the path of pure capital accumulation $\rho = (\rho^t)$ inductively by $\rho^0 = \omega^0$, and given ρ^{t-1}, set:

$$\rho^t = \max\left\{\sum_{b=1}^{n} f_b\left(z_b, \sum_{c=1}^{n} z_c\right) \middle| \sum_{c=1}^{n} z_c \leq \rho^{t-1}\right\} + \omega^t.$$

Using the path (ρ^t), we define the commodity space L of the economy by:

$$L = \ell^{\infty}_{\rho} = \{\boldsymbol{x} = (x^t) | \sup_{t \geq 0} |x^t|/\rho^t < +\infty\}.$$

The space ℓ^{∞}_{ρ} is the ℓ^{∞} weighted by ρ, or the order ideal generated by ρ (see Appendix H).

Since we assume that the total endowment vector is strictly positive, so is the path ρ. It goes without saying that our main concern is the case that $\rho \to +\infty$ as $t \to +\infty$.

The dual space of ℓ^{∞}_{ρ} is the weighted ba space ba_{ρ} defined by:

$$ba_{\rho} = \Big\{\pi : 2^{\mathbb{N}} \to \mathbb{R} \Big| \sup_{E \subset \mathbb{N}} \int_E \rho d|\pi| < +\infty, \pi(E \cup F) = \pi(E) + \pi(F) \text{ whenever } E \cap F = \emptyset\Big\},$$

and the subspace of countably additive set functions is identified with the weighted ℓ^1 space:

$$\ell^1_{\rho} = \left\{\boldsymbol{p} = (p^t) \middle| \sum_{t=0}^{\infty} |\rho^t p^t| < +\infty\right\}$$

which is the mathematically natural candidate for the price space of the economy. However, by a technical reason explained later, we will look for the equilibrium price vector from the (standard) ℓ^1 space,

$$\ell^1 = \left\{\boldsymbol{p} = (x^t) \middle| \sum_{t=0}^{\infty} |p^t| < +\infty\right\}.$$

Note that for $\boldsymbol{x} \in \ell^{\infty}_{\rho}$ and $\boldsymbol{p} \in \ell^1$, the value of the inner product $\boldsymbol{px} = \sum_{t=0}^{\infty} p^t x^t$ could be $\pm\infty$, or even does not exist. Let $\ell^{\infty}_{\rho+}$ denotes the positive orthant of ℓ^{∞}_{ρ},

$$\ell^{\infty}_{\rho+} = \{\boldsymbol{x} \in \ell^{\infty}_{\rho} \mid \boldsymbol{x} \geq \boldsymbol{0}\}.$$

The technology function $F_b : \ell^{\infty}_{\rho+} \times \ell^{\infty}_{\rho+} \to \mathbb{R}^{\infty}_{+}$ of the firm b is defined by:

$$F_b : ((z^t), (k^t)) \mapsto (y^t), \quad y^0 = 0, \quad y^{t+1} = f_b(z^t, k^t), \ t \geq 0.$$

Note that the range of the map F_b may not be contained in $\ell^{\infty}_{\rho+}$.

The consumer a has the consumption set $X_a = \ell^\infty_{\rho+}$, and the preference relation $\prec_a \subset \ell^\infty_{\rho+} \times \ell^\infty_{\rho+}$. Certainly, we have $\omega_a \in \ell^\infty_{\rho+}$. An allocation is denoted by $((\boldsymbol{x}_a), (\boldsymbol{y}_b, \boldsymbol{z}_b)) \in (\ell^\infty_{\rho+})^{m+2n}$. The list $(\ell^\infty_\rho, \prec_a, \omega_a, f_b)$ is called an infinite time horizon economy with external increasing returns, or simply an economy and denoted by $\mathcal{E}^\infty_k$.

The following is our final definition of the competitive equilibrium.

Definition 5.1. An $(m+2n+1)$-tuple $(\boldsymbol{p}, (\boldsymbol{x}_a), (\boldsymbol{y}_b, \boldsymbol{z}_b))$ consisting of a price vector $\boldsymbol{p} \in \ell^1_+$ and an allocation $((\boldsymbol{x}_a), (\boldsymbol{y}_b, \boldsymbol{z}_b)) \in (\ell^\infty_{\rho+})^{m+2n}$ is called a competitive equilibrium if and only if

(E-1) $\boldsymbol{p}\boldsymbol{x}_a \leq \boldsymbol{p}\omega_a$ and $\boldsymbol{x}_a \succsim_a \boldsymbol{x}'$ whenever $\boldsymbol{p}\boldsymbol{x}' \leq \boldsymbol{p}\omega_a$, $a = 1 \cdots m$,
(E-2) $\boldsymbol{y}_b \leq F_b(\boldsymbol{z}_b, \sum_{c=1}^n \boldsymbol{z}_c)$, and $\boldsymbol{p}(\boldsymbol{y} - \boldsymbol{z}) \leq \boldsymbol{p}(\boldsymbol{y}_b - \boldsymbol{z}_b) = 0$ for all $(\boldsymbol{y}, \boldsymbol{z})$ such that $\boldsymbol{y} \leq F_b(\boldsymbol{z}, \sum_{c=1}^n \boldsymbol{z}_c)$,
(E-3) $\sum_{a=1}^m \boldsymbol{x}_a \leq \sum_{b=1}^n (\boldsymbol{y}_b - \boldsymbol{z}_b) + \sum_{a=1}^m \omega_a$.

The economic meaning of all conditions should be clear enough. In the condition (E-1), we have $\boldsymbol{p}\omega_a < +\infty$ for $\omega_a \in \ell^\infty_+$ and we will assume that in Theorem 5.1, the endowment vector is contained in ℓ^∞_+ (the assumption (PE$_\infty$)). In the condition (E-2), by the homogeneity in the first variable of F_b, the value of the profit $\boldsymbol{p}(\boldsymbol{y}_b - \boldsymbol{z}_b) = 0$ in equilibrium. Hence, for all technologically feasible production vectors $(\boldsymbol{y}, \boldsymbol{z})$, the value $\boldsymbol{p}(\boldsymbol{y} - \boldsymbol{z})$ exists, including $-\infty$. Then, our main theorem reads as follows.

Theorem 5.1. *Suppose that an economy $\mathcal{E}^\infty_k = (\ell^\infty_\rho, \prec_a, \omega_a, f_b)$ satisfies the following assumptions,*

(CT$_\infty$) *(Continuity in Mackey topology) the preference relation $\prec_a \subset X_a \times X_a$ is open relative to $X_a \times X_a$ in the $\tau(\ell^\infty_\rho, \ell^1_\rho)$-topology,*
(CV) *(Convexity) for every $\boldsymbol{x} \in X_a$, the set $\{\boldsymbol{z} \in X_a \mid \boldsymbol{z} \succsim_a \boldsymbol{x}\}$ is convex, $a = 1 \cdots m$.*
(MT) *(Monotonicity) for every $\boldsymbol{x}, \boldsymbol{z} \in X_a$, $\boldsymbol{x} \leq \boldsymbol{z}$ and $\boldsymbol{z} \neq \boldsymbol{x}$ imply $\boldsymbol{x} \prec_a \boldsymbol{z}$, $a = 1 \cdots m$.*
(PE$_\infty$) *(Positive endowment) $\omega_a \in \ell^\infty_+$ and there exists a positive number $\gamma_a > 0$ such that:*

$$\omega^t_a \geq \gamma_a \quad \text{for all } t \in \mathbb{N}, \ a = 1 \cdots m.$$

(FCT) *(Continuous production function) the (one period) production function $f: \mathbb{R}_+ \times \mathbb{R}_+ \to \mathbb{R}_+$, $(z, k) \mapsto y = f_b(z, k)$ is continuous (in the usual topology).*

Then, there exists a competitive equilibrium $(\boldsymbol{p}, (\xi_a), (\eta_b, \zeta_b))$ for $\mathcal{E}^\infty_k$.

Proof. The proof goes on as in that of Theorem 2.1. First, we define the T-period truncated economy $\mathcal{E}^T_k$. Consider, as in the proof of Theorem 4.1, the set of sequences whose coordinates are zero after T,

$$K^T = \{\boldsymbol{x} = (x^t) \mid 0 = x^{T+1} = x^{T+2} = \cdots\} \subset L, \ T = 0, 1, \ldots.$$

The set K^T is naturally identified with $\mathbb{R}^{T+1}$, $K^T \approx \mathbb{R}^{T+1}$. Let:

$$X_a^T = X_a \cap K^T \approx \mathbb{R}_+^{T+1}, \quad a = 1 \cdots m,$$

and we have a preference relation $\prec_a^T$ on X_a^T which is the restriction of $\prec_a$ on the truncated consumption set $X_a^T, a = 1 \cdots m$. Let $\omega_a(T) \in K^T \approx \mathbb{R}^{T+1}$ be the truncated initial endowment vector which is defined as $\omega_a^t(T) = (\omega_a^0, \omega_a^1 \cdots \omega_a^T, 0, 0 \ldots)$. The technology function F_b^T is also defined as the restriction of the map F_b to $\mathbb{R}^{T+1}$,

$$F_b^T : \mathbb{R}^{T+1} \to \mathbb{R}^{T+1}, \quad (z^t, k^t) \mapsto y^{t+1} = f_b(z^t, k^t), \ t = 0 \cdots T,$$

where we set $y^0 = 0$. Then, we have obtained the T-period economy $\mathcal{E}_{\boldsymbol{k}}^T = ((X_a^T, \prec_a^T), \omega_a(T), F_b^T)$. The competitive equilibrium for the economy $\mathcal{E}_{\boldsymbol{k}}^T$ is obviously defined as in Definition 5.1. The next lemma is easy.

Lemma 5.1. *Under the assumptions of Theorem 5.1, there exists a competitive equilibrium $(\boldsymbol{p}(T), (\boldsymbol{x}_a(T)), (\boldsymbol{y}_b(T), \boldsymbol{z}_b(T)))$ for the T-period economy $\mathcal{E}_{\boldsymbol{k}}^T$, $T = 0, 1, \ldots$.*

Proof. We will show that the T-period economy $\mathcal{E}_{\boldsymbol{k}}^T$ satisfies the assumptions of Theorem 3.1. of Chap. 4. We prove that the set of feasible allocations:

$$\mathcal{F} = \left\{ ((\boldsymbol{x}_a), (\boldsymbol{y}_b, \boldsymbol{z}_b)) \in (\mathbb{R}_+^{T+1})^m \times (\mathbb{R}_+^{T+1})^{2n} \middle| \sum_{a=1}^m \boldsymbol{x}_a \right.$$
$$\left. \leq \sum_{b=1}^n (\boldsymbol{y}_b - \boldsymbol{z}_b) + \sum_{a=1}^m \omega_a, \ \boldsymbol{y}_b \leq F_b \left(\boldsymbol{z}_b, \sum_{c=1}^n \boldsymbol{z}_c \right) \right\}$$

is bounded. The other assumptions of Theorem 3.1. hold obviously. Let $((\boldsymbol{x}_a), (\boldsymbol{y}_b, \boldsymbol{z}_b))$ be a feasible allocation. Then, we have:

$$0 \leq z_b^s \leq \rho^s, \ s = 0, \ldots, T, \ b = 1 \cdots n.$$

Indeed, for $s = 0$, $0 \leq z_b^0 \leq \sum_{b=1}^n z_b^0 \leq \sum_{a=1}^m x_a^0 + \sum_{b=1}^n z_b^0 \leq \sum_{a=1}^m \omega_a^0 \leq \rho^0$. Given $0 \leq z_b^{s-1} \leq \rho^{s-1}$, it follows from the definition of ρ that:

$$0 \leq z_b^s \leq \sum_{b=1}^n z_b^s \leq \sum_{a=1}^m x_a^s + \sum_{b=1}^n z_b^s$$
$$\leq \sum_{b=1}^n f_b \left(z_b^{s-1}, \sum_{c=1}^n z_c^{s-1} \right) + \sum_{a=1}^m \omega_a^s \leq \rho^s.$$

Hence, the mathematical induction is established. Similarly, we have:

$$0 \leq y_b^s \leq \rho^s, \quad s = 0 \cdots T, \ b = 1 \cdots n,$$

$$0 \leq x_a^s \leq \rho^s, \quad s = 0 \cdots T, \ a = 1 \cdots m.$$

Hence, it follows from Theorem 3.1. of Chap. 2 that there exists a competitive equilibrium $(\boldsymbol{p}(T), (\boldsymbol{x}_a(T)), (\boldsymbol{y}_b(T)))$ for $\mathcal{E}_k^T$. □

Let $(\boldsymbol{p}(T), (\boldsymbol{x}_a(T)), (\boldsymbol{y}_b(T)))$ be an equilibrium for $\mathcal{E}_k^T$. Note that the monotonicity assumption (MT) implies that $\boldsymbol{p}(T) \geq \mathbf{0}$ for all $T \in \mathbb{N}$. Of course $\boldsymbol{p}(T) \neq \mathbf{0}$. Hence, we can normalize the equilibrium price vector $\boldsymbol{p}(T) = (p^t(T))$ as $\|\boldsymbol{p}(T)\| = \sum_{t=0}^{T} p^t(T) = 1$. Let $\pi(T) = (\boldsymbol{p}(T), 0, 0, \dots)$. Then, $\pi(T) \in \ell^1 \subset ba$ and $\pi(T) \geq \mathbf{0}$ for all $T \in \mathbb{N}$. Hence, $\|\pi(T)\| = \sum_{t=1}^{T} p^t(T) = 1$, where the norm stands for ℓ^1-norm on the sequence space.

Similarly, we define $\xi_a(T) = (\boldsymbol{x}_a(T), 0, 0 \dots)$, $a = 1 \cdots m$, $\eta_b(T) = (\boldsymbol{y}_b(T), 0, 0 \dots)$ and $\zeta_b(T) = (\boldsymbol{z}_b(T), 0, 0 \dots)$, $b = 1 \cdots n$. Then by Lemma 5.1, we have:

$$\|\xi_a(T)\|_\rho, \|\eta_b(T)\|_\rho, \|\zeta_b(T)\|_\rho \leq 1$$

for all T. Let $B = \{\boldsymbol{x} \in \ell_\rho^\infty | \ \|\boldsymbol{x}\|_\rho \leq 1\}$ and $B' = \{\pi \in ba | \ \|\pi\| \leq 1\}$. Then by Alaoglu's theorem (Theorem H3 of Appendix H), B and B' are weakly compact in the weak* topologies. Hence, by Proposition B6 of Appendix B, there exist cluster points $\pi \in ba$ and $\xi_a, \eta_b, \zeta_b \in \ell_\rho^\infty$ of the sequences $\{\pi(T)\}$ and $\{\xi_a(T)\}, \{\eta_b(T)\}, \{\zeta_b(T)\}$, respectively. Since the positive orthant of ℓ_ρ^∞ is weak* closed, it follows that $\xi_a, \eta_b, \zeta_b \in X_a = \ell_+^\infty$, $a = 1 \cdots m$, $b = 1 \cdots n$. Similarly, we have $\pi \geq \mathbf{0}$. Let $\mathbf{1} = (1, 1, \dots) \in \ell^\infty$. Since $\pi(T)\mathbf{1} = 1$ for all $T \in \mathbb{N}$, it follows that $\pi\mathbf{1} = 1$ in the limit. Hence, $\pi \neq \mathbf{0}$.

We now claim,

Lemma 5.2. *The vector $\pi \in ba$ and the allocation $((\xi_a), (\eta_b, \zeta_b))$ satisfy the following conditions.*

(*BA-1*) $\pi\boldsymbol{x} > \pi\omega_a$ *whenever* $\boldsymbol{x} \in X_a$ *and* $\xi_a \prec_a \boldsymbol{x}$, $a = 1 \cdots m$,

(*BA-2*) $\eta_b \leq F_b\left(\zeta_b, \sum_{c=1}^{n} \zeta_c\right)$ *and* $\pi(\{t+1\})y - \pi(\{t\})z \leq \pi(\{t+1\})\eta_b^{t+1} - \pi(\{t\})\zeta_b^t = 0$ *whenever* $y \leq f_b(z, \sum_{c=1}^{n} \zeta_c^t)$, $t \in \mathbb{N}$, $b = 1 \cdots n$,

(*E-3*) $\sum_{a=1}^{m} \xi_a \leq \sum_{b=1}^{n} (\eta_b - \zeta_b) + \sum_{a=1}^{m} \omega_a$.

Proof. Suppose that the condition (BA-1) does not hold. Then for some a, there exists $\boldsymbol{z} \in X_a$ such that $\xi_a \prec_a \boldsymbol{z}$ and $\pi\boldsymbol{z} \leq \pi\omega_a$. Since $\omega_a^t \geq \gamma_a > 0$ for all t and $\pi\mathbf{1} = 1$, we have $\pi\omega_a > 0$. Hence, we can take $\alpha \in (0, 1)$ such that $\xi_a \prec_a \alpha\boldsymbol{z}$ and $\pi\alpha\boldsymbol{z} < \pi\omega_a$. Let $\boldsymbol{z}(T) = (z^0, z^1 \cdots z^T, 0, 0 \dots)$ be the projection of $\boldsymbol{z}$ to $\mathbb{R}^{T+1}$. Given $\boldsymbol{q} = (q^t) \in \ell_\rho^1$, one has $\boldsymbol{q}\alpha\boldsymbol{z}(T) = \sum_{t=0}^{T} \alpha q^t z^t \to \sum_{t=0}^{\infty} \alpha q^t z^t = \boldsymbol{q}\alpha\boldsymbol{z}$ as $T \to +\infty$. Hence, $\alpha\boldsymbol{z}(T) \to \alpha\boldsymbol{z}$ in the $\sigma(\ell_\rho^\infty, \ell_\rho^1)$-topology (Proposition H2 of Appendix H). Since $\pi \geq \mathbf{0}$ and $\alpha\boldsymbol{z}(T) \leq \alpha\boldsymbol{z}$ for all T, $\pi\alpha\boldsymbol{z}(T) \leq \pi\alpha\boldsymbol{z} < \pi\omega_a$ for all T. Since ℓ_ρ^∞ is isometrically isomorphic to ℓ^∞ and ℓ_ρ^1 is also isometrically isomorphic to ℓ^1, we can apply Proposition H5 of Appendix H, and there exists a $T_0 \in \mathbb{N}$ such that $\xi_a \prec_a \alpha\boldsymbol{z}(T_0)$. Let $Q \subset ba$ and $U \subset \ell_\rho^\infty$ be neighborhoods in the weak* topologies of π and ξ_a respectively, which satisfy $\pi'\alpha\boldsymbol{z}(T_0) < \pi'\omega_a$ and $\boldsymbol{x} \prec_a \alpha\boldsymbol{z}(T_0)$ for all $\pi' \in Q$ and for all $\boldsymbol{x} \in U$. Since π and ξ_a are cluster points of the sequences $(\pi(T))$ and $(\xi_a(T))$, respectively, there exists $S \geq T_0$ such that $\pi(S) \in Q$ and $\xi_a(S) \in U$.

Therefore, $\pi(S)\alpha\mathbf{z}(T_0) < \pi(S)\omega_a = \sum_{t=0}^{S} p^t(S)\omega_a^t = \pi(S)\omega_a(S)$ and $\xi_a(S) \prec_a \alpha\mathbf{z}(T_0)$. This contradicts the fact that $\xi_a(S)$ or $\boldsymbol{x}_a(S)$ is an equilibrium consumption bundle of the consumer a in the economy $\mathcal{E}_{\boldsymbol{k}}^S$. This establishes the condition (BA-1).

Since $\eta_b(T)$ and $\zeta_b(T)$ converge in the weak* topology, they converge in the product topology. Since $\zeta_b^0(T) = z_b^0(T)$ and given $t \geq 1$, $\eta_b^{t+1}(T) \leq f_b(\zeta_b^t(T), \sum_{c=1}^n \zeta_c^t(T))$ for all $T \geq t+1$. Hence, $\eta_b^{t+1} \leq f_b(\zeta_b^t, \sum_{c=1}^n \zeta_c^t)$ in the limit.

Let $\eta = (0 \cdots 0, y^{t+1}, 0 \dots)$ and $\zeta = (0 \cdots 0, z^t, 0 \dots)$ satisfy that $y^{t+1} \leq f_b(z^t, \sum_{c=1}^n \zeta_c^t)$. We want to show that $\pi(\eta - \zeta) = \pi(\{t+1\})y^{t+1} - \pi(\{t\})z^t \leq 0$. Suppose it were the case that $\pi(\eta - \zeta) > 0$. Let $\epsilon > 0$ be given. By the continuity of f_b, we can take $\delta_b > 0$, $b = 1 \cdots n$, such that for all z_b with $|z_b - \zeta_b^t| < \delta_b$, it follows that (i) $|f_b(z^t, \sum_{c=1}^n z_c) - f_b(z^t, \sum_{c=1}^n \zeta_c^t)| < \epsilon/3$ and (ii) $y^{t+1} < f_b(z^t, \sum_{c=1}^n z_c)$ if $y^{t+1} < f_b(z^t, \sum_{c=1}^n \zeta_c^t)$. We define a set $U_b = \mathbb{R} \times \cdots \times \mathbb{R} \times (\zeta_b^t - \delta_b, \zeta_b^t + \delta_b) \times \mathbb{R} \times \dots$, $b = 1 \dots n$. Since U_b is an open neighborhood of ζ_b in the product topology, the set $V_b = U_b \cap \ell_\rho^\infty$ is an open neighborhood of ζ_b in the weak* topology. Take a neighborhood Q of π in the weak* topology such that $|\pi'\eta - \pi\eta| < \epsilon/3$ and $|\pi'\zeta - \pi\zeta| < \epsilon/3$ for all $\pi' \in Q$, and we define $\eta(T) = (0 \cdots 0, y^{t+1}(T), 0 \dots)$ by $y^{t+1}(T) = f_b(z^t, \sum_{c=1}^n \zeta_c^t(T))$ if $y^{t+1} = f_b(z^t, \sum_{c=1}^n \zeta_c^t)$ and $y^{t+1}(T) = y^{t+1}$ if $y^{t+1} < f_b(z^t, \sum_{c=1}^n \zeta_c^t)$. Then $\zeta_b(T) \in V_b$ for all b and sufficiently large T, hence we have:

$$\begin{aligned}
&\pi(T)(\eta(T) - \zeta) - \pi(\eta - \zeta)| \\
&\quad \leq |\pi(T)\eta(T) - \pi(T)\eta| + |\pi(T)\eta - \pi\eta| + |\pi(T)\zeta - \pi\zeta| \\
&\quad \leq \epsilon/3 + \epsilon/3 + \epsilon/3 = \epsilon.
\end{aligned}$$

Hence, $\eta(T) \leq F_b(\zeta, \sum_{c=1}^n \zeta_c(T))$ and $\pi(T)(\eta(T) - \zeta) > 0$ for ϵ sufficiently small. This contradicts the profit maximization of the economy $\mathcal{E}_{\boldsymbol{k}}^T$. Let $t \in \mathbb{N}$ be given. Since $p^{t+1}(T)\eta_b^{t+1}(T) - p^t(T)\zeta_b^t(T) = 0$ for all T sufficiently large, we have $\pi(\{t+1\})\eta_b^{t+1} - \pi(\{t\})\zeta_b^t = 0$ in the limit. Therefore, the condition (BA-2) is established.

Since $\sum_{a=1}^m \xi_a(T) \leq \sum_{b=1}^n (\eta_b(T) - \zeta_b(T)) + \sum_{a=1}^m \omega_a(T)$ for all T, it follows that $\sum_{a=1}^m \xi_a \leq \sum_{b=1}^n (\eta_b - \zeta_b) + \sum_{a=1}^m \omega_a$ in the limit. Hence, the condition (E-3) is proved. □

The proof of Theorem 5.1. with an equilibrium price vector in ℓ^1 will be established by the next lemma.

Lemma 5.3. *There exists a vector $\boldsymbol{p} \in \ell_+^1$ such that $(\boldsymbol{p}, (\xi_a), (\eta_b, \zeta_b))$ satisfies the following conditions.*

(E-1) $\boldsymbol{p}\xi_a \leq \boldsymbol{p}\omega_a$ *and* $\xi_a \succsim_a \mathbf{z}$ *for all* $\mathbf{z} \in X_a$ *such that* $\boldsymbol{p}\mathbf{z} \leq \boldsymbol{p}\omega_a$,
(E-2) $\boldsymbol{p}(\boldsymbol{y} - \mathbf{z}) \leq \boldsymbol{p}(\eta_b - \zeta_b) = 0$ *for all* $(\boldsymbol{y}, \mathbf{z})$ *with* $\boldsymbol{y} \leq F_b(\mathbf{z}, \sum_{c=1}^n \zeta_c)$, $b = 1 \cdots n$, *and*
(E-3) $\sum_{a=1}^m \xi_a \leq \sum_{b=1}^n (\eta_b - \zeta_b) + \sum_{a=1}^m \omega_a$.

Proof. We have already established the condition (E-3) in the previous lemma. Since $\pi \geq \mathbf{0}$, it follows from the Yosida–Hewitt theorem (Theorem H4 of Appendix H) that we can write:

$$\pi = \pi_c + \pi_p,$$

where $\pi_c \geq \mathbf{0}$ and $\pi_p \geq \mathbf{0}$ are the countably additive part and the purely finitely additive part of π, respectively.

For given $a = 1 \cdots m$, take $\mathbf{z} = (z^t) \in X_a = \ell^{\infty}_{p+}$ with $\xi_a \prec_a \mathbf{z}$ and let $\mathbf{z}(T) = (z^0, z^1 \cdots z^T, 0, 0 \ldots)$ be the projection of $\mathbf{z}$ to $\mathbb{R}^{T+1}$. Then, as in the proof of Lemma 5.2, we can show that $\xi_a \prec_a \mathbf{z}(T)$ for T sufficiently large. By Lemma 5.2, we have $\pi \mathbf{z}(T) > \pi\omega_a$. Since π_p is purely finitely additive, $\pi_p(\{0, 1, \ldots T\}) = 0$ for all T. It follows from this and $\pi_c \geq \mathbf{0}$ that:

$$\pi \mathbf{z}(T) = (\pi_c + \pi_p)\mathbf{z}(T) = \pi_c \mathbf{z}(T) \leq \pi_c \mathbf{z},$$

since $\mathbf{z}(T) \leq \mathbf{z}$. On the other hand, $\pi_p \geq \mathbf{0}$ and $\omega_a \geq \mathbf{0}$ imply that $\pi\omega_a = (\pi_c + \pi_p)\omega_a \geq \pi_c\omega_a$, hence we have: $\pi_c \mathbf{z} > \pi_c \omega_a$. Denoting $\pi_c(\{t\}) = p^t$, we get a vector $\boldsymbol{p} = (p^t)$ in ℓ^1 which satisfies,

$$(*) \;\; \xi_a \prec_a \mathbf{z} \;\text{ implies that }\; \boldsymbol{p}\mathbf{z} > \boldsymbol{p}\omega_a, \;\; a = 1 \cdots m.$$

Note that this implies that $\boldsymbol{p} \neq \mathbf{0}$. Take $(\boldsymbol{y}, \mathbf{z}) = ((y^t), (z^t)) \in \ell^{\infty}_{p+} \times \ell^{\infty}_{p+}$ such that $\boldsymbol{y} \leq F_b(\mathbf{z}, \sum_{c=1}^{n} \zeta_c)$. By Lemma 5.2, we have $p^{t+1}y^{t+1} - p^t z^t \leq 0$ for all $t \geq 0$. Setting $S(T) = \sum_{t=0}^{T} (p^{t+1}y^{t+1} - p^t z^t)$, it follows that $S(T)$ is monotonically decreasing, $0 \geq S(0) \geq S(1) \ldots$. Hence, $S(+\infty) = \sum_{t=0}^{\infty} (p^{t+1}y^{t+1} - p^t z^t)$ exists, including $-\infty$. Therefore,

$$(**) \;\; \boldsymbol{p}(\boldsymbol{y} - \mathbf{z}) \leq 0 \quad \text{whenever } \boldsymbol{y} \leq F_b\left(\mathbf{z}, \sum_{c=1}^{n} \zeta_c\right), \;\; b = 1 \cdots n.$$

Since $\eta_b \leq F_b(\mathbf{z}, \sum_{c=1}^{n} \zeta_c)$ by Lemma 5.2, we have $\boldsymbol{p}(\eta_b - \zeta_b) \leq 0$, $b = 1 \cdots n$. By the monotonicity (MT) of preferences, one can take $\mathbf{z} \in X_a = \ell^{\infty}_{p+}$ which is arbitrarily close to ξ_a such that $\xi_a \prec_a \mathbf{z}$. Thus, (*) implies that:

$$\boldsymbol{p}\xi_a \geq \boldsymbol{p}\omega_a, \quad a = 1 \cdots m.$$

$\boldsymbol{p}\xi_a$ is finite or $+\infty$ and $\boldsymbol{p}\omega_a$ is finite. Hence, $\sum_{a=1}^{m} \boldsymbol{p}(\xi_a - \omega_a) \geq 0$ is finite or $+\infty$. On the other hand, since $\sum_{a=1}^{m} (\xi_a - \omega_a) \leq \sum_{b=1}^{n} (\eta_b - \zeta_b)$ by Lemma 5.2, we have:

$$0 \leq \sum_{a=1}^{m} \boldsymbol{p}(\xi_a - \omega_a) \leq \sum_{b=1}^{n} \boldsymbol{p}(\eta_b - \zeta_b) \leq 0.$$

Consequently, we have $\sum_{a=1}^{m} \boldsymbol{p}(\xi_a - \omega_a) = 0$, hence:

$$\boldsymbol{p}\xi_a = \boldsymbol{p}\omega_a, \;\; a = 1 \cdots m.$$

This and (*) imply the condition (E-1). We also have:

$$p(\eta_b - \zeta_b) = 0, \quad b = 1 \cdots n.$$

This and (**) imply the condition (E-2). This establishes the proof. □

The first remark we have to give about Lemma 5.2 is the condition (BA-2). We could not claim that the vector $\pi \in ba$ is an equilibrium price vector, since we just showed that the firms maximized profits at π only in one period. This is in contrast to Lemma 2.2 for exchange economies, and the reason why we could not show that:

$$\pi\left(F_b\left(\zeta_b, \sum_{c=1}^{n} \zeta_c\right) - \zeta_b\right) \geq \pi\left(F_b\left(\mathbf{z}, \sum_{c=1}^{n} \zeta_c\right) - \mathbf{z}\right)$$

for every $\mathbf{z} \in \ell^{\infty}_{\rho+}$ is the following. Suppose on the contrary that:

$$\pi\left(F_b\left(\zeta_b, \sum_{c=1}^{n} \zeta_c\right) - \zeta_b\right) < \pi\left(F_b\left(\mathbf{z}, \sum_{c=1}^{n} \zeta_c\right) - \mathbf{z}\right)$$

for some $\mathbf{z} \in \ell^{\infty}_{\rho+}$ and try to deduce a contradiction. Strictly speaking, we have a problem here that the price vector π belongs to ba rather than ba_ρ, so that the profit value may not exist. We ignore this problem for the moment and assume that the value exists in order to concentrate on the problem arising from the production externalities. Then we have:

$$\pi(T)\left(F_b\left(\zeta_b, \sum_{c=1}^{n} \zeta_c\right) - \zeta_b\right) < \pi(T)\left(F_b\left(\mathbf{z}, \sum_{c=1}^{n} \zeta_c\right) - \mathbf{z}\right)$$

for some T large enough. If (the projections of) $F_b(\zeta_b, \sum_{c=1}^{n} \zeta_c)$ and $F_b(\mathbf{z}, \sum_{c=1}^{n} \zeta_c)$ are technologically feasible outputs for the finite-dimensional economy $\mathcal{E}^T_k$, we would obtain the desired contradiction. This is not so, however, and what we actually need to show is that:

$$\pi(T)\left(F_b\left(\zeta_b(T), \sum_{c=1}^{n} \zeta_c(T)\right) - \zeta_b(T)\right) < \pi(T)\left(F_b\left(\mathbf{z}, \sum_{c=1}^{n} \zeta_c(T)\right) - \mathbf{z}\right).$$

Lacking the joint continuity of the inner product $\pi\mathbf{z}$ in the weak* topology, this inequality does not follow.[1] Note that if the externalities do not exist in production, this problem does not occur under the constant returns to scale technology $F_b(\zeta_b)$.

[1]Hence, we do not have any problem if the externalities come into the utility functions in the condition (BA-1), if they are continuous in the weak* topology.

Indeed in this case, all we have to show is that

$$0 < \pi(T)(F_b(\mathbf{z}) - \mathbf{z}) \quad \text{for some } T,$$

since $\pi(T)(F_b(\zeta_b(T)) - \zeta_b(T)) = 0$ for all T. This is exactly what Bewley (1970) showed in his Theorem 3, p.525. It should be pointed out that the assumption that the technology function consists of the one-period production functions is also important. However, we can say that this assumption of the periodic structure of the production process is natural in order to study the infinite time horizon economies.

We took ℓ^1 rather than ℓ^1_ρ as the price space in Theorem 5.1. This is needed for making each consumer's income to be positive in the condition (BA-1). We had $\pi(T)\omega_a(T) > 0$ for all T. What we indeed wanted is that $\pi\omega_a > 0$ in the limit. This is ensured by the (strong) positivity of the endowment (PE_∞) and the normalization $\|\pi(T)\| = \pi(T)\mathbf{1} = 1$ for all T. If we take ℓ^1_ρ and normalize the price vector $\|\pi(T)\|_\rho = \pi(T)\rho = 1$, the assumption (PE_∞) had to be replaced by this:

$$\omega^t_a \geq \epsilon\rho^t \quad \text{for some } \epsilon > 0, \ t = 0, 1 \ldots$$

or the endowment must grow at least the same rate with the path of pure capital accumulation $\rho = (\rho^t)$, which is obviously unrealistic.

5.6. ON THE DIFFERENTIABILITY OF DEMAND

Let the commodity space L be a Banach space, and the consumption set X a convex subset of L. In order to do differential calculus on X, we assume X to be an open subset of L in the norm topology. We also assume that the consumer's preference is represented by a utility function on X,

$$u \colon X \to \mathbb{R}, \ \mathbf{x} \mapsto u(\mathbf{x}), \ a = 1 \cdots m,$$

which is of class C^2. It follows from Proposition E8 of Appendix E that for each $\mathbf{x} \in X$, there exists a linear functional $Du(\mathbf{x}) \in L^* = \mathscr{L}(L, \mathbb{R})$, and a linear map $D^2u(\mathbf{x}) \in \mathscr{L}^2(L, \mathbb{R})$ such that for every $\mathbf{h} \in L$,

$$u(\mathbf{x} + \mathbf{h}) = u(\mathbf{x}) + Du(\mathbf{x})\mathbf{h} + (1/2)D^2u(\mathbf{x})(\mathbf{h}, \mathbf{h}) + o(\|\mathbf{h}\|^3),$$

where $o(\|\mathbf{h}\|^3)/\|\mathbf{h}\|^3 \to 0$ as $\|\mathbf{h}\| \to 0$. By Proposition E6 of Appendix E, the map $D^2u(\mathbf{x})(\cdot, \cdot)$ is symmetric, or:

$$D^2u(\mathbf{x})(\mathbf{h}, \mathbf{k}) = D^2u(\mathbf{x})(\mathbf{k}, \mathbf{h}) \quad \text{for all } \mathbf{h}, \mathbf{k} \in L.$$

We will sometimes denote $D^2u(\mathbf{x})(\mathbf{h}, \cdot)$ simply by $D^2u(\mathbf{x})\mathbf{h}$ and consider it as an element of L^*. We impose the condition that the utility function is differentiably concave and nondegenerate, which we call as differentiably and regularly concave condition,

(DRC) for every $\boldsymbol{x} \in X$ and every $\boldsymbol{h} \in L$,

(i) $D^2u(\boldsymbol{x})(\boldsymbol{h},\boldsymbol{h}) \leq 0$,
(ii) there exists $\boldsymbol{h}_0 \in L$ with $D^2u(\boldsymbol{x})\boldsymbol{h}_0 \neq \boldsymbol{0}$ and $D^2u(\boldsymbol{x})(\boldsymbol{h},\boldsymbol{h}) < 0$ whenever $D^2u(\boldsymbol{x})\boldsymbol{h} \neq \boldsymbol{0}$.

Let $\boldsymbol{p} \in L^*_+$ and $w > 0$. In this section, we are concerned with the differentiability of the individual demand,

$$\phi(\boldsymbol{p}, w) = \{\boldsymbol{x} \in X \mid \boldsymbol{px} \leq w, \text{ and } u(\boldsymbol{z}) \leq u(\boldsymbol{x}) \text{ whenever } \boldsymbol{pz} \leq w\}.$$

The main result is Theorem 6.1 proposed by Araujo (1987), which states that the demand function $\phi(\boldsymbol{p}, w)$ is of class C^1 only if the commodity space L is isomorphic to a Hilbert space. The next proposition is an immediate consequence of the (infinite-dimensional version of) Kuhn–Tucker theorem (Theorem E4 of Appendix E) and it could be of independent interest.

Proposition 6.1. *$\hat{\boldsymbol{x}} \in X$ satisfies $\hat{\boldsymbol{x}} = \phi(\hat{\boldsymbol{p}}, \hat{w})$ if and only if there exists $\lambda \geq 0$ such that $Du(\hat{\boldsymbol{x}}) = \lambda\hat{\boldsymbol{p}}$ and $\hat{\boldsymbol{p}}\hat{\boldsymbol{x}} = \hat{w}$.*

Then the main result of this section now reads:

Theorem 6.1. *Suppose that $\phi(\hat{\boldsymbol{p}}, \hat{w}) \neq \emptyset$ with $\hat{w} > 0$ and there exist a neighborhood $V \subset L^*$ of $\hat{\boldsymbol{p}}$ and $\epsilon > 0$ with $\epsilon < \hat{w}$ such that $\phi(\cdot,\cdot)$ is nonempty and of class C^1 on $V \times (\hat{w} - \epsilon, \hat{w} + \epsilon)$. Then, the commodity space L is isomorphic to a Hilbert space H.*

In order to prove Theorem 6.1, we need the next lemma.

Lemma 6.1. *Let L be a Banach space. Suppose that there exists an isomorphism $T: L \to L^*$ such that:*

(i) *$T(\boldsymbol{x})(\boldsymbol{y}) = T(\boldsymbol{y})(\boldsymbol{x})$ for all $\boldsymbol{x},\boldsymbol{y} \in L$,*
(ii) *$T(\boldsymbol{x})(\boldsymbol{x}) \leq 0$ for all $\boldsymbol{x} \in L$ and with strict inequality for $\boldsymbol{x} \neq \boldsymbol{0}$.*

Then, L is isomorphic to a Hilbert space with the inner product $\langle \boldsymbol{x},\boldsymbol{y}\rangle = -T(\boldsymbol{x})(\boldsymbol{y})$. Moreover, the norm induced from this inner product is equivalent to the original norm on L.

Proof. Clearly, L is a pre-Hilbert space with the inner product $\langle \boldsymbol{x},\boldsymbol{y}\rangle = -T(\boldsymbol{x})(\boldsymbol{y})$. We will prove that L is complete with respect to the reduced norm. Let H be the completion of L with respect to the norm $\|\boldsymbol{x}\| = \sqrt{-T(\boldsymbol{x})(\boldsymbol{x})}$ (Theorem H1 of Appendix H). Then, H is a Hilbert space with the inner product $\langle \boldsymbol{x},\boldsymbol{y}\rangle$ for all $\boldsymbol{x},\boldsymbol{y} \in H$, and we have an inclusion $J: L \to H$ which is continuous because of the continuity of T. Let $J^* : H^* = H \to L^*$ be the adjoint of J defined by $J^*(\boldsymbol{x})(\boldsymbol{y}) = J(\boldsymbol{y})(\boldsymbol{x})$ or $\langle J^*(\boldsymbol{x}), \boldsymbol{y}\rangle = \langle \boldsymbol{x}, J(\boldsymbol{y})\rangle$ for all $\boldsymbol{x}, \boldsymbol{y} \in H$ (see Appendix H). We will show that the next diagram commutes this.

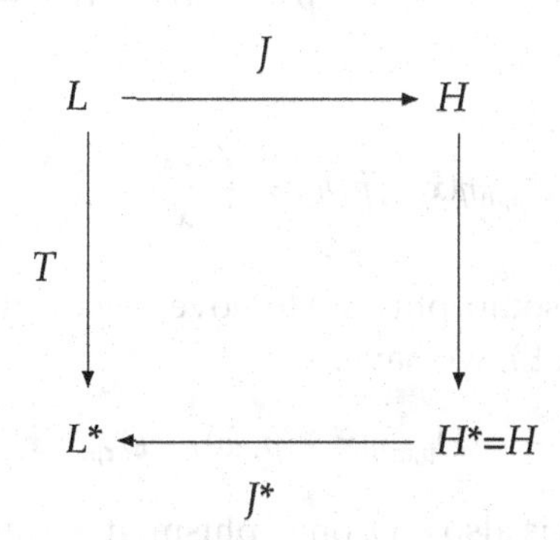

Indeed, for every $\boldsymbol{x}, \boldsymbol{y} \in L$, we have:

$$J^*(J(\boldsymbol{x}))(\boldsymbol{y}) = J(\boldsymbol{y})(J(\boldsymbol{x})) = \langle J(\boldsymbol{x}), J(\boldsymbol{y})\rangle = -T(\boldsymbol{x})(\boldsymbol{y}), \text{ or } J^* \circ J(\boldsymbol{x}) = -T(\boldsymbol{x}).$$

By assumption, T is bijection, or one-to-one and onto. It follows that J^* is onto. Since J has dense range and $\mathcal{N}(J^*) = \mathcal{R}(J)^{\perp}$ by Proposition E3 of Appendix E, it follows that $\mathcal{N}(J^*) = \{\mathbf{0}\}$, or J is one-to-one. Therefore, J^* is a bijection, hence J is also onto, since $J(L) = -(J^*)^{-1} \circ T(L) = H$. It follows from the open mapping theorem (Theorem E1 of Appendix E) that J is an isomorphism. □

We now turn to the proof of Theorem 6.1. We define a mapping:

$$\psi: X \times \mathbb{R}_{++} \to L^* \times \mathbb{R}, \quad (\boldsymbol{x}, \lambda) \mapsto (\lambda^{-1} Du(\boldsymbol{x}), \lambda^{-1} Du(\boldsymbol{x})(\boldsymbol{x})).$$

Then ψ is of class C^1. For $w > 0$, we put:

$$\lambda(\boldsymbol{p}, w) = \frac{Du(\phi(\boldsymbol{p}, w))\phi(\boldsymbol{p}, w)}{w}$$

and consider the "demand":

$$\xi(\boldsymbol{p}, w) = (\phi(\boldsymbol{p}, w), \lambda(\boldsymbol{p}, w)).$$

By Proposition 6.1, ψ is locally the right and left inverse of ξ around $(\hat{\boldsymbol{x}}, \hat{\lambda}) = \xi(\hat{\boldsymbol{p}}, \hat{w})$ and $(\hat{\boldsymbol{p}}, \hat{w})$, respectively. Since the demand function $\phi(\boldsymbol{p}, w)$ is also assumed to be of class C^1 on $V \times (\hat{w} - \epsilon, \hat{w} + \epsilon)$, it follows that $\xi(\boldsymbol{p}, w)$ is also of class C^1 on $V \times (\hat{w} - \epsilon, \hat{w} + \epsilon)$, and $D\psi(\hat{\boldsymbol{x}}, \hat{\lambda})$ is an isomorphism between $L \times \mathbb{R}$ and $L^* \times \mathbb{R}$.

Consider the map $\eta: X \times \mathbb{R} \times L^* \times \mathbb{R}_{++} \to L^* \times \mathbb{R}$ defined by:

$$(\boldsymbol{x}, \lambda, \boldsymbol{p}, w) \mapsto (Du(\boldsymbol{x}) - \lambda \boldsymbol{p}, \boldsymbol{p}\boldsymbol{x} - w).$$

Then, η is class C^1 and we have:

$$D\eta(\hat{x}, \hat{\lambda}, \hat{p}, \hat{w}) = \begin{pmatrix} D^2u(\hat{x}) & -\hat{p} & -\hat{\lambda} & 0 \\ \hat{p} & 0 & \hat{x} & -1 \end{pmatrix}.$$

Hence, it follows from:

$$D_{p,w}\eta(\hat{x}, \hat{\lambda}, \hat{p}, \hat{w}) = \begin{pmatrix} -\hat{\lambda} & 0 \\ \hat{x} & -1 \end{pmatrix}$$

that $D_{p,w}\eta(\hat{x}, \hat{\lambda}, \hat{p}, \hat{w})$ is an isomorphism. Moreover, by the implicit function theorem (Theorem E2 of Appendix E), we have:

$$D\psi(\hat{x}, \hat{\lambda}) = -(D_{p,w}\eta(\hat{x}, \hat{\lambda}, \hat{p}, \hat{w}))^{-1} D_{x,\lambda}\eta(\hat{x}, \hat{\lambda}, \hat{p}, \hat{w}).$$

Therefore, $D_{x,\lambda}\eta(\hat{x}, \hat{\lambda}, \hat{p}, \hat{w})$ is also an isomorphism. It follows from:

$$D_{x,\lambda}\eta(\hat{x}, \hat{\lambda}, \hat{p}, \hat{w}) = \begin{pmatrix} D^2u(\hat{x}) & -\hat{p} \\ \hat{p} & 0 \end{pmatrix} = \begin{pmatrix} D^2u(\hat{x}) & -\hat{\lambda}^{-1}Du(\hat{x}) \\ \hat{\lambda}^{-1}Du(\hat{x}) & 0 \end{pmatrix}$$

that $D^2u(\hat{x})$ satisfies that $D^2u(\hat{x})(h, h) < 0$ if $Du(\hat{x})h = \mathbf{0}$ with $h \neq \mathbf{0}$. Note that here we are using the differentiable and regular concavity (DRC),

$$D^2u(\hat{x})h \neq \mathbf{0} \text{ implies that } D^2u(\hat{x})(h, h) < 0.$$

Since $L_0 = \{h \in L \mid Du(\hat{x})h = 0\}$ is a Banach space of codimension 1, L will be a Hilbert space if we show that L_0 is a Hilbert space. Therefore, if we show that $D^2u(\hat{x})$ is an isomorphism from L_0 onto itself, the proof is complete by Lemma 6.1.

Since $D_{x,\lambda}\eta(\hat{x}, \hat{\lambda}, \hat{p}, \hat{w})$ is an isomorphism from $L \times \mathbb{R}$ onto $L^* \times \mathbb{R}$, $D^2u(\hat{x})$ is one-to-one and continuous as a restriction of the one-to-one and continuous map.

We show that $D^2u(\hat{x})$ is also onto. Since $D_{x,\lambda}\eta(\hat{x}, \hat{\lambda}, \hat{p}, \hat{w})$ is onto, given $(p, 0)$ in $L_0^* \times \mathbb{R}$ there exists $(h, \alpha) \in L \times \mathbb{R}$ such that $D^2u(\hat{x})(h, \cdot) - (\alpha/\lambda)Du(\hat{x}) = p(\,\cdot\,)$ and $(1/\lambda)Du(\hat{x})h = 0$. By the second equation, we have $h \in L_0$. For $k \in L_0$, one has $Du(\hat{x})k = 0$, hence $D^2u(\hat{x})(h, k) = p(k)$. This means that:

$$D^2u(\hat{x})(h, \cdot) = p(\,\cdot\,) \quad \text{on } L_0,$$

and we have proved that $D^2u(\hat{x})$ is onto. □

By Lemma 2.1, the Hilbert norm $(-D^2u(\hat{x})(x, x))^{1/2}$ is equivalent to the original one. Hence, there exists an $\epsilon > 0$ such that:

$$D^2u(\hat{x})(x, x) \leq -\epsilon \|x\|^2.$$

Conversely, Araujo (1987) showed that this inequality is sufficient for the demand function $\phi(p, w)$ being C^1 at $(\hat{p}, \hat{w})$, where $\hat{x} = \phi(\hat{p}, \hat{w})$. Then, it is easy to see that the norm $(-D^2u(\hat{x})(x, x))^{1/2}$ is a Hilbert norm on L_0, which is equivalent to the original one. Therefore, we can say that, essentially speaking, the existence of C^1 demand function at $(\hat{\boldsymbol{p}}, \hat{w})$ is equivalent to the map on $L \times L$ defined by $\langle \boldsymbol{x}, \boldsymbol{y} \rangle = -D^2 u(\hat{\boldsymbol{x}})(\boldsymbol{x}, \boldsymbol{y})$ being a Hilbert inner product on L_0, hence on L.

5.7. NOTES

Section 5.1: Peleg and Yaari (1969) worked with the commodity space $\mathbb{R}^\infty$. In order to prove the existence of equilibrium, they applied the limit theorem of the core by Debreu and Scarf (Theorem 6.1 of Chap. 2). Their technique and results were developed by Aliprantis *et al.* (1987).

Section 5.2: Theorems 2.1 and 2.3 are proposed by Bewley (1970). Bewley proved these theorems more generally, namely that the commodity space is $L^\infty(\Omega, \mathcal{A}, \mu)$, the set of essentially bounded measurable functions on (probably nonatomic) measure space $(\Omega, \mathcal{A}, \mu)$; hence, his theorems include the case of continuous time models. Theorem 2.2 is essentially put forth by Debreu (1954).

Section 5.3: An equilibrium model of the commodity differentiation based on the commodity space $ca(K)$ was initiated by Mas-Colell (1975). In order to avoid the problem arising from the unbounded marginal rate of substitution indicated in Example 4.1, he introduced the indivisible commodities into the model. Consequently, his proof is very complicated. Remarkably, he also proved the core equivalence theorem in such a model. The expositions in this section are based on the works of Jones (1984), which defined the bounded marginal rate of substitution condition and simplified Mas-Colell's proof.

Section 5.4: Mas-Colell (1986) generalized the condition of the bounded marginal rate of substitution to the concept of the proper preferences as in the text. In fact, he worked with a topological vector space of the commodities. All of the expositions in this section including the examples follow the Mas-Colell's paper in which he proved:

Proposition 7.1. *Let $\prec$ be a (norm) continuous preference relation on the order interval $[\boldsymbol{a}, \boldsymbol{b}] \subset L$ for some fixed vectors $\boldsymbol{a}, \boldsymbol{b} \in L$. Suppose that the preference relation is weakly monotone, that is, $\boldsymbol{x} \geq \boldsymbol{y}$ implies that $\boldsymbol{x} \succsim \boldsymbol{y}$. Then there exists a continuous function $u : [\boldsymbol{a}, \boldsymbol{b}] \to \mathbb{R}$ such that $\boldsymbol{x} \succsim \boldsymbol{y}$ if and only if $u(\boldsymbol{x}) \geq u(\boldsymbol{y})$.*

Hence the assumption in Theorem 4.2 that the preference relations are represented by utility functions is not necessary. Zame (1987) generalized Mas-Colell's result to production economies with competitive firms. Noguchi (1997) proved a far-reaching result of the existence of competitive equilibria of a production economy with measure spaces of the consumers and the firms.

Section 5.5: The basic reference is Suzuki (1996). As shown in this paper, there are no difficulties to extend Theorem 5.1 to the case of non-transitive preferences, since we relied on Theorem 7.1 of Chap. 2 proposed by Shafer and Sonnenschein (1975).

Section 5.6: Theorem 6.1 was proved by Araujo (1987) and the exposition of this section entirely follows Araujo's paper. He also proved that if a demand function exists (i.e., well defined) on the dual space of a Banach space of the commodities, then the commodity space has to be reflexive. Therefore, in the infinite-dimensional setting, the strong non-existence results of the demand functions are the rules, nevertheless the competitive equilibria themselves generally exist. These results seem to suggest that the concept of the "competitive equilibrium" is more solid and fundamental than that of the "demand" in general equilibrium theory.

Economies with Monopolistically Competitive Firms

Chapter

6

6.1. MONOPOLISTICALLY COMPETITIVE MARKETS

A remarkable aspect of the development of the actual markets is that on one hand, the number of consumers has grown rapidly, which is perhaps the meaning of the "extension of the markets", but on the other hand, the production activities have been concentrated on more and more small numbers of firms. In other words, as the markets have become large in the sense as given in Chap. 3, they have become more monopolistic. Lenin wrote in 1917 that:

> "The enormous growth of the industry and remarkably rapid process of concentration of production in ever-larger enterprises represent one of the most characteristic feature of capitalism. Modern censuses of production give very complete and exact data on this process.
>
> In Germany, for example, for every 1,000 industrial enterprises, large enterprises, i.e., those employing more than 50 workers, numbered three in 1882, six in 1895, and nine in 1907; and out of every 100 workers employed, this group of enterprises employed 22, 30 and 37, respectively. Concentration of production, however, is much more intense than the concentration of workers, since labour in the large enterprises is much more productive. This is shown by the figures available on steam engines and electric motors.
>
> If we take what in Germany is called industry in the broad sense of the term, that is, including commerce, transport, etc., we get the following picture: large-scale enterprises 30,588 out of a total of 3,265,623, that is to say, 0.9 per cent. These large-scale enterprises employ 5,700,000 workers out of a total of 14,400,000, that is 39.4 per cent; they use 6,660,000 steam horse power out of a total of 8,800,000, that is, 75.3 per cent and 1,200,000 kilo watts of electricity out of a total of 1,500,000, that is, 77.2 per cent.
>
> Less than one-hundredth of the total enterprises utilize more than three-fourths of the steam and electric power! Two million nine hundred and seventy thousand small enterprises (employing up to five

workers), representing 91 per cent of the total, utilize only 7 per cent of the steam and electric power. Tens of thousands of large-scale enterprises are everything; millions of small ones are nothing.

In 1907, there were in Germany 586 establishments employing one thousand and more workers. They employed nearly one-tenth (1,380,000) of the total number of workers employed in industry and utilized *almost one-third* (32 per cent) of the total steam and electric power employed. As we shall see, money capital and banks make this superiority of a handful of the largest enterprises still more overwhelming, in the most literal sense of the word, since millions of small, medium, and even some big "masters" are in fact in complete subjection to some hundreds of millionaire financiers.[1]

In another advanced country of modern capitalism, the United States of America, the growth of the concentration of production is still greater. Here statistics single out industry in the narrow sense of the word and group enterprises according to the value of their annual output. In 1904, large-scale enterprises with an annual output of one million dollars and over numbered 1,900 (out of 216,180, i.e., 0.9 per cent). These employed 1,400,000 workers (out of 5,500,000, i.e., 25.6 per cent) and their combined annual output was valued at $5,600,000,000 (out of $14,800,000,000 i.e., 38 per cent). Five years later, in 1909, the corresponding figures were: large-scale enterprises: 3,060 out of 268,491, i.e., 1.1 per cent, employing: 2,000,000 workers out of 6,600,000, i.e., 30.5 per cent, output: $9,000,000,000 out of $20,700,000,000, i.e., 43.8 per cent.

Almost half of the total production of all the enterprises of the country was carried on by a hundredth part of those enterprises! These 3,000 giant enterprises embrace 268 branches of industry. From this it can be seen that, at a certain stage of its development, concentration itself, as it were, leads right to monopoly; for a score or so of giant enterprises can easily arrive at an agreement, while on the other hand, the difficulty of competition and the tendency towards monopoly arise from the very dimensions of the enterprises. This transformation of competition into monopoly is one of the most — if not the most important — phenomena of modern capitalist economy, and we must deal with it in greater detail. . . . (Lenin (1917, pp. 2–3))"

We see that the observation which Lenin described in the above statements has been approved by the history of the 20th century. This means that the competitive equilibrium for the production economies discussed in Chap. 4 is

[1]In the following chapters of his book, Lenin also emphasized the concentration in banks and the financial sectors, which is out of focus of this monograph.

seriously insufficient for the equilibrium analysis, if we expect that the equilibrium is an approximation in one sense or another of the actual state of the economy.

Therefore, our first task is to generalize the price-taking behavior so as to incorporate the monopolistic behavior of the firms. This has been achieved by Negishi (1961). He assumed that each firm has its subjective (either perceived or expected) inverse demand $q^t = r_b(y_b^t, p^t)$, where the commodity t is assumed to be produced by the firm b, and p^t is the current market price of the commodity t. In normal situations, the firm would expect a downward-sloping inverse demand, which obeys the law of demand, and Negishi specified the function r_b to be linear with respect to the output quantity, y^t, $q^t = a_b(p^t)y_b^t + d_b(p^t)$, where $a_b(p^t) \leq 0$ and $d_b(p^t) \geq 0$ for all p^t, and they satisfy the condition:

$$\sum_a x_a^t = \sum_b y_b^t + \sum_a \omega_a^t \text{ implies that } q^t = p^t,$$

which means that in equilibrium, the firm must have the correct or consistent expectation with the prevailing market price.

In particular, if the firm has the constant expectation $q^t = p^t$, this firm is a price taker, hence Negishi's formulation of the monopolistic firms contains the competitive firms as a special case (see Figs. 6.1(a) and (b)).

Generally speaking, we believe that the technologies of the monopolistic firms exhibit at least one of the following properties: (i) the increasing returns to scale (Fig. 6.2(a)), (ii) the large setup costs (Fig. 6.2(b)), (iii) the differentiated commodities (see Section 5.3 of Chap. 5).

The essential point is that in the cases of the increasing returns (a) and the large setup costs (b), the convexity of the production sets will violate. Unfortunately, Negishi discussed the monopolistic competition with convex production sets on a finite-dimensional commodity space. Therefore, none of the above cases is contained in his model.

In the following sections, we will discuss the Negishi-type monopolistically competitive equilibria with the non-trivial setup costs following Dehez *et al.* (2003). Section 6.2 will be devoted to develop the fundamental machinery for proving the existence of the monopolistically competitive equilibria. A basic tool is the pricing rule which is a correspondence defined on the production set of each firm. It assigns to each (efficient) production vector, a subset of the normal cone of the production set at this point. Since the firms should maximize their profits at least locally even if their sets are not convex, the first-order condition at the equilibrium point is a natural condition to be stipulated. The pricing rule correspondence will also depend on the market price vector, so that Negishi's consistency condition explained above is naturally embodied to it. In order to apply the usual fixed-point argument, the correspondence has to be nonempty, compact and convex-valued, and upper

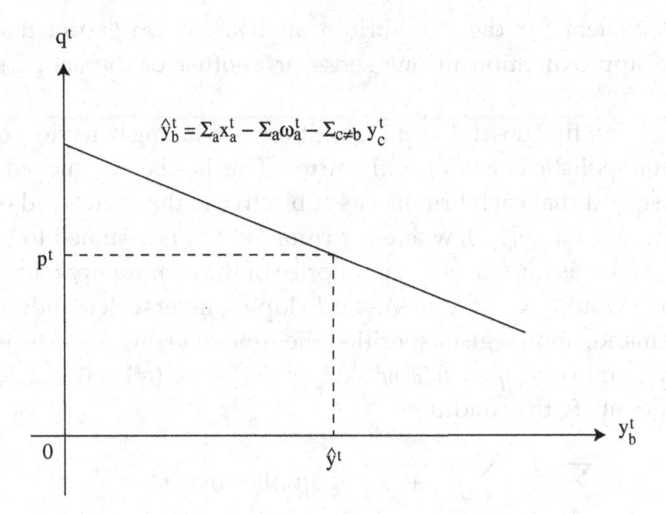

Figure 6.1(a). Perceived Demand of a Monopolist.

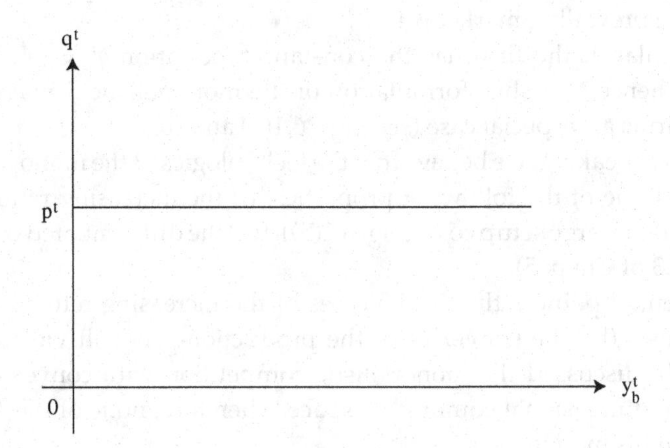

Figure 6.1(b). Perceived Demand of a Competitive Firm.

hemi-continuous. Most of our technical jobs are to guarantee these conditions on the pricing rule when the production sets are not convex.

In Section 6.3, we will discuss the monopolistically competitive equilibria for economies with convex production sets. The model is close to the original one of Negishi (1961), however, the method of the proof is different. We show that the existence of the monopolistically competitive equilibria is a straightforward consequence of the general result of Section 6.2, and this suggests the possibility to extend the theorem to non-convex production sets. Unfortunately, however, the extension is not unconditional. We present a simple example of a production set

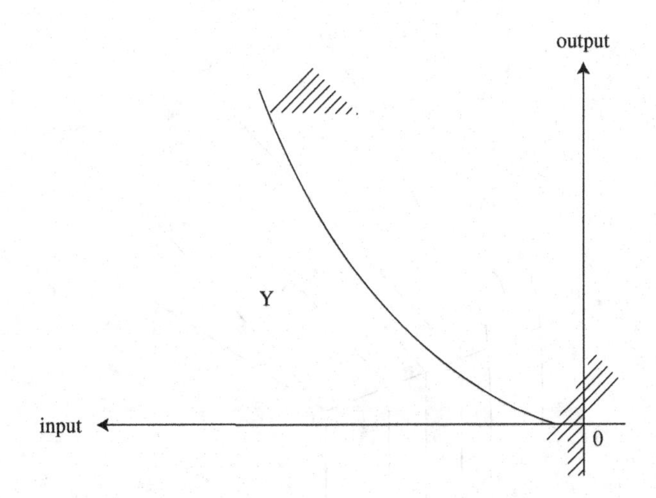

Figure 6.2(a). Increasing Returns.

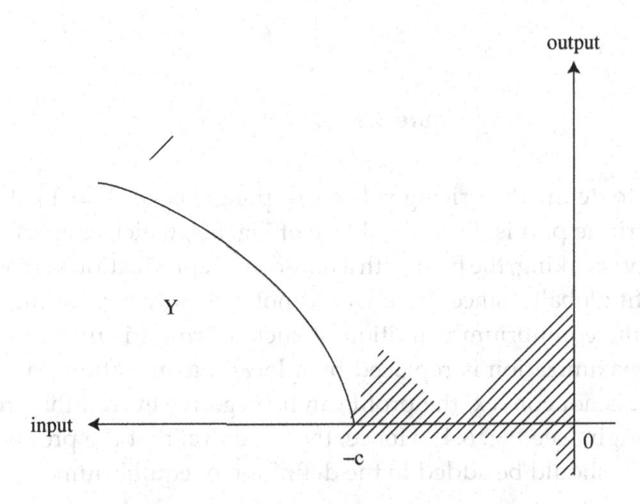

Figure 6.2(b). Setup Cost.

which exhibits the increasing returns and the upper hemi-continuous pricing rule correspondence would not exist on it.

Section 6.4 is the main body of this chapter. We will prove the existence of monopolistically competitive equilibrium with the large setup costs. On account of the non-existence example given in Section 6.3, an essential assumption to be imposed is that for each firm, a commodity which is used as the setup cost is distinguished from the other commodities. As a consequence, the non-convex production set Y is a union of two convex sets Y_1 and Y_2, $Y = Y_1 \cup Y_2$. In Fig. 6.3, the commodity z_1 is a variable input, z_2 a fixed input which represents the setup cost, and y is an output.

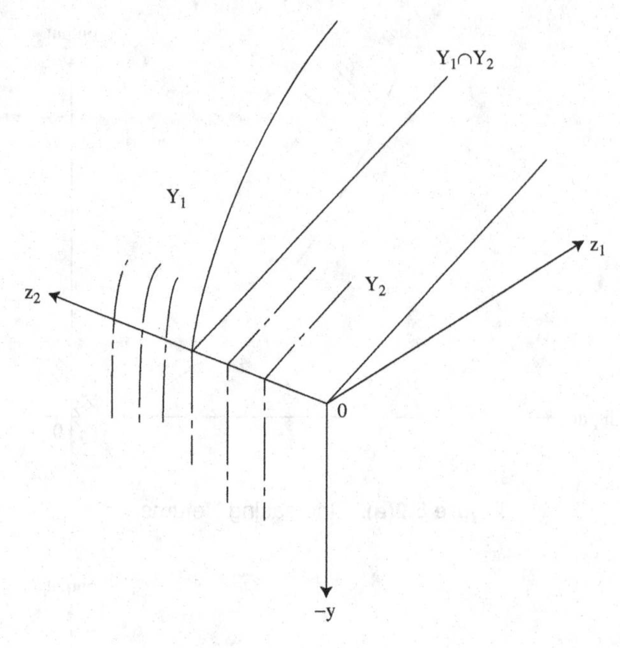

Figure 6.3. $Y = Y_1 \cup Y_2$.

In order to define the pricing rule correspondence on $Y = Y_1 \cup Y_2$, the most technically intricate part is the vertical face of Fig. 6.3, which connects to $Y_1 \cap Y_2$.

Generally speaking, the firm with a non-convex production set could not maximize the profit globally, since there would not exist any supporting hyperplane. Therefore, in the equilibrium condition of such a firm, it is unavoidable that the global profit maximization is replaced by a local maximization. Moreover, if the production set is not convex, the profit can be negative even if the production set contains the origin (see Fig. 6.4). Hence, the condition that the profits of the firms are nonnegative should be added to the definition of equilibrium.

To sum up:

Each firm has a downward-sloping expected inverse demand. In equilibrium, it maximizes the expected profit locally, and the profit is nonnegative.

6.2. EXISTENCE OF EQUILIBRIA WITH PRICING RULES

As usual, we assume that there exist ℓ commodities in the market, such that $t = 1 \cdots \ell$. The consumption sector is standard enough. There exist m consumers

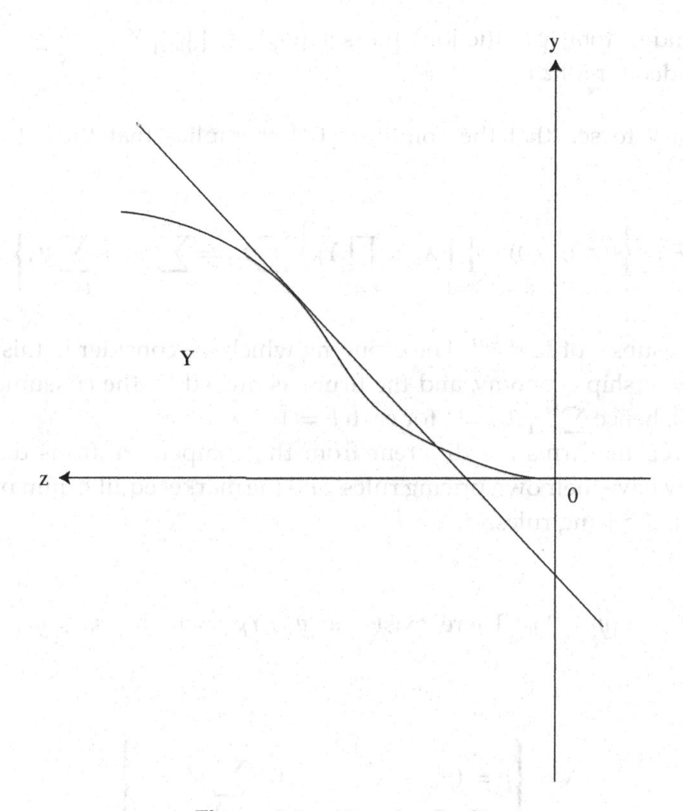

Figure 6.4. Negative Profit.

indexed by $a = 1 \cdots m$, and the consumption set X_a of a is a closed and convex subset of $\mathbb{R}^\ell$, which is bounded from below. The preference relation $\prec_a \subset X_a \times X_a$ is an irreflexive and transitive binary relation on X_a which satisfies the following conditions:

(CT) (Continuity) the set $\prec = \{(\boldsymbol{x},\boldsymbol{y}) \in X_a \times X_a \mid \boldsymbol{x} \prec_a \boldsymbol{y}\}$ is open in $X_a \times X_a$,
(CV) (Convexity) for every $\boldsymbol{x} \in X_a$, the set $\{\boldsymbol{y} \in X_a \mid \boldsymbol{y} \succsim_a \boldsymbol{x}\}$ is convex, and
(MT) (Monotonicity) if $\boldsymbol{x},\boldsymbol{y} \in X_a$ and $\boldsymbol{x} < \boldsymbol{y}$, then $\boldsymbol{x} \prec_a \boldsymbol{y}$.

The consumer a has the initial endowment vector which satisfies the minimum income (MI) condition,

(MI) for every $\boldsymbol{p} \geq \boldsymbol{0}$ with $\boldsymbol{p} \neq \boldsymbol{0}$, $\boldsymbol{p}\omega_a > \inf \boldsymbol{p}X_a$, $a = 1 \cdots m$.

There exist n firms in the economy indexed by $b = 1 \cdots n$. The production set Y_b of the firm b is a closed subset of $\mathbb{R}^\ell$ which satisfies the following conditions:

(NFP) (No free production) $Y_b \cap \mathbb{R}^\ell_+ = \{\boldsymbol{0}\}$,
(FD) (Free disposability) for every $\boldsymbol{y} \in Y_b$, $\boldsymbol{y} + \mathbb{R}^\ell_- \subset Y_b$ and

(BTP) (Bounded total production) the set $\{(\boldsymbol{y}_b) \in \prod_{b=1}^{n} Y_b \mid \boldsymbol{z} \leq \sum_{b=1}^{n} \boldsymbol{y}_b\}$ is bounded for all $\boldsymbol{z} \in \mathbb{R}^{\ell}$.

It is easy to see that the condition (BTP) implies that the set of feasible allocations:

$$\mathcal{F} = \left\{ ((\boldsymbol{x}_a), (\boldsymbol{y}_b)) \in \prod_{a=1}^{m} X_a \times \prod_{b=1}^{n} Y_b \,\middle|\, \sum_{a=1}^{m} \boldsymbol{x}_a = \sum_{a=1}^{m} \omega_a + \sum_{b=1}^{n} \boldsymbol{y}_b \right\}$$

is a bounded subset of $\mathbb{R}^{\ell(m+n)}$. The economy which we consider in this chapter is a private ownership economy, and the firm b is owned by the consumer a by the share $\theta_{ab} \geq 0$, hence $\sum_{a=1}^{m} \theta_{ab} = 1$ for each $b = 1 \cdots n$.

However, the firms are different from the competitive firms discussed in Chap. 4. They have their own pricing rules, and the market equilibrium price vector has to obey the pricing rules.

Let,

$$\partial Y_b = \{\boldsymbol{y}_b \in Y_b \mid \text{There exists no } \boldsymbol{y} \in Y_b \text{ such that } \boldsymbol{y} \gg \boldsymbol{y}_b\},$$

and

$$\Delta = \left\{ \boldsymbol{p} = (p^t) \in \mathbb{R}^{\ell} \,\middle|\, p^t \geq 0, \ \sum_{t=1}^{\ell} p^t = 1 \right\}$$

∂Y_b is the set of (weakly) efficient production plans, and Δ is the standard unit simplex. Let $\hat{\boldsymbol{p}}$ be a price vector in $\mathbb{R}_{+}^{\ell}$ and we denote $\hat{\boldsymbol{\xi}} = ((\hat{\boldsymbol{x}}_a), (\hat{\boldsymbol{y}}_b))$ be an allocation in $\mathcal{F}$. we call $\ell(m+n+1)$-tuple $(\hat{\boldsymbol{p}}, \hat{\boldsymbol{\xi}})$ a market data.

Definition 2.1. A pricing rule for the firm b is a correspondence:

$$\gamma_b: \partial Y_b \times \mathbb{R}_{+}^{\ell} \times \mathcal{F} \to \Delta, \quad (\boldsymbol{y}_b, \hat{\boldsymbol{p}}, \hat{\boldsymbol{\xi}}) \mapsto \gamma_b(\boldsymbol{y}_b \mid \hat{\boldsymbol{p}}, \hat{\boldsymbol{\xi}})$$

which assigns to each (weakly) efficient production plan $\boldsymbol{y}_b \in \partial Y_b$, a set of prices $\gamma_b(\boldsymbol{y}_b \mid \hat{\boldsymbol{p}}, \hat{\boldsymbol{\xi}}) \subset \Delta$, given the market data $(\hat{\boldsymbol{p}}, \hat{\boldsymbol{\xi}})$ satisfying:

(PR-1) the correspondence $\gamma_b: \partial Y_b \times \mathbb{R}_{+}^{\ell} \times \mathcal{F} \to \Delta$ is nonempty, compact and convex-valued, and upper hemi-continuous and

(PR-2) $\boldsymbol{p}\boldsymbol{y}_b \geq 0$ for all $\boldsymbol{p} \in \gamma_b(\boldsymbol{y}_b \mid \hat{\boldsymbol{p}}, \hat{\boldsymbol{\xi}})$.

The list $(X_a, \prec_a, \omega_a, \theta_{ab}, Y_b, \gamma_b)$ is called an economy with pricing rules and denoted by $\mathcal{E}_{\gamma}$.

The fundamental equilibrium concept of this section reads as follows.

Definition 2.2. A market data $(\hat{\boldsymbol{p}}, (\hat{\boldsymbol{x}}_a)_{a=1}^m, (\hat{\boldsymbol{y}}_b)_{b=1}^n)$ is said to consist of an equilibrium with the pricing rules if and only if

(E-1) $\hat{\boldsymbol{p}}\hat{\boldsymbol{x}}_a \leq \hat{\boldsymbol{p}}\omega_a + \sum_{b=1}^n \theta_{ab}\hat{\boldsymbol{p}}\hat{\boldsymbol{y}}_b$ and
$\hat{\boldsymbol{x}}_a \succsim_a \boldsymbol{x}$, whenever $\hat{\boldsymbol{p}}\boldsymbol{x} \leq \hat{\boldsymbol{p}}\omega_a + \sum_{b=1}^n \theta_{ab}\hat{\boldsymbol{p}}\hat{\boldsymbol{y}}_b$, $a = 1 \cdots m$,
(E-2) $\hat{\boldsymbol{p}} \in \gamma_b(\hat{\boldsymbol{y}}_b \mid \hat{\boldsymbol{p}}, \hat{\boldsymbol{\zeta}})$, $b = 1 \cdots n$, and
(E-3) $\sum_{a=1}^m \hat{\boldsymbol{x}}_a \leq \sum_{b=1}^n \hat{\boldsymbol{y}}_b + \sum_{a=1}^m \omega_a$.

The next result is a basis of our discussion of the monopolistically competitive equilibria.

Theorem 2.1. *Suppose that an economy $\mathcal{E}_\gamma = (X_a, \prec_a, \omega_a, \theta_{ab}, Y_b, \gamma_b)$ satisfies continuity (CT), the convexity (CV) and the minimum income condition (MI) for every $a = 1 \cdots m$, and for every $b = 1 \cdots nB$, the free disposability (FD) and no free production (NFP) conditions hold. Finally, we assume that the condition of the bounded total production (BTP). Then, there exists a competitive equilibrium $(\hat{\boldsymbol{p}}, \hat{\boldsymbol{x}}_1 \cdots \hat{\boldsymbol{x}}_m)$ for $\mathcal{E}_\gamma$.*

Proof. Since the set of feasible allocations $\mathcal{F}$ is nonempty and bounded by the assumption (BTP), the feasible consumption sets and feasible production sets which are defined as the projections of $\mathcal{F}$ to X_a and Y_b, respectively are bounded. Therefore, there exists a closed cube K in $\mathbb{R}^\ell$ with the length $k > 0$, centered at the origin and containing in its interior, the feasible consumption and production plans, $\mathcal{F} \subset$ interior K. Then, we define:

$$\hat{X}_a = X_a \cap K, \quad a = 1 \cdots m,$$
$$\dot{Y}_b = (Y_b + k\mathbf{1}) \cap \mathbb{R}_{++}^\ell, \quad \text{where } \mathbf{1} = (1 \cdots 1),$$
$$\hat{Y}_b = \text{closure } \dot{Y}_b \text{ in } \mathbb{R}_{+}^\ell, \quad b = 1 \cdots n.$$

Let g_b denote the projection of points in $\mathbb{R}_+^\ell \setminus \{0\}$ on the unit simplex Δ, and we define for each $\boldsymbol{v} \in \Delta$,

$$\lambda_b(\boldsymbol{v}) = \sup\{\lambda > 0 \mid \lambda \boldsymbol{v} \in \hat{Y}_b\},$$
$$h_b(\boldsymbol{v}) = \lambda_b(\boldsymbol{v})\boldsymbol{v},$$
$$\partial\hat{Y}_b = \{h_b(\boldsymbol{v}) \in \hat{Y}_b \mid \boldsymbol{v} \in \Delta\}, \quad b = 1 \cdots n.$$

Obviously, $h_b(\boldsymbol{v})$ is weakly efficient (up to $k\mathbf{1}$), or $h_b(\boldsymbol{v}) - k\mathbf{1} \in \partial Y_b$ for each $\boldsymbol{v} \in \Delta$, and if $\boldsymbol{y}$ is a feasible production vector, then $\boldsymbol{y} + k\mathbf{1} \gg \mathbf{0}$ (see Fig. 6.5). □

We now prove:

Lemma 2.1. $g_b\colon \partial\hat{Y}_b \to \Delta$ *is a homeomorphism between $\partial\hat{Y}_b$ and Δ, which satisfies that $g_b(\boldsymbol{y}) \gg \mathbf{0}$ if and only if $\boldsymbol{y} \gg \mathbf{0}$.*

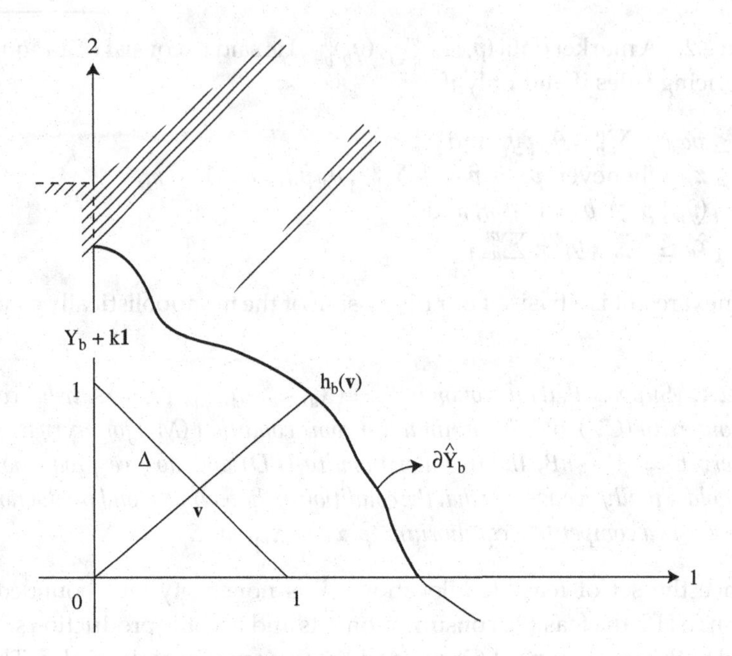

Figure 6.5. Lemma 2.1.

Proof. First note that the mappings g_b: $\partial\hat{Y}_b \to \Delta$ and h_b: $\Delta \to \partial\hat{Y}_b$ are the inverse mapping of each other, or $h_b \circ g_b$ and $g_b \circ h_b$ are identities on $\partial\hat{Y}_b$ and Δ, respectively. By definition, we can write g_b: $\partial\hat{Y}_b \to \Delta$ as

$$g_b(\boldsymbol{y}) = \left(\frac{1}{\sum_{t=1}^{\ell} y^t}\right)\boldsymbol{y} \quad \text{for } \boldsymbol{y} = (y^t) \in \partial\hat{Y}_b.$$

From this expression, it is immediately known that $g_b(\boldsymbol{y}) \gg \mathbf{0}$ if and only if $\boldsymbol{y} \gg \mathbf{0}$, and g_b is continuous, since $\sum_{t=1}^{\ell} y^t > 0$ for every $\boldsymbol{y} \in \partial\hat{Y}_b$. Suppose that for $\boldsymbol{y} = (y^t) \in \partial\hat{Y}_b$ and $\mathbf{z} = (z^t) \in \partial\hat{Y}_b$, $g_b(\boldsymbol{y}) = g_b(\mathbf{z})$. Then, we have:

$$\boldsymbol{y} = \left(\frac{\sum_{t=1}^{\ell} y^t}{\sum_{t=1}^{\ell} z^t}\right)\mathbf{z},$$

hence $\sum_{t=1}^{\ell} y^t = \sum_{t=1}^{\ell} z^t$, since otherwise $\boldsymbol{y} \gg \mathbf{z}$ or $\mathbf{z} \gg \boldsymbol{y}$, contradicting the definition of $\partial\hat{Y}_b$. Therefore, $\boldsymbol{y} = \mathbf{z}$ and the map g_b is one-to-one. It is onto, since for every $\boldsymbol{v} \in \partial\hat{Y}_b$, there exists a vector, $g_b^{-1}(\boldsymbol{v}) = h_b(\boldsymbol{v}) \in \partial\hat{Y}_b$, by definition of $\partial\hat{Y}_b$. Since Δ is closed and g_b is continuous, it follows that $g_b^{-1}(\Delta) = \partial\hat{Y}_b$ is closed. Since $\partial\hat{Y}_b$ is bounded, it is compact. Therefore, by Proposition B7 of Appendix B, g_b is a homeomorphism between Δ and $\partial\hat{Y}_b$. □

Let us then define a function η_b on Δ by $\eta_b(\boldsymbol{v}) = g_b^{-1}(\boldsymbol{v}) - k\mathbf{1}$. The free disposability (FD) ensures that $\eta_b(\boldsymbol{v}) \in \partial Y_b$ for all $\boldsymbol{v} \in \Delta$. Moreover, if $\eta_b(\boldsymbol{b})$ is a feasible production vector, then $g_b^{-1}(\boldsymbol{v}) \gg \mathbf{0}$, hence $\boldsymbol{v} \gg \mathbf{0}$. Let the budget relation $\beta_a : \Delta^{n+1} \to \hat{X}_a$ and the quasi-demand relation $\hat{\phi}_a : \Delta^{n+1} \to \hat{X}_a$ be defined by:

$$\beta_a(\boldsymbol{p}, \boldsymbol{v}_1 \cdots \boldsymbol{v}_n) = \left\{ \boldsymbol{x} \in \hat{X}_a \,\middle|\, \boldsymbol{p}\boldsymbol{x} \le \boldsymbol{p}\omega_a + \sum_{b=1}^{n} \theta_{ab} \boldsymbol{p} \eta_b(\boldsymbol{v}_b) \right\},$$

and

$$\hat{\phi}_a(\boldsymbol{p}, \boldsymbol{v}_1 \cdots \boldsymbol{v}_n) = \begin{cases} \{\boldsymbol{x} \in \hat{X}_a \mid \boldsymbol{x} \in \beta_a(\boldsymbol{p}, \boldsymbol{v}_1 \cdots \boldsymbol{v}_n) \text{ and } \boldsymbol{x} \succsim_a \boldsymbol{z} \text{ for all } \boldsymbol{z} \in \beta_a(\boldsymbol{p}, \boldsymbol{v}_1 \cdots \boldsymbol{v}_n)\} \\ \qquad\qquad \text{if } \inf \boldsymbol{p}\hat{X}_a < \boldsymbol{p}\omega_a + \sum_{b=1}^{n} \theta_{ab} \boldsymbol{p} \eta_b(\boldsymbol{v}_b), \\ \{\boldsymbol{x} \in \hat{X}_a \mid \boldsymbol{p}\boldsymbol{x} = \inf \boldsymbol{p}\hat{X}_a\} \qquad \text{otherwise,} \end{cases}$$

respectively. By Proposition 3.3 given in Chap. 3, the quasi-demand correspondence $\hat{\phi}_a$ is nonempty, compact and convex-valued, and upper hemi-continuous correspondence.

For each $b = 1 \cdots n$, we define the correspondence ψ_b: $\Delta^3 \to \Delta$ by:

$$\psi_b(\boldsymbol{p}, \boldsymbol{q}_b, \boldsymbol{v}_b) = \left(\frac{\max\{0, v_b^t + p^t - q_b^t\}}{\sum_{t=1}^{\ell} \max\{0, v_b^t + p^t - q_b^t\}} \right).$$

Obviously, $\sum_{t=1}^{\ell} \max\{0, v_b^t + p^t - q_b^t\} \ge 1$, hence ψ_b is a continuous function. Here, $\boldsymbol{p}$ denotes a "market price" as opposed to $\boldsymbol{q}_b$, which denotes a "producer price".

The market-price vector $\boldsymbol{p}$ is determined through the standard market correspondence μ: $\prod_{a=1}^{m} \hat{X}_a \times \Delta^n \to \Delta$ defined by:

$$\mu((\boldsymbol{x}_a), (\boldsymbol{v}_b)) = \left\{ \boldsymbol{p} \in \Delta \,\middle|\, \boldsymbol{r} \left(\sum_{a=1}^{m} \boldsymbol{x}_a - \sum_{b=1}^{n} \eta(\boldsymbol{v}_b) - \sum_{a=1}^{m} \omega_a \right) \right.$$
$$\left. \le \boldsymbol{p} \left(\sum_{a=1}^{m} \boldsymbol{x}_a - \sum_{b=1}^{n} \eta_b(\boldsymbol{v}_b) + \sum_{a=1}^{m} \omega_a \right) \text{ for all } \boldsymbol{r} \in \Delta \right\}.$$

The continuity of η_b ensures that the correspondence μ is upper hemi-continuous with non-empty, compact, and convex values.

For each b, the producer price $\boldsymbol{q}_b$ is determined through the correspondence τ: $\Delta^2 \times \prod_{a=1}^{m} \hat{X}_a \to \Delta$ which is defined by:

$$\tau_b(\boldsymbol{v}_b, \boldsymbol{p}, (\boldsymbol{x}_a)) = \gamma_b(\eta_b(\boldsymbol{v}_b) \mid \boldsymbol{p}, (\boldsymbol{x}_a), (\eta_b(\boldsymbol{v}_b))).$$

Since the correspondences γ_b are upper hemi-continuous with non-empty compact and convex values, and η_bs are continuous functions, it follows from Proposition D1 of Appendix D that the correspondence τ_b is upper hemi-continuous whose values are nonempty, compact, and convex.

Then, we can define a fixed-point mapping which is simply the direct product of these correspondences,

$$\begin{aligned}&\Phi(\boldsymbol{p},(\boldsymbol{q}_b),(\boldsymbol{v}_b),(\boldsymbol{x}_a))\\&\quad=\mu((\boldsymbol{x}_a),(\boldsymbol{v}_b))\times\prod_{b=1}^{n}\tau_b(\boldsymbol{v}_b,\boldsymbol{p},(\boldsymbol{x}_a))\times\prod_{b=1}^{n}\psi_b(\boldsymbol{p},\boldsymbol{q}_b,\boldsymbol{v}_b)\times\prod_{a=1}^{m}\hat{\phi}_a(\boldsymbol{p},(\boldsymbol{v}_b)).\end{aligned}$$

By the Kakutani's fixed-point theorem (Theorem D1 of Appendix D), the correspondence Φ has a fixed point. Let $(\hat{\boldsymbol{p}},(\hat{\boldsymbol{q}}_b),(\hat{\boldsymbol{v}}_b),(\hat{\boldsymbol{x}}_a))$ be the fixed point and we define $\hat{\boldsymbol{y}}_b=\eta_b(\hat{\boldsymbol{v}}_b)$ and $\hat{\boldsymbol{z}}=\sum_{a=1}^{m}\hat{\boldsymbol{x}}_a-\sum_{a=1}^{m}\omega_a-\sum_{b=1}^{n}\hat{\boldsymbol{y}}_b$.

Then, $\hat{\boldsymbol{y}}_b\in\partial Y_b$ for all b and the following conditions hold.

$$\begin{aligned}&\hat{\boldsymbol{v}}_b=\psi_b(\hat{\boldsymbol{p}},\hat{\boldsymbol{q}}_b,\hat{\boldsymbol{v}}_b),\quad b=1\cdots n,\\&\hat{\boldsymbol{x}}_a\in\hat{\phi}_a(\hat{\boldsymbol{p}},(\hat{\boldsymbol{v}}_b)),\quad a=1\cdots m,\\&\boldsymbol{p}\hat{\boldsymbol{z}}\leq\hat{\boldsymbol{p}}\hat{\boldsymbol{z}}\quad\text{for all }\boldsymbol{p}\in\Delta,\end{aligned}$$

and

$$\hat{\boldsymbol{q}}_b\in\tau(\hat{\boldsymbol{v}}_b,\hat{\boldsymbol{p}},(\hat{\boldsymbol{x}}_a)),\quad b=1\cdots n.$$

Let us define $\hat{\sigma}_b=\sum_{t=1}^{\ell}\max\{0,\hat{v}_b^t+\hat{p}^t-\hat{q}_b^t\}$. Then, $\hat{\boldsymbol{v}}_b=\psi_b(\hat{\boldsymbol{p}},\hat{\boldsymbol{q}}_b,\hat{\boldsymbol{v}}_b)$ implies that for each b,

$$\hat{\sigma}_b\hat{v}_b^t\geq\hat{v}_b^t+\hat{p}^t-\hat{q}_b^t,\quad t=1\cdots\ell,$$

with the exact equality whenever $\hat{v}_b^t>0$. Multiplying both sides of the above inequalities by $\hat{v}_b^t$ and summing over t, we have:

$$(\hat{\sigma}_b-1)\hat{\boldsymbol{v}}_b\hat{\boldsymbol{v}}_b=(\hat{\boldsymbol{p}}-\hat{\boldsymbol{q}}_b)\hat{\boldsymbol{v}}_b,$$

where $\hat{\sigma}_b\geq1$ and $\hat{\boldsymbol{v}}_b\hat{\boldsymbol{v}}_b\geq1/\ell$. Therefore, it follows that:

$$(\hat{\boldsymbol{p}}-\hat{\boldsymbol{q}}_b)\hat{\boldsymbol{v}}_b\geq0,\quad b=1\cdots n.$$

By the definition of g_b, there exists $\hat{\lambda}_b>0$ such that:

$$\hat{\boldsymbol{v}}_b=\hat{\lambda}_b(\hat{\boldsymbol{y}}_b+k\mathbf{1}),\quad b=1\cdots n.$$

Using the fact that $(\hat{\boldsymbol{p}}-\hat{\boldsymbol{q}}_b)\mathbf{1}=0$, we obtain $(\hat{\boldsymbol{p}}-\hat{\boldsymbol{q}}_b)\hat{\boldsymbol{v}}_b=\hat{\lambda}_b(\hat{\boldsymbol{p}}-\hat{\boldsymbol{q}}_b)\hat{\boldsymbol{y}}_b$, which combined with $(\hat{\boldsymbol{p}}-\hat{\boldsymbol{q}}_b)\hat{\boldsymbol{v}}_b\geq0$ implies that:

$$\hat{\boldsymbol{p}}\hat{\boldsymbol{y}}_b\geq\hat{\boldsymbol{q}}_b\hat{\boldsymbol{y}}_b,\quad b=1\cdots n.$$

Since $\hat{q}_b \in \gamma_b(\hat{y}_b \mid \hat{p}, (\hat{x}_a), (\hat{y}_b))$, we have $\hat{q}_b\hat{y}_b \geq 0$, hence $\hat{p}\hat{y}_b \geq 0$ for all b, and the minimum income assumption (MI) ensures that for every a,

$$\hat{p}\omega_a + \sum_{b=1}^{n} \theta_{ab}\hat{p}\hat{y}_b > \inf pX_a.$$

From the definition of $\hat{\phi}_a$, we see that the budget inequalities apply for all a,

$$\hat{p}\hat{x}_a \leq \hat{p}\omega_a + \sum_{b=1}^{n} \theta_{ab}\hat{p}\hat{y}_b, \quad a = 1 \cdots m.$$

Summing over all a, we get $\hat{p}\hat{z} \leq 0$. Since $p\hat{z} \leq \hat{p}\hat{z}$ for all $p \in \Delta$, setting $p = e_s = (\delta_s^t)$, $s = 1 \ldots \ell$, where $\delta_s^t = 1$ when $s = t$, and $\delta_s^t = 0$ for $s \neq t$, it follows that $\hat{z} \leq 0$, and therefore $((\hat{x}_a), (\hat{y}_b)) \in \mathcal{F}$. Consequently, we have $g_b(\hat{y}_b) \gg 0$ and therefore $\hat{v}_b \gg 0$ by Lemma 2.1. Then, the inequalities:

$$\hat{\sigma}_b\hat{v}_b^t \geq \hat{v}_b^t + \hat{p}^t - \hat{q}_b^t, \quad t = 1 \cdots \ell,$$

yield $\hat{\sigma}_b = 1$ and $\hat{p}^t = \hat{q}_b^t$ for all $t = 1 \cdots \ell$ and for all $b = 1 \cdots n$. The equilibrium condition (E-2), or $\hat{p} \in \gamma_b(\hat{y}_b \mid \hat{p}, (\hat{x}_a), (\hat{y}_b))$ is therefore established on $\hat{Y}_b$. Since $\hat{p}\omega_a + \sum_{b=1}^{n} \theta_{ab}\hat{p}\hat{y}_b > \inf pX_a$ and $\hat{x}_a \in \hat{\phi}_a(\hat{p}, (\hat{v}_b))$ for all a, the condition (E-1) on $\hat{X}_a$ follows from the definition of $\hat{\phi}_a$. The usual limiting argument as in Theorem 1.2 or Theorem 3.1 of Chap. 4 will verify the conditions (E-1) and (E-2) on X_a and Y_b, respectively. Finally we have already showed the condition (E-3), or $\hat{z} \leq 0$. Hence, the proof of Theorem 2.1 is complete. □

6.3. MONOPOLISTICALLY COMPETITIVE EQUILIBRIA UNDER CONVEX TECHNOLOGIES

The consumption sector is the same as the one discussed in Section 6.2. The consumer a ($= 1 \cdots m$) is characterized by the consumption set X_a, which is a convex and closed subset of $\mathbb{R}^\ell$ and it is assumed to be bounded from below. The consumer a has also the initial endowment vector $\omega_a \in \mathbb{R}^\ell$. We assume that the minimum income condition:

(MI) for every $p \geq 0$ with $p \neq 0$, $p\omega_a > \inf pX_a$, $a = 1 \cdots m$

holds. As usual, the preference relation $\prec_a \subset X_a \times X_a$ is a irreflexive and transitive binary relation on X_a, which satisfies the continuity (CT), the convexity (CV), and the monotonicity (MT).

For the production sector, we assume that each firm produces only one commodity: there exist ℓ firms in the economy and the firm t is assumed to produce the commodity t ($= 1 \cdots \ell$). This specification is just for simplicity and not necessary (see Section 6.5).

Recall that a pair $(\hat{\boldsymbol{p}}, (\hat{\boldsymbol{x}}_a), (\hat{\boldsymbol{y}}_t))$ of a price vector $\hat{\boldsymbol{p}}$ and a feasible allocation $\hat{\xi} = ((\hat{\boldsymbol{x}}_a), (\hat{\boldsymbol{y}}_t)) \in \mathcal{F}$ is called a market data. The firm t is assumed to have a perceived demand function $p_t(\boldsymbol{y} \mid \hat{\boldsymbol{p}}, \hat{\xi})$ defined as:

$$p_t^t(\boldsymbol{y} = (y^t) \mid \hat{\boldsymbol{p}}, \hat{\xi}) = a_t(\hat{\boldsymbol{p}}, \hat{\xi}) y^t + d_t(\hat{\boldsymbol{p}}, \hat{\xi}),$$
$$p_t^s(\boldsymbol{y} \mid \hat{\boldsymbol{p}}, \hat{\xi}) = \hat{p}^s \quad \text{for } s \neq t,$$

where $a_t(\hat{\boldsymbol{p}}, \hat{\xi})$ is a real-valued function on $\mathbb{R}_+^\ell \times \mathcal{F}$ such that $a_t(\hat{\boldsymbol{p}}, \hat{\xi}) \leq 0$ for all $(\hat{\boldsymbol{p}}, \hat{\xi}) \in \mathbb{R}_+^\ell \times \mathcal{F}$.

The economic meaning of the perceived demand function is that the firm t expects to be able to sell the quantity y^t at the price p_t^t when the market data is $(\hat{\boldsymbol{p}}, \hat{\xi})$. Note that the firm t behaves competitively in the market of the commodities other than t. In other words, it is a price taker in the market of the input commodities such as factors or labors. This assumption is not necessary (see Section 6.5). However, the firm which behaves monopolistically as a seller, but competitively as a buyer seems to be intuitively natural.

The second term $d_t(\hat{\boldsymbol{p}}, \hat{\xi})$ is assumed to ensure that:

(CN) (Consistency with observations) $p_t^t(\hat{\boldsymbol{y}}_t \mid \hat{\boldsymbol{p}}, \hat{\xi}) \equiv \hat{p}^t, t = 1 \cdots \ell$.

Furthermore, we require that the perceived demand function is homogeneous of degree 1 with respect to price and it is continuous,

(HG) (Homogeneity and continuity)

$$\begin{aligned} p_t^t(\boldsymbol{y} \mid \lambda\hat{\boldsymbol{p}}, \hat{\xi}) &= a_t(\lambda\hat{\boldsymbol{p}}, \hat{\xi}) y^t + d_t(\lambda\hat{\boldsymbol{p}}, \hat{\xi}) \\ &= \lambda a_t(\hat{\boldsymbol{p}}, \hat{\xi}) y^t + \lambda d_t(\hat{\boldsymbol{p}}, \hat{\xi}) = \lambda p_t^t(\boldsymbol{y} \mid \hat{\boldsymbol{p}}, \hat{\xi}) \end{aligned}$$

for all $\lambda \geq 0$, and $a_t((\hat{\boldsymbol{p}}, \hat{\xi})$ and $d_t(\hat{\boldsymbol{p}}, \hat{\xi})$ are continuous.

The expected profit function of the firm t is then given by:

$$\pi_t(\hat{\boldsymbol{p}}, \hat{\xi}) = a_t(\hat{\boldsymbol{p}}, \hat{\xi})(y_t^t)^2 + d_t(\hat{\boldsymbol{p}}, \hat{\xi}) y_t^t - \sum_{s \neq t} \hat{p}^s y_t^s,$$

and from this, one obtains:

$$\begin{aligned} \frac{\partial \pi_t(\hat{\boldsymbol{p}}, \hat{\xi})}{\partial y_t^t} &= 2a_t(\hat{\boldsymbol{p}}, \hat{\xi}) y_t^t + d_t(\hat{\boldsymbol{p}}, \hat{\xi}) \\ &= a_t(\hat{\boldsymbol{p}}, \hat{\xi}) y_t^t + p_t(\boldsymbol{y}_t \mid \hat{\boldsymbol{p}}, \hat{\xi}) \equiv q^t. \end{aligned}$$

By free disposability (FD), one obtains $q^t \geq 0$.

Let $\boldsymbol{q} = (\hat{p}^1 \cdots \hat{p}^{t-1}, q^t, \hat{p}^{t+1} \cdots \hat{p}^\ell)$. The first-order condition at $\boldsymbol{y}_t \in Y_t$ is that $\boldsymbol{q}\boldsymbol{y} \leq \boldsymbol{q}\boldsymbol{y}_t$ for all $\boldsymbol{y} \in Y_t$ sufficiently close to $\boldsymbol{y}_t$, or $\boldsymbol{q}$ is normal to Y_t at $\boldsymbol{y}_t$.

Let $N(\boldsymbol{y}, Y)$ be the normal cone of a set Y at $\boldsymbol{y} \in \partial Y$ which is defined by:

$$N(\boldsymbol{y}, Y) = \{\boldsymbol{q} \in \mathbb{R}^{\ell} \mid \boldsymbol{qz} \leq \boldsymbol{qy} \text{ for all } \boldsymbol{z} \in Y\}.$$

If the set Y is convex, then $N(\boldsymbol{y}, Y) \neq \emptyset$ by the separation hyperplane theorem (Theorem A1 of Appendix A). $N(\boldsymbol{y}, Y)$ is a closed and convex cone with the vertex at the origin.

Suppose that the production set Y_t is convex and satisfies the free disposability (FD). Then, for each $\boldsymbol{y}_t \in \partial Y_t$, $N(\boldsymbol{y}_t, Y_t) \cap \Delta \neq \emptyset$. Let,

$$\hat{\boldsymbol{q}} = (\hat{q}^t) \in N(\boldsymbol{y}_t, Y_t) \cap \Delta.$$

Then by the homogeneity (HG), the first-order condition is written as:

$$\begin{aligned} p_t^t(\boldsymbol{y} \mid \lambda\hat{\boldsymbol{p}}, \hat{\boldsymbol{\xi}}) + a_t(\lambda\hat{\boldsymbol{p}}, \hat{\boldsymbol{\xi}})y_t^t &= \nu_\lambda \hat{q}_t, \\ p_t^s(\boldsymbol{y} \mid \lambda\hat{\boldsymbol{p}}, \hat{\boldsymbol{\xi}}) &= \nu_\lambda \hat{q}^s \quad (s \neq t). \end{aligned}$$

Summing over $s = 1 \cdots \ell$, we have:

$$\sum_{s=1}^{\ell} p_t^s(\boldsymbol{y} \mid \hat{\boldsymbol{p}}, \hat{\boldsymbol{\xi}}) + a_t(\hat{\boldsymbol{p}}, \hat{\boldsymbol{\xi}})y^t = \frac{\nu_\lambda}{\lambda} \sum_{s=1}^{\ell} \hat{q}_s = \frac{\nu_\lambda}{\lambda}.$$

In order to obtain $\sum_{s=1}^{\ell} p_t^s(\boldsymbol{y} \mid \hat{\boldsymbol{p}}, \hat{\boldsymbol{\xi}}) = 1$, it is necessary that:

$$\frac{\nu_\lambda}{\lambda} = 1 + a_t(\hat{\boldsymbol{p}}, \hat{\boldsymbol{\xi}})y^t.$$

This yields the first-order conditions in the normalized form:

$$\begin{aligned} p_t^t(\boldsymbol{y} \mid \hat{\boldsymbol{p}}, \hat{\boldsymbol{\xi}}) &= (1 + a_t(\hat{\boldsymbol{p}}, \hat{\boldsymbol{\xi}})y_t^t)\hat{q}^t - a_t(\hat{\boldsymbol{p}}, \hat{\boldsymbol{\xi}})y_t^t, \\ p_t^s(\boldsymbol{y} \mid \lambda\hat{\boldsymbol{p}}, \hat{\boldsymbol{\xi}}) &= (1 + a_t(\hat{\boldsymbol{p}}, \hat{\boldsymbol{\xi}})y_t^t)\hat{q}^s \quad (s \neq t). \end{aligned}$$

These equations can be written as:

$$p_t^s(\boldsymbol{y} \mid \hat{\boldsymbol{p}}, \hat{\boldsymbol{\xi}}) = (1 + a_t(\hat{\boldsymbol{p}}, \hat{\boldsymbol{\xi}})y_t^t)\hat{q}^t - \delta_t^s a_t(\hat{\boldsymbol{p}}, \hat{\boldsymbol{\xi}})y_t^t, \quad s = 1 \cdots \ell,$$

where $\delta_s^t = 1$ for $s = t$, and $\delta_s^t = 0$ for $s \neq t$.

For every $t (= 1 \cdots \ell)$, we define the pricing rule $\gamma_t \colon \partial Y_t \times \Delta \times \mathcal{F} \to \mathbb{R}_+^{\ell}$ by:

$$\gamma_t(\boldsymbol{y} \mid \hat{\boldsymbol{p}}, \hat{\boldsymbol{\xi}}) = \left\{ \boldsymbol{p} = (p^s) \in \mathbb{R}_+^{\ell} \,\middle|\, \text{for some } \hat{\boldsymbol{q}} = (\hat{q}^s) \in N(\boldsymbol{y}_t, Y) \cap \Delta, \right.$$
$$\left. p^s = \hat{q}^s \max\{0, 1 + a_t(\hat{\boldsymbol{p}}, \hat{\boldsymbol{\xi}})y_t^t\} - \frac{\delta_t^s a_t(\hat{\boldsymbol{p}}, \hat{\boldsymbol{\xi}})y_t^t}{\max\{1, -a_t(\hat{\boldsymbol{p}}, \hat{\boldsymbol{\xi}})y_t^t\}} \quad s = 1 \cdots \ell \right\}.$$

Then, we can prove that:

Lemma 3.1. *Under the assumptions that $a_t(\hat{\boldsymbol{p}}, \hat{\boldsymbol{\xi}})$ is continuous and the production set Y_t is convex, the pricing rule γ_t satisfies the conditions (PR-1) and (PR-2). Moreover, γ_t satisfies that the first-order condition for the profit maximization, or else some of the prices are nonpositive (in fact, zero).*

Proof. Let $\boldsymbol{p} = (p^t) \in \gamma_t(\boldsymbol{y} \mid \hat{\boldsymbol{p}}, \hat{\boldsymbol{\xi}})$. Then, by definition, there exists $\hat{q} \in N(\boldsymbol{y}_t, Y_t) \cap \Delta$ such that:

$$p^s = \hat{q}^s \max\{0, 1 + a_t(\hat{\boldsymbol{p}}, \hat{\boldsymbol{\xi}})y_t^t\} - \frac{\delta_t^s a_t(\hat{\boldsymbol{p}}, \hat{\boldsymbol{\xi}})y_t^t}{\max\{1, -a_t(\hat{\boldsymbol{p}}, \hat{\boldsymbol{\xi}})y_t^t\}}, \quad s = 1 \cdots \ell.$$

If $1 + a_t(\hat{\boldsymbol{p}}, \hat{\boldsymbol{\xi}})y_t^t > 0$, then $1 > -a_t(\hat{\boldsymbol{p}}, \hat{\boldsymbol{\xi}})y_t^t$. Thus, we have:

$$p^s = (1 + a_t(\hat{\boldsymbol{p}}, \hat{\boldsymbol{\xi}})y_t^t)\hat{q}^s - \delta_t^s a_t(\hat{\boldsymbol{p}}, \hat{\boldsymbol{\xi}})y_t^t \geq 0, \quad s = 1 \cdots \ell.$$

Summing over s,

$$\sum_{s=1}^{\ell} p^s = (1 + a_t(\hat{\boldsymbol{p}}, \hat{\boldsymbol{\xi}})y_t^t) \sum_{s=1}^{\ell} \hat{q}^s - a_t(\hat{\boldsymbol{p}}, \hat{\boldsymbol{\xi}})y_t^t = 1.$$

Similarly, if $1 + a_t(\hat{\boldsymbol{p}}, \hat{\boldsymbol{\xi}})y_t^t \leq 0$, then $1 \leq -a_t(\hat{\boldsymbol{p}}, \hat{\boldsymbol{\xi}})y_t^t$. Thus,

$$p^s = \frac{-\delta_t^s a_t(\hat{\boldsymbol{p}}, \hat{\boldsymbol{\xi}})y_t^t}{-a_t(\hat{\boldsymbol{p}}, \hat{\boldsymbol{\xi}})y_t^t} = \delta_t^s \geq 0, \quad s = 1 \cdots \ell,$$

therefore, $\sum_{s=1}^{\ell} p^s = 1$.

For given $(\hat{\boldsymbol{p}}, \hat{\boldsymbol{\xi}}) \in \Delta \times \mathcal{F}$, it is easy to see that the correspondence $N(\cdot, Y_t)$ is closed. Hence, the correspondence $N(\cdot, Y_t) \cap \Delta$ is upper hemi-continuous and compact-valued by Proposition D2 of Appendix D. Therefore, the correspondence γ_t is compact-valued. It is convex-valued, since $N(\cdot, Y_t) \cap \Delta$ is convex-valued. It is non-empty valued by construction.

Since $\max\{0, 1 + a_t(\hat{\boldsymbol{p}}, \hat{\boldsymbol{\xi}})y_t^t\}$ is continuous for $(\boldsymbol{y}, \hat{\boldsymbol{p}}, \hat{\boldsymbol{\xi}})$, the first term of p^s is upper hemi-continuous, as the product of two upper hemi-continuous and compact-valued correspondences by Proposition D6 of Appendix D. Since the second term is obviously continuous, the correspondence $\gamma_t(\boldsymbol{y} \mid \hat{\boldsymbol{p}}, \hat{\boldsymbol{\xi}})$ is upper hemi-continuous.

The profit function $\pi_t(\boldsymbol{y} \mid \hat{\boldsymbol{p}}, \hat{\boldsymbol{\xi}})$ is calculated as:

$$\begin{aligned}
\pi_t(\boldsymbol{y} \mid \hat{\boldsymbol{p}}, \hat{\boldsymbol{\xi}}) &= \boldsymbol{p}_t(\boldsymbol{y} \mid \hat{\boldsymbol{p}}, \hat{\boldsymbol{\xi}})\boldsymbol{y}_t \\
&= \sum_{s=1}^{\ell} \left(\hat{q}^s y_t^s \max\{0, 1 + a_t(\hat{\boldsymbol{p}}, \hat{\boldsymbol{\xi}})y_t^t\} - \frac{\delta_t^s a_t(\hat{\boldsymbol{p}}, \hat{\boldsymbol{\xi}})y_t^t y_t^s}{\max\left\{1, -a_t(\hat{\boldsymbol{p}}, \hat{\boldsymbol{\xi}})y_t^t\right\}} \right) \\
&\geq \max\left\{0, 1 + a_t(\hat{\boldsymbol{p}}, \hat{\boldsymbol{\xi}})y_t^t\right\} \hat{\boldsymbol{q}}\hat{\boldsymbol{y}}_t \geq 0,
\end{aligned}$$

since $a_t(\hat{\boldsymbol{p}}, \hat{\boldsymbol{\xi}}) \leq 0$ and $N(\boldsymbol{y}_t, Y_t)\boldsymbol{y}_t \geq 0$. This shows that $\pi_t(\boldsymbol{y} \mid \hat{\boldsymbol{p}}, \hat{\boldsymbol{\xi}}) \geq 0$.

Finally, when $1+a_t(\hat{p},\hat{\xi})y_t^t > 0$, then the pricing rule implements the first-order condition:

$$p_t^s(y \mid \hat{p},\hat{\xi}) = (1 + a_t(\hat{p},\hat{\xi})y_t^t)\hat{q}^t - \delta_t^s a_t(\hat{p},\hat{\xi})y_t^t \geq 0, \quad s = 1 \cdots \ell.$$

When $1 + a_t(\hat{p},\hat{\xi})y_t^t \leq 0$, then:

$$p_t^s(y \mid \hat{p},\hat{\xi}) = 0 \quad \text{for } s \neq t.$$

This proves Lemma 3.1. □

We now state the definition of the monopolistically competitive equilibrium. Since we will assume that the production sets Y_t are convex for all $t = 1 \cdots \ell$, the profit maximizations are global and the maximized profits are ensured to be nonnegative at equilibria.

Definition 3.1. A market data $(\hat{p},(\hat{x}_a)_{a=1}^m,(\hat{y}_t)_{t=1}^\ell)$ is said to consist of a monopolistically competitive equilibrium if and only if

(E-1) $\hat{p}\hat{x}_a \leq \hat{p}\omega_a + \sum_{t=1}^\ell \theta_{at}\hat{p}\hat{y}_t$ and
$\hat{x}_a \succsim_a x$ whenever $\hat{p}x \leq \hat{p}\omega_a + \sum_{t=1}^\ell \theta_{at}\hat{p}\hat{y}_t$, $a = 1 \cdots m$,
(E-2) $p_t(y \mid \hat{p},\hat{\xi})y \leq \hat{p}\hat{y}_t$ for all $y \in Y_t$, $t = 1 \cdots \ell$, and
(E-3) $\sum_{a=1}^m \hat{x}_a \leq \sum_{t=1}^\ell \hat{y}_t + \sum_{a=1}^m \omega_a$.

The existence of the equilibrium is immediately obtained by using Lemma 3.1 and the result of the previous section.

Theorem 3.1. *Under the assumptions of the continuity (CT), the convexity (CV), the monotonicity (MT), and the minimum income condition (MI), for every $a = 1 \cdots m$, and for every $t = 1 \ldots \ell$, suppose that the free disposability (FD) and no free production (NFP) conditions hold and Y_t is convex. Finally, we assume that the condition of the bounded total production (BTP). Then there exists a monopolistically competitive equilibrium.*

Proof. Follows from Theorem 2.1 and Lemma 3.1. □

Theorem 3.1 generalizes Theorem 1.2 given in Chap. 4 in a way that it implements the monopolistic behavior of the firms. However, it is not completed satisfactory, since it assumes the convexity of the production sets. As pointed out in Section 6.1, the realistic monopolistically competitive firms would take production activities under non-convex production technologies. The next example, however, indicates that not all non-convex production sets are compatible with the pricing rule verifying (PR-1) and (PR-2) with the first-order condition as given in Lemma 3.1.

Example 3.1. There are two commodities, an output y and an input z with the prices p and r, respectively. The production set is defined by:

$$Y = \{(y,z) \in \mathbb{R}_+ \times \mathbb{R}_- \mid y \leq \max\{0, -z - c\}\},$$

(see Fig. 6.6).

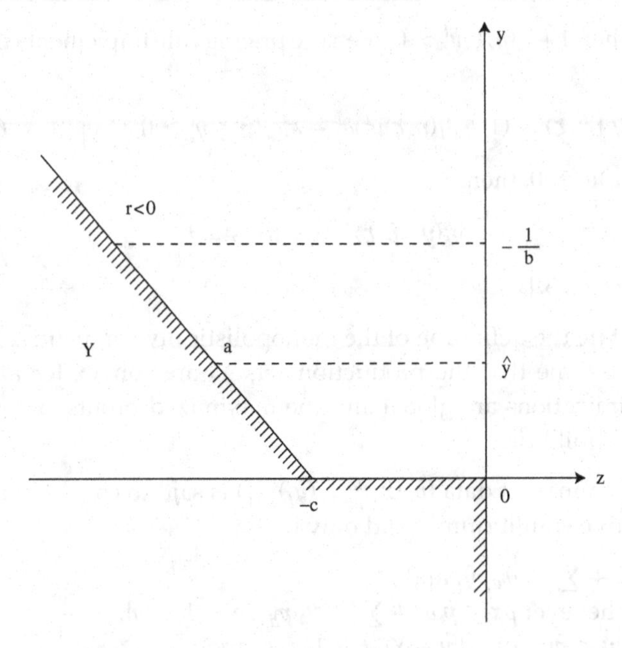

Figure 6.6. Example 3.1.

Note that $y \leq 0$ when $z \geq -c$, and $y \leq -z - c$ when $z \leq -c$. The production set Y represents a technology which includes a setup cost c.

The perceived inverse demand function of the firm is defined by:

$$p(y, z \mid \hat{\boldsymbol{p}}, \hat{\boldsymbol{\xi}}) = \hat{p} + b(y - \hat{y}), \quad b < 0,$$
$$r(y, z \mid \hat{\boldsymbol{p}}, \hat{\boldsymbol{\xi}}) = \hat{r},$$

given $\hat{\boldsymbol{p}} = (\hat{p}, \hat{r})$ and $\hat{\boldsymbol{\xi}} = ((\hat{\boldsymbol{x}}_a), (\hat{y}, \hat{z}))$. Then, the profit function is calculated as:

$$\pi(y, z \mid \hat{\boldsymbol{p}}, \hat{\boldsymbol{\xi}}) = \begin{cases} \hat{r}z & \text{if } z \geq -c, \\ (\hat{p} + b(y - \hat{y}))y - \hat{r}(y + c) & \text{otherwise.} \end{cases}$$

The first-order condition for the profit maximization, given $z \leq -c$ is then given by:

$$\frac{d\pi}{dy} = p(y, z \mid \hat{\boldsymbol{p}}, \hat{\boldsymbol{\xi}}) + by - \hat{r} = 0, \quad p = \hat{r} - by.$$

This yields the normalized price vector $\boldsymbol{p} = (p, r)$ given by:

$$p = \frac{1 - by}{2}, \quad r = \frac{1 + by}{2}.$$

The corresponding profits are given by:

$$\pi = \left(\frac{1 - by}{2}\right) y - \left(\frac{1 + by}{2}\right)(y + c) = -by^2 - \frac{1 + by}{2} c.$$

Therefore, we see that:

$$\pi \geq 0 \quad \text{if and only if } \hat{y} = \frac{bc - \sqrt{b^2c^2 - 8bc}}{4b} \leq y \leq -\frac{1}{b}.$$

For example, if $b = -1$, then $0 \leq \hat{y} \leq 1$ for $c \leq 1$.

We wish to construct, on the boundary ∂Y of Y, a pricing rule which is upper hemi-continuous (PR-1), which yields non-negative profits (PR-2), and such that $p, r \geq 0$ or else $pr \leq 0$ (see Lemma 3.1). We will show that a contradiction arises at the point $(\hat{y}, -\hat{y} - c)$, labeled a in Fig. 6.6.

For $y \in [\hat{y}, -1/b]$, the first-order condition imposes the unique prices given by $p(\hat{y}) = (1 - b\hat{y})2^{-1}$ and $r(\hat{y}) = (1 + b\hat{y})2^{-1}$. In particular, $p(\hat{y}), r(\hat{y}) > 0$ for $\hat{y} < -1/b$. On the other hand, for $0 \leq y < \hat{y}$, the first-order condition yields negative profits, hence the conditions of Lemma 3.1 impose $pr \leq 0$ with $p \geq 0$ and $r \leq 0$.

Let $y_n \to \hat{y}$ as $n \to +\infty$ and $y^n < \hat{y}$ for all n. Take $(p_n, r_n) \in \gamma(y_n)$, so that $p^n \geq 0$ and $r_n \leq 0$. If $\gamma(y)$ is upper hemi-continuous, there exists a (not necessarily unique) limit $(\bar{p}, \bar{r})$ such that $\bar{p} \geq 0$ and $\bar{r} \leq 0$. Since $\gamma(y)$ is upper hemi-continuous, compact and convex-valued at $\hat{y}$, it follows that $(p_\alpha, r_\alpha) = \alpha(p(\hat{y}), r(\hat{y})) + (1-\alpha)(\bar{p}, \bar{r}) \in \gamma(\hat{y})$ for all $\alpha \in [0, 1]$. But then, $(p_\alpha, r_\alpha) \gg (0, 0)$ for α close to 1. For $\alpha \neq 1$, we have a strictly positive vector $(p_\alpha, r_\alpha) \in \gamma(\hat{y})$ which is different from $(p(\hat{y}), (\hat{y}))$, hence (p_α, r_α) does not obey the first-order condition. This is a contradiction and we cannot construct a pricing rule with the desired properties in this example. □

Example 3.1 explains why we introduce the restriction that fixed inputs and variable inputs are disjoint sets of commodities.

6.4. MONOPOLISTICALLY COMPETITIVE EQUILIBRIA WITH FIXED COSTS

In this section, we will prove the existence of monopolistically competitive equilibria for a class of technologies with fixed costs. On account of the counter example given in Section 6.3, we will assume that for each firm, fixed inputs are distinct from variable inputs or outputs, so that the production set is the union of two convex sets, and one of them contains the origin.

As usual, there exist ℓ (≥ 3) commodities in the economy. For definitiveness and simplicity, we assume that the commodity $t(= 1 \cdots \ell)$ is produced by one firm, hence the firm is indexed by $t(= 1 \cdots \ell)$. Each firm uses a single, fixed input commodity with a fixed investment threshold, say $c_t > 0$.

More specifically, let $f(\neq t)$ be the index of the fixed input to the firm t, hence $v \in \boldsymbol{v} \equiv \{1 \cdots \ell\} \backslash \{f, t\}$ are the indices of the variable inputs.

For each t, we re-arrange the coordinates of $\boldsymbol{y}_t \in \mathbb{R}^\ell$ and denote $\boldsymbol{y}_t = (y_t^f, (y_t^v), y_t^t)$. Sometimes, we denote $\boldsymbol{y}_t^v = (y_t^v) \in \mathbb{R}^{\ell-2}$. We assume that the production set Y_t of the firm t is written in the form:

$$Y_t = Y_t^1 \cup Y_t^2,$$
$$Y_t^1 = \{\boldsymbol{y}_t = (y_t^f, \boldsymbol{y}_t^v, y_t^t) \mid y_t^f \leq -c_t, y_t^t \leq f_t(\boldsymbol{y}_t^v)\},$$
$$Y_t^2 = \{\boldsymbol{y}_t = (y_t^f, \boldsymbol{y}_t^v, y_t^t) \mid -c_t \leq y_t^f \leq 0, (\boldsymbol{y}_t^v, y_t^t) \in \hat{Y}_t^{v,t}\},$$

where $f_t: \mathbb{R}_-^{\ell-2} \to \mathbb{R}_+$ is a monotonically decreasing, continuous, and concave function such that $f_t(\boldsymbol{0}) = 0$, $\hat{Y}_t^{v,t}$ is a closed and convex subset of $\mathbb{R}^{\ell-1}$ such that $\hat{Y}_t^{v,t} + \mathbb{R}_-^{\ell-1} \subset \hat{Y}_t^{v,t}$, $\hat{Y}_t^{v,t} \cap \mathbb{R}_+^{\ell-1} = \{0\}$ (see Fig. 6.3).

We also assume that $y^t < f_t(\boldsymbol{y}^v)$ for all $(\boldsymbol{y}^v, y^t) \in \hat{Y}_t^{v,t}$. This reflects a fact that a production plan which is feasible without fixed investment remains feasible with fixed investment.

The set of weakly efficient production plans or the boundary ∂Y_t of Y_t,

$$\partial Y_t = \{\boldsymbol{y}_t \in Y_t \mid \text{There exists no } \boldsymbol{z} \in Y_t \text{ such that } \boldsymbol{y}_t \ll \boldsymbol{z}\}$$

can be described as follows in terms of $Y_t^1, \hat{Y}_t^{v,t}$, and their boundaries $\partial Y_t^1, \partial\hat{Y}_t^{v,t} \subset \mathbb{R}^{\ell-1}$,

$$\partial Y_t = \{\boldsymbol{y}_t = (y_t^f, \boldsymbol{y}_t^v, y_t^t) \mid y_t^f < -c_t, y_t^t = f_t(\boldsymbol{y}_t^v)\},$$
$$\cup \{\boldsymbol{y}_t = (y_t^f, \boldsymbol{y}_t^v, y_t^t) \mid y_t^f = -c_t, y_t^t \leq f_t(\boldsymbol{y}_t^v), (\boldsymbol{y}_t^v, y_t^t) \notin \hat{Y}_t^{v,t} \backslash \partial\hat{Y}_t^{v,t}\},$$
$$\cup \{\boldsymbol{y}_t = (y_t^f, \boldsymbol{y}_t^v, y_t^t) \mid -c_t < y_t^f < 0, (\boldsymbol{y}_t^v, y_t^t) \in \partial\hat{Y}_t^{v,t}\}, \text{ and}$$
$$\cup \{\boldsymbol{y}_t = (y_t^f, \boldsymbol{y}_t^v, y_t^t) \mid y_t^f = 0, (\boldsymbol{y}_t^v, y_t^t) \in \hat{Y}_t^{v,t}\}.$$

In order to define a pricing rule $\gamma_t: \partial Y_t \times \Delta \times \mathcal{F} \to \Delta$, we can rely on Lemma 3.1 for the first and third sets of the union defining ∂Y_t, but we have to extend this specification so as to cover the second and fourth sets preserving the upper hemicontinuity at $y_t^f = -c_t$ and 0.

The definition of the monopolistically competitive equilibrium with fixed costs reads as follows.

Definition 4.1. A market data $(\hat{\boldsymbol{p}}, (\hat{\boldsymbol{x}}_a)_{a=1}^m, (\hat{\boldsymbol{y}}_t)_{t=1}^\ell)$ is said to consist of a monopolistically competitive equilibrium (with fixed costs) if and only if

(E-1) $\hat{\boldsymbol{p}}\hat{\boldsymbol{x}}_a \leq \hat{\boldsymbol{p}}\omega_a + \sum_{t=1}^\ell \theta_{at}\hat{\boldsymbol{p}}\hat{\boldsymbol{y}}_t$ and
$\hat{\boldsymbol{x}}_a \succsim_a \boldsymbol{x}$ whenever $\hat{\boldsymbol{p}}\boldsymbol{x} \leq \hat{\boldsymbol{p}}\omega_a + \sum_{t=1}^\ell \theta_{at}\hat{\boldsymbol{p}}\hat{\boldsymbol{y}}_t$, $a = 1 \cdots m$,
(E-2) there exists a neighborhood U of $\hat{\boldsymbol{y}}_t$ such that:
$p_t(\boldsymbol{y}|\hat{\boldsymbol{p}}, \hat{\zeta})\boldsymbol{y} \leq \hat{\boldsymbol{p}}\hat{\boldsymbol{y}}_t$ for all $\boldsymbol{y} \in U \cap Y_t$, and $\hat{\boldsymbol{p}}\hat{\boldsymbol{y}}_t \geq 0$, $t = 1 \cdots \ell$, and
(E-3) $\sum_{a=1}^m \hat{\boldsymbol{x}}_a \leq \sum_{t=1}^\ell \hat{\boldsymbol{y}}_t + \sum_{a=1}^m \omega_a$.

The conditions (E-1) and (E-3) are standard, and need no explanations. The only difference between Definitions 3.1 and 4.1 is the condition (E-2). Since the production sets are not convex, we can expect at best the local profit maximization for each firm. The nonnegativity of the profits is not obvious, and we add it into the definition.

The fundamental existence theorem of this chapter now reads as:

Theorem 4.1. *Under the assumptions of the continuity (CT), the convexity (CV), the monotonicity (MT), and the minimum income condition (MI) for every $a \in A$, and for every $t = 1 \cdots \ell$, the production set Y_t is defined as above. Finally, we assume that the condition of the bounded total production (BTP). Then, there exists a monopolistically competitive equilibrium.*

Proof. The proof of Theorem 4.1 relies on:

Lemma 4.1. *Under the assumptions that $a_t(\hat{\boldsymbol{p}}, \hat{\boldsymbol{\xi}})$, which was defined in Section 6.3, is continuous and the production set Y_t is as defined as above, there exists a pricing rule $\gamma_t \colon \partial Y_t \times \Delta \times \mathcal{F} \to \Delta$ satisfying the conditions (PR-1) and (PR-2). Moreover, γ_t satisfies that the first-order condition for the profit maximization, or else some of the prices are zero (nonpositive).*

Proof. The proof is constructive. A desired pricing rule is defined successively for all $\boldsymbol{y}_t = (y_t^f, \boldsymbol{y}_t^v, y_t^t) \in \partial Y_t$ such that:

1. $-c_t < y_t^f < 0$;
2. $y_t^f < -c_t$;
3. $y_t^f = -c_t$; and
4. $y_t^f = 0$.

This proof applies to an arbitrary firm, so we omit the subscript t. Similarly, we often omit explicit reference to $(\hat{\boldsymbol{p}}, \hat{\boldsymbol{\xi}})$.

We write $N^1(\boldsymbol{y}_t^v, y_t^t)$ for the normal cone to the set $\hat{Y}^1 = \{(\boldsymbol{z}^v, z^t) \in \mathbb{R}^{\ell-1} \mid z^t \leq f(\boldsymbol{z}^v)\}$ at $(\boldsymbol{y}_t^v, y_t^t)$ and similarly, $N^2(\boldsymbol{y}_t^v, y_t^t)$ for the normal cone to the set $\hat{Y}^{v,t} \subset \mathbb{R}^{\ell-1}$ (see the definition of $Y_t = Y_t^1 \cup Y_t^2$) at $(\boldsymbol{y}_t^v, y_t^t)$. Let $\gamma^i(\boldsymbol{y}_t^v, y_t^t)$ $(i = 1, 2)$ be the pricing rules defined in Lemma 3.1 with $\bar{q} \in N^i(\boldsymbol{y}_t^v, y_t^t) \cap \Delta(\subset \mathbb{R}^{\ell-1})$.

1. When $-c < y^f < 0$, then $(\boldsymbol{y}^v, y^t) \in \partial\hat{Y}^{v,t}$. We set $p^f = 0$ and $(\boldsymbol{p}^v, p^t) \in \gamma^2(\boldsymbol{y}^v, y^t)$, or:

$$\begin{aligned}\gamma(\boldsymbol{y} = (y^f, \boldsymbol{y}^v, y^t) \mid -c < y^f < 0, (\boldsymbol{y}^v, y^t) \in \partial\hat{Y}^{v,t}) \\ = \{\boldsymbol{p} = (p^f, \boldsymbol{p}^v, p^t) \in \Delta \mid p^f = 0, (\boldsymbol{p}^v, p^t) \in \gamma^2(\boldsymbol{y}^v, y^t)\}.\end{aligned}$$

Since Lemma 3.1 applies to $\gamma^2(\boldsymbol{y}^v, y^t)$, it also applies to $\gamma(y^f, \boldsymbol{y}^v, y^t)$.

2. When $y^f < -c$, then $(\boldsymbol{y}^v, y^t) \in \partial\hat{Y}^1$, and we define:

$$\gamma(\boldsymbol{y} = (y^f, \boldsymbol{y}^v, y^t) \mid y^f < -c, (\boldsymbol{y}^v, y^t) \in \partial\hat{Y}^1)$$
$$= \{\boldsymbol{p} = (p^f, \boldsymbol{p}^v, p^t) \in \Delta \mid p^f = 0, (\boldsymbol{p}^v, p^t) \in \gamma^1(\boldsymbol{y}^v, y^t)\}.$$

Again Lemma 3.1 applies to $\gamma(y^f, \boldsymbol{y}^v, y^t)$ as defined above.

3. When $y^f = -c$, then $(\boldsymbol{y}^v, y^t) \notin \hat{Y}^{v,t} \backslash \partial\hat{Y}^{v,t}$, and $(y^f, \boldsymbol{y}^v, y^t)$ is efficient if and only if $y^t = f(\boldsymbol{y}^v)$. Otherwise, the first-order conditions cannot be satisfied at strictly positive price vectors, and $\boldsymbol{y}$ cannot be an equilibrium production plan when the preferences are (strictly) monotone (MT). We then set $p^f = 0$ when $y^t < f(\boldsymbol{y}^v)$ and extend the rule appropriately. In order to do so, it is convenient to distinguish three subcases.

3.1 Suppose that $y^t = f(\boldsymbol{y}^v)$. Then, we set $p^f \in [0, (1/c)(\boldsymbol{p}^v\boldsymbol{y}^v + p^t y^t)]$ and $(\boldsymbol{p}^v, p^t) = (1 - p^f)(\hat{\boldsymbol{p}}^v, \hat{p}^t)$ with $(\hat{\boldsymbol{p}}^v, \hat{p}^t) \in \gamma^1(\boldsymbol{y}^v, y^t)$. This is equivalent to that for some $(\hat{\boldsymbol{p}}^v, \hat{p}^t) \in \gamma^1(\boldsymbol{y}^v, y^t)$, we set:

$$p^f \in \left[0, \frac{\hat{\boldsymbol{p}}^v\boldsymbol{y}^v + \hat{p}^t y^t}{c + \hat{\boldsymbol{p}}^v\boldsymbol{y}^v + \hat{p}^t y^t}\right]$$

and $(\boldsymbol{p}^v, p^t) = (1 - p^f)(\hat{\boldsymbol{p}}^v, \hat{p}^t)$. This defines the correspondence:

$$\gamma(\boldsymbol{y} = (y^f, \boldsymbol{y}^v, y^t) \mid y^f = -c, y^t = f(\boldsymbol{y}^v))$$
$$= \Bigg\{\boldsymbol{p} = (p^f, \boldsymbol{p}^v, p^t) \in \Delta \mid \text{for some } (\hat{\boldsymbol{p}}^v, \hat{p}^t) \in \gamma^1(\boldsymbol{y}^v, y^t),$$
$$p^f \in \left[0, \frac{\hat{\boldsymbol{p}}^v\boldsymbol{y}^v + \hat{p}^t y^t}{c + \hat{\boldsymbol{p}}^v\boldsymbol{y}^v + \hat{p}^t y^t}\right], (\boldsymbol{p}^v, p^t) = (1 - p^f)(\hat{\boldsymbol{p}}^v, \hat{p}^t)\Bigg\},$$

which is upper hemi-continuous by Proposition D1 of Appendix D and nonempty, compact, and convex-valued. In order to verify the convex values, take $\boldsymbol{p}_i = (p_i^f, \boldsymbol{p}_i^v, p_i^t) \in \gamma(\boldsymbol{y} \mid y^f = -c, y^t = f(\boldsymbol{y}^v))$, $i = 1, 2$. That is,

$$(\hat{\boldsymbol{p}}_i^v, \hat{p}_i^t) \in \gamma^1(\boldsymbol{y}^v, y^t),$$
$$0 \le p_i^f \le \frac{\hat{\boldsymbol{p}}_i^v\boldsymbol{y}^v + \hat{p}_i^t y^t}{c + \hat{\boldsymbol{p}}_i^v\boldsymbol{y}^v + \hat{p}_i^t y^t},$$

and

$$(\boldsymbol{p}_i^v, p_i^t) = (1 - p_i^f)(\hat{\boldsymbol{p}}_i^v, \hat{p}_i^t), \quad i = 1, 2.$$

We define $\boldsymbol{p}_\lambda = \lambda \boldsymbol{p}_1 + (1-\lambda)\boldsymbol{p}_2 = (\lambda p_1^f + (1-\lambda)p_2^f, \lambda \boldsymbol{p}_1^{\boldsymbol{v}} + (1-\lambda)\boldsymbol{p}_2^{\boldsymbol{v}}, \lambda p_1^t + (1-\lambda)p_2^t) \equiv (p_\lambda^f, \boldsymbol{p}_\lambda^{\boldsymbol{v}}, p_\lambda^t)$. Set:

$$\nu_\lambda = \frac{\lambda(1-p_1^f)}{\lambda(1-p_1^f)+(1-\lambda)(1-p_2^f)} = \frac{\lambda(1-p_1^f)}{1-p_\lambda^f},$$
$$1-\nu_\lambda = \frac{(1-\lambda)(1-p_2^f)}{1-p_\lambda^f},$$

and

$$(\hat{\boldsymbol{p}}_\lambda^{\boldsymbol{v}}, \hat{p}_\lambda^t) = \nu_\lambda(\hat{\boldsymbol{p}}_1^{\boldsymbol{v}}, \hat{p}_1^t) + (1-\nu_\lambda)(\hat{\boldsymbol{p}}_2^{\boldsymbol{v}}, \hat{p}_2^t) \in \gamma^1(\boldsymbol{y}^{\boldsymbol{v}}, y^t).$$

Then, we have:

$$(\boldsymbol{p}_\lambda^{\boldsymbol{v}}, p_\lambda^t) = (1-p_\lambda^f)(\hat{\boldsymbol{p}}_\lambda^{\boldsymbol{v}}, \hat{p}_\lambda^t),$$

and

$$\begin{aligned}
0 \le p_\lambda^f &= \lambda p_1^f + (1-\lambda)p_2^f \\
&\le (1/c)\{(\lambda \boldsymbol{p}_1^{\boldsymbol{v}}\boldsymbol{y}^{\boldsymbol{v}} + \lambda p_1^t y^t + (1-\lambda)\boldsymbol{p}_2^{\boldsymbol{v}}\boldsymbol{y}^{\boldsymbol{v}} + (1-\lambda)p_2^t y^t\} \\
&= (1/c)\{(\lambda \boldsymbol{p}_1^{\boldsymbol{v}} + (1-\lambda)\boldsymbol{p}_2^{\boldsymbol{v}})\boldsymbol{y}^{\boldsymbol{v}} + (\lambda p_1^t + (1-\lambda)p_2^t)y^t\} \\
&= (1/c)(\boldsymbol{p}_\lambda^{\boldsymbol{v}}\boldsymbol{y}^{\boldsymbol{v}} + p_\lambda^t y^t).
\end{aligned}$$

Therefore, it follows that:

$$p_\lambda^f \in \left[0, \frac{\hat{\boldsymbol{p}}_\lambda^{\boldsymbol{v}}\boldsymbol{y}^{\boldsymbol{v}} + \hat{p}_\lambda^t y^t}{c + \hat{\boldsymbol{p}}_\lambda^{\boldsymbol{v}}\boldsymbol{y}^{\boldsymbol{v}} + \hat{p}_\lambda^t y^t}\right].$$

This shows that $\boldsymbol{p}_\lambda \in \gamma(\boldsymbol{y} \mid y^f = -c, y^t = f(\boldsymbol{y}^{\boldsymbol{v}}))$.

Since $p^f \le (1/c)(\boldsymbol{p}^{\boldsymbol{v}}\boldsymbol{y}^{\boldsymbol{v}} + p^t y^t)$, $\boldsymbol{p}\boldsymbol{y} \ge 0$ for each $\boldsymbol{p} = (p^f, \boldsymbol{p}^{\boldsymbol{v}}, p^t) \in \gamma(\boldsymbol{y} \mid \hat{\boldsymbol{p}}, \hat{\zeta})$, which verifies the condition (PR-2). Hence, Lemma 3.1 applies to this case.

3.2 Suppose that $y^t < f(\boldsymbol{y}^{\boldsymbol{v}})$ and $(\boldsymbol{y}^{\boldsymbol{v}}, y^t) \in \partial\hat{Y}^{\boldsymbol{v},t}$. Then, we define:

$$\begin{aligned}
\gamma(\boldsymbol{y} &= (y^f, \boldsymbol{y}^{\boldsymbol{v}}, y^t) \mid y^f = -c, (\boldsymbol{y}^{\boldsymbol{v}}, y^t) \in \partial\hat{Y}^{\boldsymbol{v},t}) \\
&= \{\boldsymbol{p} = (p^f, \boldsymbol{p}^{\boldsymbol{v}}, p^t) \in \Delta \mid p^f = 0, (\boldsymbol{p}^{\boldsymbol{v}}, p^t) \in \gamma^2(\boldsymbol{y}^{\boldsymbol{v}}, y^t)\}.
\end{aligned}$$

Again Lemma 3.1 applies.

3.3 For the case that $y^t < f(\boldsymbol{y}^{\boldsymbol{v}})$ and $(\boldsymbol{y}^{\boldsymbol{v}}, y^t) \notin \partial\hat{Y}^{\boldsymbol{v},t}$, the construction of the pricing rule is more intricate.

Given $\boldsymbol{y} = (-c, \boldsymbol{y}^v, y^t)$, we take $d_1 > 0$ and $d_2 > 0$ such that:

$$y^1 + d_1 = f(\boldsymbol{y}^v + d_1\mathbf{1}), \quad (\boldsymbol{y}^v - d_2\mathbf{1}, y^t - d_2) \in \partial\hat{Y}^{v,t},$$

where $\mathbf{1} = (1\ldots1) \in \mathbb{R}^{\ell-2}$. Such d_1 and d_2 exist, since Y is closed, the set $\{\boldsymbol{z} \in Y |\ \boldsymbol{z} \geq \boldsymbol{y}\}$ is bounded, and $\mathbb{R}^{\ell-1}_{-} \subset \hat{Y}^{v,t}$. They are unique by the definition of ∂Y. Clearly they are continuous in $(\boldsymbol{y}^v, y^t)$. We then define:

$$\begin{aligned}
&\gamma(\boldsymbol{y} = (y^f, \boldsymbol{y}^v, y^t) |\ y^f = -c, y^t < f(\boldsymbol{y}^v), (\boldsymbol{y}^v, y^t) \notin \partial\hat{Y}^{v,t}) \\
&= \left\{ \boldsymbol{p} = (p^f, \boldsymbol{p}^v, p^t) \in \Delta \,\middle|\, p^f = 0, (\boldsymbol{p}^v, p^t) = \frac{d_2}{d_1 + d_2}(\hat{\boldsymbol{p}}^v_1, \hat{p}^t_1) + \frac{d_1}{d_1 + d_2}(\hat{\boldsymbol{p}}^v_2, \hat{p}^t_2) \right. \\
&\quad \text{for some } (\hat{\boldsymbol{p}}^v_1, \hat{p}^t_1) \in \gamma^1(\boldsymbol{y}^v + d_1\mathbf{1}, y^t + d_1) \text{ such that } y^t + d_1 = f(\boldsymbol{y}^v + d_1\mathbf{1}), \\
&\quad \text{and for some } (\hat{\boldsymbol{p}}^v_2, \hat{p}^t_2) \in \gamma^2(\boldsymbol{y}^v - d_2\mathbf{1}, y^t - d_2) \\
&\quad \left. \text{such that } (\boldsymbol{y}^v - d_2\mathbf{1}, y^t - d_2) \in \partial\hat{Y}^{v,t} \right\}
\end{aligned}$$

That is to say, $\gamma(\boldsymbol{y} = (y^f, \boldsymbol{y}^v, y^t) |\ y^f = -c, y^t < f(\boldsymbol{y}^v), (\boldsymbol{y}^v, y^t) \notin \partial\hat{Y}^{v,t})$ is a set of vectors $(0, \boldsymbol{p}^v, p^t)$, where $(\boldsymbol{p}^v, p^t)$ is a convex combination of elements of $\gamma_1(\boldsymbol{y}^v + d_1\mathbf{1}, y^t + d_1)$ and $\gamma_2(\boldsymbol{y}^v - d_2\mathbf{1}, y^t - d_2)$ for d_1 and d_2 as defined above. Obviously, if $(\boldsymbol{y}^v_n, y^t_n) \to (\hat{\boldsymbol{y}}^v, \hat{y}^t)$ with $\hat{y}^t = f(\hat{\boldsymbol{y}}^v)$, then $\gamma(-c, \boldsymbol{y}^v_n, y^t_n) \to \{\boldsymbol{p} = (p^f, \boldsymbol{p}^v, p^t) \in \Delta \mid p^f = 0, (\boldsymbol{p}^v, p^t) \in \gamma^1(\hat{\boldsymbol{y}}^v, \hat{y}^t)\}$, and if $(\boldsymbol{y}^v_n, y^t_n) \to (\tilde{\boldsymbol{y}}^v, \tilde{y}^t)$ with $(\tilde{\boldsymbol{y}}^v, \tilde{y}^t) \in \partial\hat{Y}^{v,t}$, then $\gamma(-c, \boldsymbol{y}^v_n, y^t_n) \to \{\boldsymbol{p} = (p^f, \boldsymbol{p}^v, p^t) \in \Delta \mid p^f = 0, (\boldsymbol{p}^v, p^t) \in \gamma^2(\tilde{\boldsymbol{y}}^v, \tilde{y}^t)\}$.

Hence, by Proposition D5 of Appendix D, the correspondence is:

$$\boldsymbol{y} = (y^f, \boldsymbol{y}^v, y^t) \mapsto \gamma(\boldsymbol{y} \mid y^f = -c, y^t < f(\boldsymbol{y}^v), (\boldsymbol{y}^v, y^t) \notin \partial\hat{Y}^{v,t})$$

is upper hemi-continuous. It is clearly nonempty and compact-valued. It is also convex-valued, as the set of convex combinations of elements from two convex sets. Therefore, (PR-1) is verified. In order to verify (PR-2), note that by Lemma 3.1, it follows that $\hat{\boldsymbol{p}}^v_i(\boldsymbol{y}^v + \sigma_i d_i\mathbf{1}) + \hat{p}^t_i(y^t + \sigma_i d_i) \geq 0$ for $(\hat{\boldsymbol{p}}^v_i, \hat{p}^t_i) \in \gamma^i(\boldsymbol{y}^v + \sigma_i d_i\mathbf{1}, y^t + \sigma_i d_i)$, $i = 1, 2$, where $\sigma_1 = 1$ and $\sigma_2 = -1$. Hence, we have:

$$\begin{aligned}
\boldsymbol{p}\boldsymbol{y} &= \boldsymbol{p}^v\boldsymbol{y}^v + p^t y^t \\
&= \frac{d_2}{d_1 + d_2}\left(\hat{\boldsymbol{p}}^v_1(\boldsymbol{y}^v + d_1\mathbf{1} - d_1\mathbf{1}) + \hat{p}^t_1(y^t + d_1 - d_1)\right) \\
&\quad + \frac{d_1}{d_1 + d_2}\left(\hat{\boldsymbol{p}}^v_2(\boldsymbol{y}^v - d_2\mathbf{1} + d_2\mathbf{1}) + \hat{p}^t_2(y^t - d_2 + d_2)\right) \\
&\geq \frac{-d_1 d_2}{d_1 + d_2} + \frac{d_1 d_2}{d_1 + d_2} = 0.
\end{aligned}$$

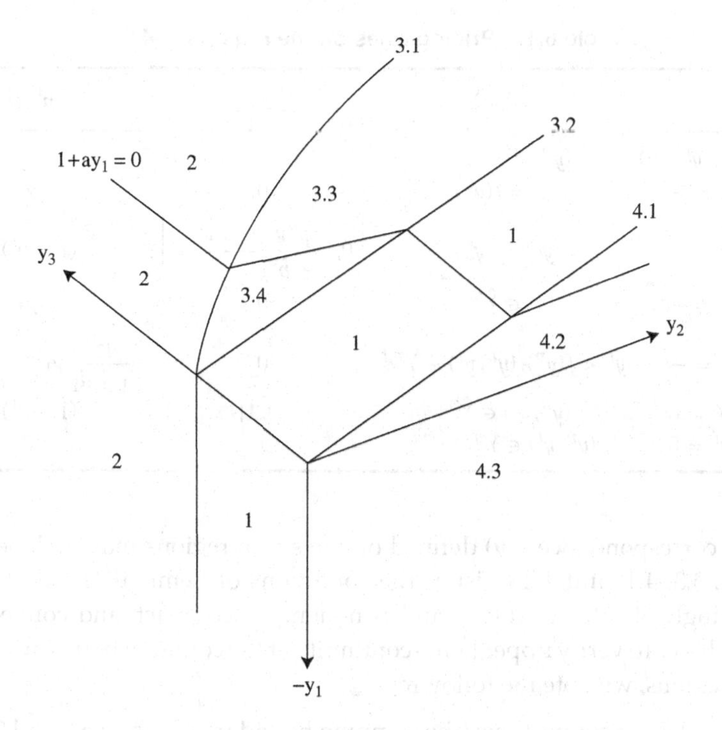

Figure 6.7. Lemma 4.1.

Therefore, (PR-2) is verified, and on account of $p^f = 0$, Lemma 4.1 holds good for $\gamma(\boldsymbol{y} = (y^f, \boldsymbol{y}^v, y^t)|\ y^f = -c, y^t < f(\boldsymbol{y}^v), (\boldsymbol{y}^v, y^t) \notin \partial\hat{Y}^{v,t})$.

It may also noted that the Case 3.2 is a special case of Case 3.3, which could as well have been defined for $(\boldsymbol{y}^v, y^t) \in \partial\hat{Y}^{v,t}$.

4. When $y^f = 0, (\boldsymbol{y}^v, y^t) \in \hat{Y}^{v,t}$, it is again convenient to distinguish the two subcases.

4.1 For $(\boldsymbol{y}^v, y^t) \in \partial\hat{Y}^{v,t}$, let:

$$\begin{aligned}\gamma(\boldsymbol{y} &= (y^f, \boldsymbol{y}^v, y^t) \mid y^f = 0, (\boldsymbol{y}^v, y^t) \in \partial\hat{Y}^{v,t}) \\ &= \{\boldsymbol{p} = (p^f, \boldsymbol{p}^v, p^t) \in \Delta \mid p^f \in [0,1], (\boldsymbol{p}^v, p^t) = (1-p^f)(\hat{\boldsymbol{p}}^v, \hat{p}^t) \\ &\qquad \text{for some } (\hat{\boldsymbol{p}}^v, \hat{p}^t) \in \gamma^2(\boldsymbol{y}^v, y^t)\}.\end{aligned}$$

By the argument spelled out from Case 3.1, Lemma 3.1 applies to this correspondence.

4.2 For $(\boldsymbol{y}^v, y^t) \notin \partial\hat{Y}^{v,t}$, let:

$$\begin{aligned}\gamma(\boldsymbol{y} &= (y^f, \boldsymbol{y}^v, y^t)|\ y^f = 0, (\boldsymbol{y}^v, y^t) \in \hat{Y}^{v,t}\backslash\partial\hat{Y}^{v,t}) \\ &= \{\boldsymbol{p} = (p^f, \boldsymbol{p}^v, p^t) \in \Delta \mid p^f = 1, (\boldsymbol{p}^v, p^t) = (0,0)\}.\end{aligned}$$

Lemma 3.1 now applies trivially.

Table 6.1. Pricing rules on the regions 1–4.2.

Case	y^f	(y^v, y^t)	p^f	(p^v, p^t)
1	$-c < y^f < 0$	$(y^v, y^t) \in \hat{Y}^{v,t}$	0	γ_2
2	$y^f < -c$	$y^f = f(y^v)$	0	γ_1
3.1	$y^f = -c$	$y^f = f(y^v)$	$\left[0, \frac{\hat{p}^v y^v + \hat{p}^t y^t}{c + \hat{p}^v y^v + \hat{p}^t y^t}\right]$	$(1 - p^f)\gamma_1$
3.2	$y^f = -c$	$y^t \in \hat{Y}^{v,t}$	0	γ_2
3.3	$y^f = -c$	$y^f < f(y^v), (y^v, y^t) \notin \hat{Y}^{v,t}$	0	$\frac{d_2}{d_1 + d_2}\gamma_1 + \frac{d_1}{d_1 + d_2}\gamma_2$
4.1	$y^f = 0$	$(y^v, y^t) \in \hat{Y}^{v,t}$	$[0, 1]$	$(1 - p^f)\gamma_2$
4.2	$y^f = 0$	$(y^v, y^t) \in \hat{Y}^{v,t} \setminus \hat{Y}^{v,t}$	1	0

The correspondence $\gamma(y)$ defined on the seven regions that are labelled 1, 2, 3.1, 3.2, 3.3, 4.1, and 4.2 satisfies the conditions of Lemma 3.1 (see Table 6.1). Accordingly, it satisfies (PR-2) and is nonempty, compact, and convex-valued everywhere. To verify upper hemi-continuity at the common boundaries of these seven regions, we note the following:

(1) The relevant connections (the common boundaries) concern 1 and 3.2, 1 and 4.1, 2 and 3.1, 3.1 and 3.3, 3.2 and 3.3, and 4.1 and 4.2;
(2) regarding 1 and 3.2, the pricing rules are identical;
(3) 1 connects to 4.1 for $p^f = 0$ in 4.1;
(4) 2 connects to 3.1 for $p^f = 0$ in 3.1;
(5) 3.1 connects to 3.3, hence also to 3.2 for $p^f = 0$ in 3.1 with $d_1 = 0$ in 3.3;
(6) 3.2 connects to 3.3 with $d_2 = 0$ in 3.3 and
(7) 4.1 connects to 4.2 for $p^f = 1$ in 4.1.

This completes the proof of Lemma 4.1 (see Fig. 6.7). □

The proof of Theorem 4.1 follows from Theorem 2.1 and Lemma 4.1. □

6.5. NOTES

Section 6.2: Theorem 2.1 is proposed to Dehez and Dreze (1988a).

Section 6.3: As explained in Introduction, an equilibrium theory in which firms behave monopolistically based on their perceived demand function was initiated by Negishi (1961). Theorem 3.1 was first published in this paper. His proof did not use the pricing rule approach. As a consequence, Negishi's proof is simpler than ours, but difficult to see the path to extend to the case of non-convex production sets.

Section 6.4: Theorem 4.1 was proposed by Dehez *et al.* (2003). They did not distinguish the input and output commodities, and formulated perceived demand map of the firm j as a vector-valued function:

$$p_j(\boldsymbol{y}_j \mid \hat{\boldsymbol{p}}, \hat{\zeta}) = H(\hat{\boldsymbol{p}}, \hat{\zeta})\boldsymbol{y}_j + K(\hat{\boldsymbol{p}}, \hat{\zeta}) \quad j = 1 \cdots n$$

where $H(\hat{\boldsymbol{p}}, \hat{\zeta})$ is a negative semi-definite $\ell \times \ell$-matrix, and $K(\hat{\boldsymbol{p}}, \hat{\zeta})$ is an ℓ-vector.

After the paper (1988a) was published, Dehez and Dreze pursued the study of the monopolistic competition of the Negishi type. Indeed in p. 214, they wrote:

> "Of course, increasing returns provide a natural invitation to model monopolistic competition along the lines suggested by Negishi (1961); like a number of others, we are currently investigating that possibility..."

They had reached essentially the equilibrium concept given in Definition 4.1 when the author of this book joined the research collaboration.

CONVEX SETS AND FUNCTIONS

Appendix **A**

We start with the notion of a set as a primitive concept, and denote it, such as $X = \{x, y, z, \dots\}$ or $X = \{x \mid \text{the properties which } x \text{ has}\}$. '$x \in X'$' means that x is an element of the set X, or x is contained in X. Suppose that if $x \in X$, then $x \in Y$. Then, we say that X is a subset of Y or the set Y includes the set X, and denote it by $X \subset Y$. The union $X \cup Y$ and the intersection $X \cap Y$ are defined by:

$$X \cup Y = \{x \mid x \in X \text{ or } x \in Y\} \quad \text{and} \quad X \cap Y = \{x \mid x \in X \text{ and } x \in Y\},$$

respectively. The empty set $\emptyset$ is the set which contains nothing, and we postulate $\emptyset \subset X$ for every set X. Let X be a subset of a set Y. The complement of X in Y is defined by $Y \backslash X = \{x \in Y \mid x \notin X\}$. Sometimes, we denote it by X^c when the set Y is understood.

Let X and Y be the sets. The set of ordered pairs $\{(x, y) \mid x \in X, \ y \in Y\}$ is called the product of X and Y, and denoted by $X \times Y$. The product of n sets, $X_1 \cdots X_n$ is defined inductively by $\prod_{i=1}^{n} X_i = (\prod_{i=1}^{n-1} X_i) \times X_n$. The subset $\mathcal{R}$ of the product $X \times Y$ is called a binary relation or simply a relation of X to Y. $(x, y) \in \mathcal{R}$ is often written as $x\mathcal{R}y$. If $X = Y$, then the binary relation $\mathcal{R} \subset X \times Y$ is called a relation on X.

A relation $\equiv$ on X which satisfies
(Reflexivity) $x \equiv x$ for every $x \in X$,
(Symmetricity) $x \equiv y$ implies $y \equiv x$ for all $x, y \in X$,
(Transitivity) $x \equiv y$ and $y \equiv z$ imply $x \equiv z$ for all $x, y, z \in X$,

is called an equivalence relation. A relation $\succsim$ on X which satisfies the reflexivity and transitivity is called a pre-order relation. The preference relation introduced in Section 2.2 is an example of the pre-order relations. A pre-order relation $\geq$ which satisfies:

$$(\text{Anti-symmetricity}) \ x \geq y \quad \text{and} \quad y \geq x \text{ imply } x = y$$

is called the order relation.

Let X and Y be sets and $f \subset X \times Y$ a relation of X to Y. If for every $x \in X$, there exists a unique $y \in Y$ such that $(x, y) \in f$, we call the relation f a map(ping) or a function from X to Y, and denote it as $f : X \to Y,\ x \mapsto y$. We often denote the function as simply $y = f(x)$. A function $f : X \to Y$ is said to be injective or one-to-one if $x \neq y$ implies $f(x) \neq f(y)$. It is called to be surjective or onto if for every $y \in Y$, there exists an $x \in X$ with $y = f(x)$. When a map f is one-to-one and onto, it is called to be bijective and the bijective map is also called a bijection. The set $f(X) = \{y \in Y |\ f(x) = y \text{ for some } x \in X\}$ is called the image of a map f, and for a set $Z \subset Y$, the set $f^{-1}(Z) = \{x \in X |\ f(x) \in Z\}$ is called the inverse image of f. A map f is injective if and only if $f^{-1}(\{y\})$ is a one point set or $\emptyset$ for every $y \in Y$, and it is surjective if and only if $f(X) = Y$.

The ℓ-times product of the real line $\mathbb{R}$, $\mathbb{R} \times \cdots \times \mathbb{R}$ is denoted as $\mathbb{R}^\ell$ and called the ℓ-dimensional Euclidean space. Hence, an element of $\mathbb{R}^\ell$ is an ℓ-tuple of real numbers $\boldsymbol{x} = (x^1 \cdots x^\ell), x^t \in \mathbb{R}, t = 1 \cdots \ell$. The elements of $\mathbb{R}^\ell$ are also called ℓ-vectors or simply vectors. The sum and the scalar product of vectors are defined respectively by:

$$(x^1 \cdots x^\ell) + (y^1 \cdots y^\ell) = (x^1 + y^1 \cdots x^\ell + y^\ell)$$

and

$$\alpha(x^1 \cdots x^\ell) = (\alpha x^1 \cdots \alpha x^\ell),$$

for all $\boldsymbol{x} = (x^1 \cdots x^\ell), \boldsymbol{y} = (y^1 \cdots y^\ell) \in \mathbb{R}^\ell$ and for all $\alpha \in \mathbb{R}$. By these operations, the space $\mathbb{R}^\ell$ becomes a vector space (see mathematical Appendix E for the definition of the vector spaces). The space $\mathbb{R}^\ell$ has the order relation $\geq$ on it, which is defined by:

$$(x^1 \cdots x^\ell) \geq (y^1 \cdots y^\ell) \quad \text{if and only if} \quad x^t \geq y^t \ \text{ for all } \ t = 1 \cdots \ell.$$

We will use the following related notations: $\boldsymbol{x} > \boldsymbol{y}$ if and only if $\boldsymbol{x} \geq \boldsymbol{y}$ and $\boldsymbol{x} \neq \boldsymbol{y}$. $\boldsymbol{x} \gg \boldsymbol{y}$ if and only if $x^t > y^t$ for all $t = 1 \cdots \ell$.

The positive orthant of $\mathbb{R}^\ell$ is defined by $\mathbb{R}^\ell_+ = \{\boldsymbol{x} \in \mathbb{R}^\ell \mid \boldsymbol{x} \geq 0\}$. Similarly we define $\mathbb{R}^\ell_- = -\mathbb{R}^\ell_+ = \{\boldsymbol{x} \in \mathbb{R}^\ell | \boldsymbol{x} \leq 0\}$.

A subset X of $\mathbb{R}^\ell$ is said to be convex if,

$$\text{for every } \ \boldsymbol{x}, \boldsymbol{y} \in X \ \text{ and } \quad \text{for every } \ 0 \leq \theta \leq 1, \theta \boldsymbol{x} + (1-\theta)\boldsymbol{y} \in X.$$

It is easy to see that if $X_1 \cdots X_m \subset X$ are convex, then $\prod_{i=1}^m X_i$, $\sum_{i=1}^m X_i$ and $\cap_{i=1}^m X_i$ are also convex.

Let S be a subset of $\mathbb{R}^\ell$. The convex hull of the set S is defined by the intersection of all convex sets which contain S, and it is denoted by coS. It is the smallest convex subset which contains S.

The next theorem is known as the separation hyperplane theorem.

Theorem A1 (Minkowski). *Let X and Y be non-empty convex subsets of $\mathbb{R}^\ell$. If $X \cap Y = \emptyset$, then there exists a vector $\boldsymbol{p} \in \mathbb{R}^\ell$ with $\boldsymbol{p} \neq \boldsymbol{0}$ such that $\boldsymbol{px} \leq \boldsymbol{py}$ for every $\boldsymbol{x} \in X$ and every $\boldsymbol{y} \in Y$.*

Knaster, Kuratowski, and Mazurkiewicz proved the following theorem known as K–K–M lemma.

Theorem A2 (K-K-M). *Let* $A = \{a_0, a_1 \cdots a_r\}$ *be a set of* $r+1$ *points in* $\mathbb{R}^\ell$. *Let* $\{S_0, S_1 \cdots S_r\}$ *be a set of* $r+1$ *closed subsets of* $\mathbb{R}^\ell$, *and* $I = \{0, 1 \cdots r\}$. *Suppose that for all* $J \subset I$, *the convex hull of* $\{a_i\}_{i \in J}$ *is contained in* $\cup_{i \in J} S_i$, *that is,* $co\{a_i\}_{i \in J} \subset \cup_{i \in J} S_i$. *Then* $\cap_{i \in I} S_i \neq \emptyset$.

Shapley (1973) generalized K-K-M lemma as follows. Let $A = \{1, 2, \ldots, N\}$ be a finite set and $\mathcal{A} = \{C \subset A \mid C \neq \emptyset\}$ be the set of non-empty subset of A. We say that a family $\mathcal{B}$ of subsets of A is balanced if there exist non-negative weights w_C for $C \in \mathcal{B}$ such that $\sum_{C \in \mathcal{B}_a} w_C = 1$ for all $a \in A$, where $\mathcal{B}_a = \{C \in \mathcal{B} \mid a \in C\}$ is the family of sets in $\mathcal{B}$ which contain $a \in A$.

Let $\boldsymbol{e}(a) = (0 \cdots 0, 1, 0 \cdots 0)$ be the a-th elementary unit vector, which has all 0 components but 1 at the a-th coordinate. For each $C \in \mathcal{B}$, the set $\Delta^C = co\{\boldsymbol{e}(a) \mid a \in C\}$ is called the unit simplex spanned by C.

The generalized K-K-M lemma now reads:

Theorem A3. *Let* $\{F_C\}_{C \in \mathcal{A}}$ *be a family of closed subsets of* Δ^A, *indexed by the members of* $\mathcal{A}$, *such that for every* $D \in \mathcal{A}$, $\Delta^D \subset \cup\{F_C \mid C \subset D\}$. *Then, there exists a balanced family* $\mathcal{B}$ *for which* $\cap\{F_C \mid C \in \mathcal{B}\} \neq \emptyset$.

A real-valued function $f: X \to \mathbb{R}$, where $X \subset \mathbb{R}^\ell$ is a convex set, is said to be concave if for every $\boldsymbol{x}, \boldsymbol{y} \in X$ and every $0 \leq \theta \leq 1$,

$$\theta f(\boldsymbol{x}) + (1-\theta) f(\boldsymbol{y}) \leq f(\theta \boldsymbol{x} + (1-\theta)\boldsymbol{y}).$$

f is called convex if the function $-f$ is concave.

Theorem A4. *A concave function is continuous on its relative interior of the domain.*

We conclude this section by the fundamental theorem of non-linear programming.

Theorem A5 (Kuhn–Tucker). *Let* $f, g_1 \cdots g_m$ *be real-valued concave functions defined on a convex set X in* $\mathbb{R}^\ell$. *Suppose that the Slater's condition holds, namely that:*

$$\textit{there exists an } \boldsymbol{x}_0 \in X \textit{ such that } g_j(\boldsymbol{x}_0) > 0, \ j = 1 \cdots m.$$

Let $\hat{\boldsymbol{x}}$ *be a point which achieves a maximum of* $f(\boldsymbol{x})$ *on X subject to* $g_j(\boldsymbol{x}) \geq 0, \ j = 1 \cdots m$. *Then, there exists a non-negative vector* $\hat{\boldsymbol{\lambda}} = (\hat{\lambda}_1 \cdots \hat{\lambda}_m) \in \mathbb{R}^m_+$ *such that:*

$$f(\boldsymbol{x}) + \hat{\boldsymbol{\lambda}} g(\boldsymbol{x}) \leq f(\hat{\boldsymbol{x}}) + \hat{\boldsymbol{\lambda}} g(\hat{\boldsymbol{x}}) \leq f(\hat{\boldsymbol{x}}) + \boldsymbol{\lambda} g(\hat{\boldsymbol{x}})$$

for every $\boldsymbol{x} \in X$ *and every non-negative* $\boldsymbol{\lambda} \in \mathbb{R}^m_+$.

In the above inequalities, we set $g(\boldsymbol{x}) = (g_1(\boldsymbol{x}) \cdots g_m(\boldsymbol{x}))$. In Theorem A5, the function $\mathcal{L}(\boldsymbol{x}, \boldsymbol{\lambda}) = f(\boldsymbol{x}) + \boldsymbol{\lambda} g(\boldsymbol{x})$ is called a Lagrangian and the point $(\hat{\boldsymbol{x}}, \hat{\boldsymbol{\lambda}}) \in X \times \mathbb{R}^m_+$ is called a saddle point of the Lagrangian $\mathcal{L}(\boldsymbol{x}, \boldsymbol{\lambda})$.

Appendix

B

ELEMENTS OF GENERAL TOPOLOGY

Let X be a set. A family $\mathcal{G}$ of subsets of X is called a topology on X if it satisfies:

(t-1) $X, \emptyset \in \mathcal{G}$;
(t-2) if $G_1, G_2 \in \mathcal{G}$, then $G_1 \cap G_2 \in \mathcal{G}$; and
(t-3) if $G_\lambda \in \mathcal{G}$, for all $\lambda \in \Lambda$, where Λ is an arbitrary set, then $\cup_{\lambda \in \Lambda} G_\lambda \in \mathcal{G}$.

In this case, we call $(X, \mathcal{G})$ a topological space and the family $\mathcal{G}$ a topology on X. We often call X itself a topological space when the topology $\mathcal{G}$ is understood. Each member of $\mathcal{G}$ is called an open set. A subset F of X is said to be closed if its complement $X \backslash F$ is open. The smallest closed set containing a set $S \subset X$ is called the closure of S and denoted by closure S or simply clS. This means that if F is closed and $S \subset F$, then $clS \subset F$. The closure of S is characterized by the intersection of all closed sets containing S. It is clear that for every subset S of X, $S \subset clS$ and S is closed if and only if $S = clS$. A point x of the set clS is called a point of closure of S. It is characterized such that every open set U containing x has non-empty intersection with X, that is, if U is open and $x \in U$, then $U \cap X \neq \emptyset$.

An open set U containing $x \in X$ is also called a(n open) neighborhood of x. A point x of a subset S of X is said to be an interior point of S if the set S contains a neighborhood of x. The set of all interior points of the set S is called the interior of S and denoted by interior S or $intS$. For every subset S of X, $clS \backslash intS$ is called the boundary of S and denoted by $bdryS$.

A directed system is a set Λ together with a relation[1] $\succ \subset \Lambda \times \Lambda$ satisfying the following conditions,

(i) if $(\lambda, \mu) \in \succ$ and $(\mu, \nu) \in \succ$, then $(\lambda, \nu) \in \succ$ and
(ii) if $\lambda, \mu \in \Lambda$, then there exists a $\nu \in \Lambda$ such that $(\nu, \lambda) \in \succ$ and $(\nu, \mu) \in \succ$.

Note that the condition (i) is nothing but the transitivity. As usual, we denote $(\lambda, \mu) \in \succ$ by $\lambda \succ \mu$. A net on a topological space X is a mapping of a directed

[1]This relation should not be confused with the preference relation.

system $(\Lambda, \succ)$ to X. We usually denote by x_λ the value of the net at $\lambda \in \Lambda$ and by $\{x_\lambda\}$ the net itself. A map from $\mathbb{N}$ to X is called a sequence. The sequence $\{x_n\}$ is an example of the net with a directed set $(\mathbb{N}, \geq)$ (in fact, the net is a generalization of the sequence!).

A point $x \in X$ is said to be the limit of a net $\{x_\lambda\}$ or the net $\{x_\lambda\}$ converges to x if for every open neighborhood U of x, there is a $\lambda_0 \in \Lambda$ such that $x_\lambda \in U$ for all $\lambda \succ \lambda_0$. In this case, we denote $x_\lambda \to x$ or $\lim_\lambda x_\lambda = x$. A point $x \in X$ is called a cluster point of the net $\{x_\lambda\}$ if for every neighborhood U of x and for every $\lambda \in \Lambda$, there exists a $\mu \succ \lambda$ such that $x_\mu \in U$.

Points of closure of a set $S \subset X$ is characterized by the net.

Proposition B1. *A point $x \in X$ is a point of closure of a set S if and only if it is the limit of a net $\{x_\lambda\}$ in S.*

A subset D of X is said to be dense in X if $X \subset clD$. A topological space X is separable if and only if there exists a countable dense subset of X.

A subset S of a topological space $(X, \mathcal{G})$ is called a subspace if S is endowed with the topology $\mathcal{G}_S$ whose open sets are the intersections with S of the open sets in X.

Proposition B2. *Every open (closed) set in the subspace S of $(X, \mathcal{G})$ is open (closed) in X if and only if S is open (closed) in X.*

A mapping f of a topological space X to a topological space Y is said to be continuous if for every open set G of Y, its inverse image of f, $f^{-1}(G)$ is open in X. In terms of the net, we can say that:

Proposition B3. *Let X and Y be topological spaces. A mapping $f: X \to Y$ is continuous if and only if for every $x \in X$ and for every net $\{x_\lambda\}$ on X converging to x, the net $\{f(x_\lambda)\}$ on Y converges to $f(x)$.*

We have derived the concepts of convergence of a net, the limit and the cluster point, and so on from the concept of the topology. On the contrary, it is known that once the concepts of a net and its convergence are given, we can construct the unique topology such that the convergence of a net with respect to this topology coincides with the originally given convergence. In this sense, the ideas of the topology and the net are equivalent. For more details, see Notes (Appendix I).

A continuous and bijective mapping from a topological space $(X, \mathcal{G})$ to a topological space $(Y, \mathcal{G}')$ whose inverse f^{-1} is also continuous is called a homeomorphism. In this case, the set X and the set Y are said to be homeomorphic. Homeomorphic spaces are considered to be identical from the topological point of view.

A subset K of a topological space is said to be compact if for every family of open sets $\{G_\alpha\}$, $\alpha \in A$, where A is an arbitrary set, such that $K \subset \cup_\alpha G_\alpha$, there exists a finite subfamily $\{G_{\alpha_i}\}$ $i = 1 \cdots n$ such that $K \subset \cup_{i=1}^n G_{\alpha_i}$. This definition of compactness is equivalent to the definition that K is compact if and only if for every family $\{F_\alpha\}$ of closed subsets of K such that every finite subfamily $\{F_{\alpha_i}\}$, $i = 1 \cdots n$, has non-empty intersection $\cap_{i=1}^n F_{\alpha_i} \neq \emptyset$, its intersection is nonempty, $\cap_\alpha F_\alpha \neq \emptyset$.

Proposition B4. *Every closed subset F of a compact space is compact.*

The compact set is also characterized by the net.

Proposition B5. *Let X be a topological space. Then, X is compact if and only if every net in X has a cluster point.*

Since the sequence is a less general concept than the net, we get a weaker result related to the compactness.

Proposition B6. *Let X be a topological space. Then, every sequence in X has a cluster point in x.*

A topological space X is called Hausdorff if and only if for every $x, y \in X$ with $x \neq y$, there are disjoint open sets U and V such that $x \in U$ and $y \in V$.

Proposition B7. *Let f be a continuous mapping of a compact space* $(X, \mathcal{G})$ *to a topological space* $(Y, \mathcal{G}')$. *Then f(X) is compact. Furthermore, if Y is Hausdorff and f is one-to-one and onto (or bijective), then f is an homeomorphism.*

Let X and Y be compact spaces and $f: X \to Y$ a continuous bijection. Then, every closed subset F of X is compact by Proposition B5. Therefore, $f(F)$ is compact by Proposition B6, hence it is closed. This shows that $f^{-1} : Y \to X$ is also continuous. Thus, we have obtained:

Proposition B8. *A continuous and bijective map f of a compact space* $(X, \mathcal{G})$ *to a compact space* $(Y, \mathcal{G}')$ *is an homeomorphism.*

Let $\{X_\alpha, \mathcal{G}_\alpha\}, \alpha \in A$ be a family of topological spaces, where A is an arbitrary set, and let $X = \prod_{\alpha \in A} X_\alpha$ be the product. Consider the subsets B of X of the form $B = \prod_{\alpha \in A} G_\alpha$, where $G_\alpha \in \mathcal{G}_\alpha$ and $G_\alpha = X_\alpha$ except for a finite number of α. Let $\mathcal{G}$ be the family of subsets of X which are unions of the sets B. Then, it is easy to see that $\mathcal{G}$ satisfies (t-1), (t-2), and (t-3) on top of this section. The topological space $(X, \mathcal{G})$ is called the product space and $\mathcal{G}$ the product topology.

Theorem B1 (Tychonoff). *The product space* $X = \prod_{\alpha \in A} X_\alpha$ *is compact if each* X_α *is compact.*

A topological space X is said to be locally compact if for every point $x \in X$ has a compact neighborhood.

Theorem B2 (Alexandroff). *Let $(X, \mathcal{G})$ be a locally compact space. Then, there exists a compact space $(X', \mathcal{G}')$ such that X is homeomorphic to X' whose complement consists of exactly one point.*

The space X' in the above theorem is unique up to an homeomorphism. The space X' is called the Alexandroff compactification or the one-point compactification and denoted by $X' = X \cup \{\infty\}$.

METRIC SPACES

Appendix

C

Let X be a set. A non-negative real-valued function d on $X \times X$,

$$d\colon X \times X \to \mathbb{R}_+, \quad (x,y) \mapsto d(x,y)$$

is called a metric on X if it satisfies the following properties. For all x, y and $z \in X$,

(m-1) $d(x,y) \geq 0$ and $d(x,y) = 0$ if and only if $x = y$;
(m-2) $d(x,y) = d(y,x)$; and
(m-3) $d(x,y) \leq d(x,z) + d(z,y)$.

The pair consisting of the set X and a metric d on X is called a metric space. We often call X itself the metric space if the metric is understood. Let $x \in X$ and $\epsilon > 0$. The set $B(x,\epsilon) = \{z \in X \mid d(x,z) < \epsilon\}$ is called an open ball or simply ball with center x and radius ϵ. The family $\mathcal{G}$ of subsets G of X such that for all $x \in G$, $B(x,\epsilon) \subset G$ for some $\epsilon > 0$ satisfies (t-1),(t-2), and (t-3) of Section B, hence the pair $(X, \mathcal{G})$ is a topological space.

Examples. Let $X = \mathbb{R}^\ell$, the ℓ-dimensional Euclidean space. We can consider three metrics on $\mathbb{R}^\ell$.

$$d_1(\boldsymbol{x},\boldsymbol{y}) = |x^1 - y^1| + \cdots + |x^\ell - y^\ell|,$$

$$d_2(\boldsymbol{x},\boldsymbol{y}) = \sqrt{(x^1 - y^1)^2 + \cdots + (x^\ell - y^\ell)^2},$$

and

$$d_\infty(\boldsymbol{x},\boldsymbol{y}) = \max\{|x^t - y^t| | 1 \leq t \leq \ell\}.$$

They look different from each other, but all these three metrics define the same topology on $\mathbb{R}^\ell$, because we have:

$$d_\infty(\boldsymbol{x},\boldsymbol{y}) \leq d_2(\boldsymbol{x},\boldsymbol{y}) \leq \sqrt{\ell} d_\infty(\boldsymbol{x},\boldsymbol{y}),$$

and

$$d_\infty(\boldsymbol{x},\boldsymbol{y}) \leq d_1(\boldsymbol{x},\boldsymbol{y}) \leq \ell d_\infty(\boldsymbol{x},\boldsymbol{y}),$$

so that any ball with respect to d_2 is contained in a ball with respect to d_∞ and so on. We denote $d_2(\boldsymbol{x},\boldsymbol{0}) = \|\boldsymbol{x}\|$, and call it as the length or the norm of $\boldsymbol{x}$.

On the metric spaces, much of the topological discussions go smoothly in terms of sequences. For example, a sequence $\{x_n\}$ in a metric space (X,d) converges to $x \in X$ if and only if $d(x_n, x) \to 0$ as $n \to \infty$. Moreover by Theorem B2, a mapping f of a metric space (X,d) to a metric space (Y,d') is continuous (at x) if $f(x_n) \to f(x)$ whenever $x_n \to x$, and we can easily prove:

Proposition C1. *A subset F of a metric space (X,d) is closed if and only if for every sequence $\{x_n\}$ in F with $x_n \to x$, it follows that $x \in F$.*

Let S be a subset of a metric space (X,d) and d_S be the restriction of d to S. Clearly, d_S is a metric on the set S and the topology on S derived from the metric d_S coincides with the sub-space topology. Hence, we call (S,d_S) a subspace of (X,d). Recall that a metric space X is separable if there exists a countable and dense subset D such that $X \subset clD$.

Proposition C2. *Every subspace of a separable metric space is separable.*

A sequence $\{x_n\}$ in a metric space (X,d) is called a Cauchy sequence if for every $\epsilon > 0$, there exists an $n \in \mathbb{N}$ such that $d(x_p, x_q) \leq \epsilon$ for all $p, q \geq n$. A metric space (X,d) is said to be complete if every Cauchy sequence converges to a point in X.

Proposition C3. *Every compact metric space is complete and separable.*

Let $\{x_n\}$ be a sequence in a topological space X. Recall that a point $x \in X$ is called a cluster point of the sequence $\{x_n\}$ if for every neighborhood U of x and for every n, there exists an $N \geq n$ such that $x_N \in U$. On metric spaces, Proposition B5 can be stated in terms of sequences, rather than nets.

Proposition C4. *Let (X,d) be a metric space. Then X is compact if and only if every sequence $\{x_n\}$ in X has a cluster point.*

If $\{x_n\}$ is an infinite sequence, we say that $\{x_{n_i}\}$ is a subsequence when n_i is a monotone mapping from $\mathbb{N}$ to $\mathbb{N}$, that is, the mapping $i \mapsto n_i$ satisfies that $i < j$ implies $n_i < n_j$. Let x be a cluster point of a sequence $\{x_n\}$ in a metric space X. Then, we have a subsequence $\{x_{n_i}\}$ such that $x_{n_i} \to x$. For this, consider the ball $B(x,(1+i)^{-1}),\ i \in \mathbb{N}$. Since x is a cluster point, there exists $x_{n_i} \in B(x,(1+i)^{-1})$ for every i. Then $\{x_{n_i}\}$ is the desired subsequence, because the radius of the balls converges to zero. The converse is easily verified; if every sequence in X has a converging subsequence, X has a cluster point. Hence, from Proposition C4, we have:

Proposition C5. *Let (X,d) be a metric space. Then X is compact if and only if for every sequence in X, there exists a subsequence converging to a point of X.*

Let $S \subset \mathbb{R}^\ell$ be a subset of the ℓ-dimensional Euclidean space $\mathbb{R}^\ell$. The set S is said to be bounded if there exists an $M > 0$ such that $S \subset B(\mathbf{0}, M)$, namely that S is contained in a ball which is large enough. We can derive easily from Proposition C5 that:

Proposition C6. *Let $K \subset \mathbb{R}^\ell$ be a subset of the ℓ-dimensional Euclidean space. Then K is compact if and only if it is closed and bounded.*

Let S, T be metric spaces and consider the real-valued function on $S \times T$,

$$f: S \times T \to \mathbb{R}, \quad (x, y) \mapsto f(x, y).$$

When f is continuous with respect to the product topology of $S \times T$, we sometimes say that f is bi-continuous or joint continuous, emphasizing the difference between the continuity for each variable separately. In terms of sequences, f is joint continuous if and only if $(x_n, y_n) \to (x, y)$ implies that $f(x_n, y_n) \to f(x, y)$. This is different from the separate continuity for each variable which says that $y \in T$, $x_n \to x \in S$ implies that $f(x_n, y) \to f(x, y)$ and for each $x \in S$, $(x, y_n) \to (x, y)$ implies that $f(x, y_n) \to f(x, y)$. The joint continuity implies the separate continuity for each variable, but not vice versa.

The inner product on $\mathbb{R}^\ell$ is an example of a map with two variables: $(\boldsymbol{p}, \boldsymbol{x}) \mapsto \boldsymbol{px}$, and it is joint continuous. Indeed, let $(\boldsymbol{p}_n, \boldsymbol{x}) \to (\boldsymbol{p}, \boldsymbol{x})$. Then, we have:

$$\begin{aligned} |\boldsymbol{p}_n\boldsymbol{x}_n - \boldsymbol{px}| &= |\boldsymbol{p}_n\boldsymbol{x}_n - \boldsymbol{p}_n\boldsymbol{x} + \boldsymbol{p}_n\boldsymbol{x} - \boldsymbol{px}| \\ &\le |\boldsymbol{p}_n\boldsymbol{x}_n - \boldsymbol{p}_n\boldsymbol{x}| + |\boldsymbol{p}_n\boldsymbol{x} - \boldsymbol{px}| \\ &\le \|\boldsymbol{p}_n\|\,\|\boldsymbol{x}_n - \boldsymbol{x}\| + \|\boldsymbol{p}_n - \boldsymbol{p}\|\,\|\boldsymbol{x}\| \to 0. \end{aligned}$$

In the above inequalities, we used the Cauchy–Schwartz inequality: $|\boldsymbol{xy}| \le \|\boldsymbol{x}\|\,\|\boldsymbol{y}\|$. In Appendix H, we will see that the problem of joint continuity of the inner product is more subtle on infinite-dimensional spaces.

Appendix

D

CONTINUITY OF CORRESPONDENCES

Let X and Y be topological spaces. A correspondence ϕ is a relation $\phi \subset X \times Y$ such that for every $x \in X$, there exists $y \in Y$ such that $(x, y) \in \phi$. For every $x \in X$, we denote the set $\{y \in Y \mid (x, y) \in \phi\}$ by $\phi(x)$ and call the value of ϕ at $x \in X$. It is a non-empty subset of Y. If the value $\phi(x)$ is a singleton, $\phi(x) = \{y\}$, then the correspondence is nothing but a mapping. Hence, we will often denote the correspondence $\phi \subset X \times Y$ as:

$$\phi: X \to Y, \quad x \mapsto \phi(x).$$

Definition D1. A correspondence $\phi : X \to Y$ between topological spaces X and Y is said to be upper hemi-continuous (u.h.c) at $x \in X$ if for every open subset V of Y which contains $\phi(x)$, there exists a neighborhood U of x such that $\phi(z) \subset V$ for every $z \in U$. The correspondence ϕ is called upper hemi-continuous if it is u.h.c. at every $x \in X$.

Let $\phi : X \to Y$ and $\psi : Y \to Z$ be correspondences. Then we can define the composition of ϕ and ψ, $\psi \circ \phi : X \to Z$ by $\psi \circ \phi(x) = \{z \in Z | \ z \in \psi(y) \text{ for some } y \in \phi(x)\}$. We can immediately deduce from the definition D1 that:

Proposition D1. *Let the correspondences $\phi: X \to Y$ and $\psi: Y \to Z$ be u.h.c. Then, the composition $\psi \circ \phi: X \to Z$ is also u.h.c.*

In the following Propositions and Theorems of Appendix D, the sets X, Y, and other sets are assumed to be metric spaces unless otherwise specified. The correspondence $\phi: X \to Y$ is said to be closed if its graph $\{(x, y) \in X \times Y \mid y \in \phi(x)\}$ is a closed subset of $X \times Y$.

Proposition D2. *Let ϕ and ψ be the correspondences of X to Y such that $\phi(x) \cap \psi(x) \neq \emptyset$ for all $x \in X$. Suppose that the following conditions: either* (i) *or* (ii) *holds:*

(i) *ϕ and ψ are closed valued and u.h.c at $\bar{x} \in X$,*
(ii) *ϕ is closed, ψ is u.h.c at $\bar{x}$ and $\psi(\bar{x})$ is compact.*

Then, the correspondence $x \mapsto \phi(x) \cap \psi(x)$ is u.h.c at $\bar{x}$.

As a corollary, it follows that the closed correspondence $\phi: X \to Y$ is u.h.c if Y is compact. Moreover, in this case, ϕ is compact-valued. For let $y_n \in \phi(x)$ for all n and $y_n \to y$. Since $(x, y_n) \to (x, y)$ and ϕ is closed, we have $y \in \phi(x)$. Hence, $\phi(x)$ is a closed subset of Y. Since every closed subset of a compact set is compact, $\phi(x)$ is also compact.

Proposition D3. *Let the correspondence ϕ of X to Y be compact-valued and u.h.c. Then, the image $\phi(K)$ of a compact set is compact.*

The next proposition characterizes the u.h.c correspondences by sequences.

Proposition D4. *A compact-valued correspondence ϕ of X to Y is u.h.c at $x \in X$ if and only if for every sequence $\{x_n\}$ converging to x and every sequence $\{y_n\}$ with $y_n \in \phi(x_n)$, there exists a converging subsequence y_{n_q} with the limit y such that $y \in \phi(x)$.*

Proposition D5. *Let $\phi_1 \cdots \phi_m$ be compact-valued and u.h.c correspondences of X to $\mathbb{R}^\ell$. Then, we have that the convex hull, $x \mapsto co\phi(x)$, and the sum, $x \mapsto \sum_{i=1}^{m} \phi_i(x)$ are also compact-valued and u.h.c.*

Proposition D6. *Let $\phi_i : X \to Y_i$ be compact valued and u.h.c correspondences of X to Y_i, $i = 1 \cdots m$. Then the product, $x \mapsto \prod_{i=1}^{m} \phi_i(x)$ is also compact-valued and u.h.c.*

The next classical theorem plays no doubt a fundamental role in the mathematical economics.

Theorem D1 (Kakutani). *Let X be a compact and convex subset of $\mathbb{R}^\ell$. If the correspondence ϕ of X to X is convex-valued and u.h.c, there exists a point $x \in X$ such that $x \in \phi(x)$.*

The point x such that $x \in \phi(x)$ is called a fixed point of ϕ.

Definition D2. A correspondence $\phi: X \to Y$ between topological spaces X and Y is said to be lower hemi-continuous (l.h.c) at $x \in X$ if for every open set G of Y such that $\phi(x) \cap G \neq \emptyset$, there exists a neighborhood U of x such that $\phi(z) \cap G \neq \emptyset$ for every $z \in U$. The correspondence ϕ is called lower hemi-continuous if it is l.h.c. at every $x \in X$.

If the correspondence ϕ is u.h.c and l.h.c (at x), it is said to be continuous (at x). The next theorem is used to guarantee the upper hemi-continuity of the individual demand correspondences or the supply correspondences of competitive firms.

Theorem D2 (Berge). *Let β be a compact-valued and continuous correspondence from X to Y, and let $f: X \times Y \to \mathbb{R}$ be a continuous function. Then, we have:*

(i) *the function $m: X \to \mathbb{R}, x \mapsto \max\{f(x,y) \mid y \in \beta(x)\}$ is continuous;*
(ii) *the correspondence $x \mapsto \{y \in \beta(x) \mid f(x,y) = m(x)\}$ is compact-valued and upper hemi-continuous.*

Let X be a metric space and $\{S_n\}$ a sequence of subsets of X.

Definition D3. A topological limes inferior $Li(S_n)$ is a subset of X such that $x \in Li(S_n)$ if and only if for every neighborhood U of x, there is an integer N such that $U \cap S_n \neq \emptyset$ for all $n \geq N$.

A topological limes superior $Ls(S_n)$ is a subset of X such that $x \in Ls(S_n)$ if and only if for every neighborhood U of x, there are infinitely many n such that $U \cap S_n \neq \emptyset$.

The next proposition is immediate from the definitions.

Proposition D7. *For every sequence $(F_n)_{n \in \mathbb{N}}$ of subsets of X, the following properties hold.*

(i) *$L_i(F_n)$ and $L_s(F_n)$ are closed (possibly empty) and $Li(S_n) \subset Ls(S_n)$.*
(ii) *$x \in L_i(F_n)$ if and only if there exists an integer $N \in \mathbb{N}$ and a sequence $(x_n)_{n \in \mathbb{N}}$ with $x_n \in F_n$ for all $n \geq N$ and $x_n \to x$.*
(iii) *$x \in L_s(F_n)$ if and only if there exists a subsequence (F_{n_q}) from which one can choose an element $x_{n_q} \in F_{n_q}$ for every n_q such that $x_{n_q} \to x$.*

The next proposition is often useful to examine the lower hemi-continuous of a correspondence.

Proposition D8. *If a correspondence $\phi: X \to Y$ is l.h.c at x, then it follows that $\phi(x) \subset Li(\phi(x))$ for every sequence $\{x_n\}$ converging to x. Conversely, if $\phi(x) \subset Ls(\phi(x))$ for every sequence $\{x_n\}$ converging to x, then the correspondence ϕ is l.h.c at x.*

From Proposition D7, we can easily obtain:

Proposition D9. *Let $\phi_1 \cdots \phi_m$ be l.h.c correspondences of X to $\mathbb{R}^\ell$. Then, we have the convex hull, $x \mapsto co\phi(x)$, and the sum, $x \mapsto \sum_{i=1}^m \phi_i(x)$ are also l.h.c.*

Proposition D10. *Let $\phi_i: X \to Y_i$ be l.h.c correspondences of X to Y_i, $i = 1 \cdots m$. Then, the product, $x \mapsto \prod_{i=1}^m \phi_i(x)$ is l.h.c.*

Let $\Delta = \{\boldsymbol{p} \in \mathbb{R}^\ell \mid p^t \geq 0, \sum_{t=1}^\ell p^t = 1\}$ be the unit simplex. Clearly, Δ is compact and convex subset of $\mathbb{R}^\ell$. The next theorem is a fundamental lemma which is used to prove the existence of equilibrium.

Theorem D3 (Gale–Nikaido). *Let X be a compact and convex subset of $\mathbb{R}^\ell$, and P a closed and convex subset of Δ. If $Z: P \to X$ is an upper hemi-continuous correspondence such that for every $\boldsymbol{p} \in P$, $Z(\boldsymbol{p})$ is a non-empty convex set and satisfies $\boldsymbol{p}Z(\boldsymbol{p}) \leq 0$ for every $\boldsymbol{p} \in P$. Then, there exists a $\boldsymbol{p}^* \in P$ such that $\boldsymbol{q}Z(\boldsymbol{p}^*) \leq 0$ for every $\boldsymbol{q} \in P$. Furthermore, if $P = \Delta$, then there exists a $\boldsymbol{p}^* \in \Delta$ such that $Z(\boldsymbol{p}^*) \leq \boldsymbol{0}$.*

Proof. We define the map $M: X \to P$ by:

$$M(\boldsymbol{z}) = \{\boldsymbol{p} \in P \mid \boldsymbol{pz} \geq \boldsymbol{qz} \text{ for every } \boldsymbol{q} \in P\},$$

and then we can construct the correspondence,

$$M \times Z : P \times X \to P \times X, \quad (\boldsymbol{p},\boldsymbol{z}) \mapsto M(\boldsymbol{z}) \times Z(\boldsymbol{p}).$$

First, we shall show that the correspondence M is upper hemi-continuous. As a closed subset of a compact set Δ, P is compact. Hence, it is enough to show that M is closed by Proposition D2. Take a sequence $\{(\boldsymbol{p}_n,\boldsymbol{z}_n)\}$ with $\boldsymbol{p}_n \in M(\boldsymbol{z}_n)$ for all n and $(\boldsymbol{p}_n,\boldsymbol{z}_n) \to (\boldsymbol{p},\boldsymbol{z})$. We want to show that $\boldsymbol{p} \in M(\boldsymbol{z})$. Suppose not. Then $\boldsymbol{pz} < \boldsymbol{qz}$ for some $\boldsymbol{q} \in P$. Since $(\boldsymbol{p}_n,\boldsymbol{z}_n) \to (\boldsymbol{p},\boldsymbol{z})$, we have $\boldsymbol{p}_n\boldsymbol{z}_n \to \boldsymbol{pz}$, hence $\boldsymbol{p}_n\boldsymbol{z}_n < \boldsymbol{q}\boldsymbol{z}_n$ for n large enough. This contradicts $\boldsymbol{p}_n \in M(\boldsymbol{z}_n)$. This proves that M is upper hemi-continuous. Hence, so is the product map $M \times Z$ by Proposition D6. Therefore, $M \times Z$ is upper hemi-continuous correspondence from a compact and convex set $X \times \Delta$ to itself. By the Kakutani's fixed-point theorem (Theorem D1), there exists a fixed-point $(\boldsymbol{p}^*,\boldsymbol{z}^*) \in M(\boldsymbol{z}^*) \times Z(\boldsymbol{p}^*)$, or $\boldsymbol{p}^* \in M(\boldsymbol{z}^*)$ and $\boldsymbol{z}^* \in Z(\boldsymbol{p}^*)$. By the definition of the correspondence M, we have:

$$\boldsymbol{p}^*\boldsymbol{z}^* \geq \boldsymbol{q}\boldsymbol{z}^* \quad \text{for every } \boldsymbol{q} \in P.$$

This proves the first part of the theorem. Now let $P = \Delta$. We have $\boldsymbol{p}^*\boldsymbol{z}^* \leq 0$, since $\boldsymbol{z}^* \in Z(\boldsymbol{p}^*)$. Taking $\boldsymbol{q} = (1,0\ldots0),(0,1\ldots0),\ldots,(0\cdots0,1)$ in the above inequality, one obtains $\boldsymbol{z}^t \leq 0$ for $t = 1\cdots\ell$. This proves the latter part of this theorem. □

Theorem D3 is strengthened such that the exact equality between the supply and the demand holds when the Walras law is met by the exact equality and the boundary condition (see below) is additionally met.

Let $int\Delta = \{\boldsymbol{p} \in \mathbb{R}^\ell \mid \sum_{t=1}^{\ell} p^t = 1,\ p^t > 0, t = 1\cdots\ell\}$ be the interior of the simplex.

Theorem D4. *Let Z be a correspondence of $int\Delta$ into $\mathbb{R}^\ell$ which satisfies the following properties:*

(i) *for every strictly positive price vector $\boldsymbol{p} \gg \boldsymbol{0}$, $\boldsymbol{p}Z(\boldsymbol{p}) = 0$;*
(ii) *the correspondence Z is compact and convex-valued, bounded from below and u.h.c; and*
(iii) *if a sequence $\{\boldsymbol{p}_n\}$ in $int\Delta$ converges to $\boldsymbol{p} \in bdry\Delta = \{\boldsymbol{p} = (p^t) \in \Delta \mid p^t = 0 \text{ for some } t\}$, then $inf\{\sum_{t=1}^{\ell} z^t \mid \boldsymbol{z} = (z^t) \in Z(\boldsymbol{p})\} > 0$ for n large enough.*

Then, there exists a vector $\boldsymbol{p}^* \gg \boldsymbol{0}$ such that $\boldsymbol{0} \in Z(\boldsymbol{p}^*)$.

Proof. For $n = 1, 2, \ldots$, we define:

$$\Delta_n = \left\{ \boldsymbol{p} = (p^t) \in \mathbb{R}^\ell \;\middle|\; \sum_{t=1}^{\ell} p^t = 1,\ p^t \geq 1/n,\ t = 1 \cdots \ell \right\}.$$

By Proposition D3, the set $Z(\Delta_n)$ is compact. Hence, we can apply Theorem D3 and there exist vectors $\boldsymbol{p}_n \in \Delta_n$ and $\mathbf{z}_n \in \mathbb{R}^\ell$ such that: (1) $\mathbf{z}_n \in Z(\boldsymbol{p}_n)$ and (2) $\boldsymbol{q}\mathbf{z}_n \leq 0$, for every $\boldsymbol{q} \in \Delta_n$. It remains to show that $\mathbf{z}_n = \mathbf{0}$, for some n. Since Δ is compact, we can assume that $\boldsymbol{p}_n \to \boldsymbol{p} \in \Delta$. We now claim that $\boldsymbol{p} \gg \mathbf{0}$. Otherwise, it would follow from property (iii) that $\sum_{t=1}^{\ell} z_n^t > 0$, since $\mathbf{z}_n \in Z(\boldsymbol{p}_n)$. Setting $\boldsymbol{q} = (1/\ell \cdots 1/\ell)$ in the property (2), one obtains $\sum_{t=1}^{\ell} z_n^t \leq 0$, a contradiction. Finally, $\boldsymbol{p} \gg \mathbf{0}$ implies that $\mathbf{z}_n = \mathbf{0}$ for n large enough. Indeed, let N be such that Δ_N contains $\boldsymbol{p}$ in its interior. By property (i), we have $\boldsymbol{p}_n\mathbf{z}_n = 0$. Since $\boldsymbol{p}_n \in int\Delta_N$ for n large enough, it follows from property (2) that $\mathbf{z}_n = \mathbf{0}$. □

Note that in Theorem D4, we restrict the domain of the correspondence Z to the interior of the price simplex, and the range of Z is not restricted to a compact set.

DIFFERENTIAL CALCULUS AND MANIFOLDS

Appendix

E

Definition E1. A set L is called a vector space over $\mathbb{R}$ if we have functions $+: L \times L \to L$ and $\cdot: \mathbb{R} \times L \to L$ which satisfy the following conditions.

(i) $\boldsymbol{x}+\boldsymbol{y}=\boldsymbol{y}+\boldsymbol{x}$ for all $\boldsymbol{x},\boldsymbol{y} \in L$;
(ii) $(\boldsymbol{x}+\boldsymbol{y})+\boldsymbol{z}=\boldsymbol{x}+(\boldsymbol{y}+\boldsymbol{z})$ for all $\boldsymbol{x},\boldsymbol{y},\boldsymbol{z} \in L$;
(iii) there exists a vector $\mathbf{0}$ in L such that $\boldsymbol{x}+\mathbf{0}=\boldsymbol{x}$, for all $\boldsymbol{x} \in L$;
(iv) $\lambda \cdot (\boldsymbol{x}+\boldsymbol{y}) = \lambda \cdot \boldsymbol{x} + \lambda \cdot \boldsymbol{y}$ for all $\lambda \in \mathbb{R}$ and all $\boldsymbol{x},\boldsymbol{y} \in L$;
(v) $(\lambda+\mu)\cdot \boldsymbol{x} = \lambda \cdot \boldsymbol{x} + \mu \cdot \boldsymbol{x}$ for all $\lambda, \mu \in \mathbb{R}$ and all $\boldsymbol{x} \in L$;
(vi) $\lambda(\mu \cdot \boldsymbol{x}) = (\lambda\mu)\cdot \boldsymbol{x}$ for all $\lambda, \mu \in \mathbb{R}$ and all $\boldsymbol{x} \in L$; and
(vii) $0 \cdot \boldsymbol{x} = \mathbf{0}$, and $1 \cdot \boldsymbol{x} = \boldsymbol{x}$, for all $\boldsymbol{x} \in L$.

The function $+$ is called the addition and the function $\cdot$ is called the multiplication by scalars.[1] Note that the element $\mathbf{0}$ in (iii) is unique, for if $\mathbf{0}'$ also satisfies (iii), then $\mathbf{0} = \mathbf{0} + \mathbf{0}' = \mathbf{0}' + \mathbf{0} = \mathbf{0}'$. The element $(-1)\boldsymbol{x}$ is called the negative of $\boldsymbol{x}$ and written $-\boldsymbol{x}$. One has $\boldsymbol{x} + (-\boldsymbol{x}) = 1\boldsymbol{x} + (-1)\boldsymbol{x} = (1-1)\boldsymbol{x} = 0\boldsymbol{x} = \mathbf{0}$.

Definition E2. A non-negative real-valued function $\|\cdot\|$ defined on a vector space L is called a norm if it satisfies:

(n-1) $\|\boldsymbol{x}\| = 0$ if and only if $\boldsymbol{x} = \mathbf{0}$,
(n-2) $\|\boldsymbol{x}+\boldsymbol{y}\| \le \|\boldsymbol{x}\| + \|\boldsymbol{y}\|$ for all $\boldsymbol{x},\boldsymbol{y} \in L$; and
(n-3) $\|\lambda \boldsymbol{x}\| = |\lambda| \|\boldsymbol{x}\|$ for all $\lambda \in \mathbb{R}$ and for all $\boldsymbol{x},\boldsymbol{y} \in L$.

A normed vector space becomes a metric space, hence a topological space with the norm topology if we define a metric d by $d(\boldsymbol{x},\boldsymbol{y}) = \|\boldsymbol{x}-\boldsymbol{y}\|$. When a normed vector space is complete in this metric, it is called a Banach space.

[1] In the following, we will often omit the dot for the scalar multiplication.

Example 1. Let $L = \mathbb{R}^\ell$, the ℓ-dimensional Euclidean space. We can consider three norms on $\mathbb{R}^\ell$.

$$\|\boldsymbol{x}\|_1 = |x^1| + \cdots + |x^\ell|,$$
$$\|\boldsymbol{x}\|_2 = \sqrt{(x^1)^2 + \cdots + (x^\ell)^2},$$
$$\|\boldsymbol{x}\|_\infty = \max\{|x^t||1 \leq t \leq \ell\}.$$

We have already observed in Appendix C that these norms are equivalent in the sense that they induce the same topology on the set L. The finite-dimensional spaces are extended to the infinite-dimensional spaces.

Example 2. The set of sequences $\boldsymbol{x} = (x^t)$ satisfying $\sum_{t=0}^{\infty} |x^t|^p < +\infty$ for $1 \leq p < +\infty$ is a Banach space with the norm:

$$\|x\|_p = \left(\sum_{t=0}^{\infty} |x^t|^p\right)^{1/p}.$$

It is called the ℓ^p spaces which are defined by:

$$\ell^p = \left\{\boldsymbol{x} = (x^t) \middle| \sum_{t=0}^{\infty} |x^t|^p < +\infty\right\}.$$

In particular, for $p = 1, 2$, the norms of ℓ^1 and the ℓ^2 spaces are clearly generalizations of the first and the second norms of Example 1 to infinite-dimensional spaces of sequences. The third norm of Example 1 is extended to the norm $\|\boldsymbol{x}\|_\infty = \sup_{t\geq 0}|x^t|$ for a sequence $\boldsymbol{x} = (x^t)$. The space with this norm is called the ℓ^∞ space and defined by:

$$\ell^\infty = \{\boldsymbol{x} = (x^t)|\sup_{t\geq 0}|x^t| < +\infty\}.$$

We will discuss more about these spaces in Appendix H. An important class of the Banach spaces is the Hilbert space which possess the inner product between any two vectors of the space.

Definition E3. A pre-Hilbert space is a vector space L together with an inner product $(\boldsymbol{x}, \boldsymbol{y}) \mapsto \langle \boldsymbol{x}, \boldsymbol{y} \rangle$ which satisfies:

(ip-1) $\langle \boldsymbol{x}, \boldsymbol{y} \rangle = \langle \boldsymbol{y}, \boldsymbol{x} \rangle$ for all $\boldsymbol{x}, \boldsymbol{y} \in L$;
(ip-2) $\langle \boldsymbol{x} + \boldsymbol{y}, \boldsymbol{z} \rangle = \langle \boldsymbol{x}, \boldsymbol{z} \rangle + \langle \boldsymbol{y}, \boldsymbol{z} \rangle$ for all $\boldsymbol{x}, \boldsymbol{y}, \boldsymbol{z} \in L$;
(ip-3) $\langle \lambda \boldsymbol{x}, \boldsymbol{y} \rangle = \lambda \langle \boldsymbol{x}, \boldsymbol{y} \rangle$ for all $\boldsymbol{x}, \boldsymbol{y} \in L$, and for all $\lambda \in \mathbb{R}$; and
(ip-4) $\langle \boldsymbol{x}, \boldsymbol{x} \rangle \geq 0$ and $\langle \boldsymbol{x}, \boldsymbol{x} \rangle = 0$ if and only if $\boldsymbol{x} = \boldsymbol{0}$.

For a pre-Hilbert space L, we can define a "norm" $\|\boldsymbol{x}\|$ by $\|\boldsymbol{x}\| = (\langle \boldsymbol{x},\boldsymbol{x}\rangle)^{1/2}$ for $\boldsymbol{x} \in L$. Then it follws that:

Proposition E1 (The Cauchy–Schwartz inequality). *For all $\boldsymbol{x},\boldsymbol{y}$ in a pre-Hilbert space L, it follows that $|\langle \boldsymbol{x},\boldsymbol{y}\rangle| \leq \|\boldsymbol{x}\| \|\boldsymbol{y}\|$. The equality holds if and only if $\boldsymbol{x} = \lambda\boldsymbol{y}$ for some $\lambda \in \mathbb{R}$, or $\boldsymbol{x} = \boldsymbol{0}$.*

Let L be a pre-Hilbert space. For any $\boldsymbol{x},\boldsymbol{y} \in L$, we have by (ip-1), (ip-2), and Proposition E1,

$$\begin{aligned}\|\boldsymbol{x}+\boldsymbol{y}\|^2 &= \langle \boldsymbol{x}+\boldsymbol{y},\boldsymbol{x}+\boldsymbol{y}\rangle \\ &= \langle \boldsymbol{x},\boldsymbol{x}\rangle + \langle \boldsymbol{x},\boldsymbol{y}\rangle + \langle \boldsymbol{y},\boldsymbol{x}\rangle + \langle \boldsymbol{y},\boldsymbol{y}\rangle \\ &\leq \|\boldsymbol{x}\|^2 + 2|\langle \boldsymbol{x},\boldsymbol{y}\rangle| + \|\boldsymbol{y}\|^2 \\ &\leq \|\boldsymbol{x}\|^2 + 2\|\boldsymbol{x}\|\|\boldsymbol{y}\| + \|\boldsymbol{y}\|^2 = (\|\boldsymbol{x}\| + \|\boldsymbol{y}\|)^2,\end{aligned}$$

hence $\|\boldsymbol{x}+\boldsymbol{y}\| \leq \|\boldsymbol{x}\| + \|\boldsymbol{y}\|$. Therefore, if L is a pre-Hilbert space, then the map $\|\cdot\|: L \to \mathbb{R}_+$ defined by $\|\boldsymbol{x}\| = \langle \boldsymbol{x},\boldsymbol{x}\rangle^{1/2}$ is indeed a norm on L. A pre-Hilbert space is called a Hilbert space if it is complete in this norm.

In Example 2, the space ℓ^2 is a Hilbert space which has an inner product defined by $\langle \boldsymbol{x},\boldsymbol{y}\rangle = \sum_{t=0}^{\infty} x^t y^t$ for $\boldsymbol{x} = (x^t)$ and $\boldsymbol{y} = (y^t)$ of ℓ^2.

Let E and F be Banach spaces. A map $\phi: E \to F$ is said to be linear if:

$$\phi(a\boldsymbol{x} + b\boldsymbol{y}) = a\phi(\boldsymbol{x}) + b\phi(\boldsymbol{y})$$

for every $\boldsymbol{x},\boldsymbol{y} \in E$ and every $a, b \in \mathbb{R}$. The idea of linearity is generalized to that of multilinear map. Let $E_1 \cdots E_n, F$ be Banach spaces. A map $\psi : E_1 \times \cdots \times E_n \to F$ is said to be n multilinear if $\psi(\boldsymbol{x}_1 \cdots \boldsymbol{x}_n)$ is linear in each variable separately. For instance, the linearity in the first variable means that:

$$\psi(a\boldsymbol{x}_1 + b\boldsymbol{y}_1 \cdots \boldsymbol{x}_n) = a\psi(\boldsymbol{x}_1 \cdots \boldsymbol{x}_n) + b\psi(\boldsymbol{y}_1 \cdots \boldsymbol{x}_n).$$

The n multilinear map ψ is continuous if and only if there exists $M > 0$ such that:

$$\|\psi(\boldsymbol{x}_1 \cdots \boldsymbol{x}_n)\| \leq M\|\boldsymbol{x}_1\| \cdots \|\boldsymbol{x}_n\| \quad \text{for all } \boldsymbol{x}_k \in E_k,\ k = 1 \cdots n.$$

The space of continuous n multilinear maps of $E_1 \cdots E_n$ to F is denoted by $\mathscr{L}(E_1 \cdots E_n, F)$. When $E_1 = \cdots = E_n = E$, it is denoted by $\mathscr{L}^n(E,F)$. When $n = 1$, we usually write $\mathscr{L}^1(E,F) = \mathscr{L}(E,F)$. This is nothing but the space of linear maps from E to F. In particular, if $F = \mathbb{R}$, then $\mathscr{L}(E,\mathbb{R})$ is often denoted as L^* and called the dual space of L.

The space $\mathscr{L}^n(E,F)$ is obviously linear space and we can endow a norm on it which is called the operator norm and defined by:

$$\|\psi\| = \sup\left\{\frac{\|\psi(x_1\cdots x_n)\|}{\|x_1\|\ldots\|x_n\|}\middle| x_1\cdots x_n \neq 0\right\}.$$

We can prove easily that the space $\mathscr{L}^n(E,F)$ is complete if F is complete. Moreover, we can show that $\mathscr{L}(E,\mathscr{L}^{n-1}(E,F)) = \mathscr{L}^n(E,F)$. Indeed, let $\phi \in \mathscr{L}(E,\mathscr{L}^{n-1}(E,F))$. Since $\phi(x_1) \in \mathscr{L}^{n-1}(E,F)$, we can define a map $\psi: E^n \to F$ by $\psi(x_1\cdots x_n) = \phi(x_1)(x_2\cdots x_n)$. Obviously, the map $\phi \mapsto \psi$ is bijective and linear, and it is easily verified that $\|\phi\| = \|\psi\|$, hence $\mathscr{L}(E,\mathscr{L}^{n-1}(E,F))$ and $\mathscr{L}^n(E,F)$ are isomorphic, which means:

Definition E4. Banach spaces E and F are said to be isomorphic if there exists a bijective (namely one-to-one and onto) and continuous map f from E to F whose inverse f^{-1} is also continuous.

Two isomorphic Banach spaces are considered to be the same space. Hence, if E and F are isomorphic, we often write $E \approx F$ or even $E = F$.

Let E and F be Banach spaces, and $\phi: E \to F$ is a map from E to F. We say that the map ϕ is open if and only if $\phi(U)$ is open in F whenever U is open in E.

Theorem E1 (Open-mapping theorem). *Let E and F be Banach spaces and suppose $\phi \in \mathscr{L}(E,F)$ is onto. Then ϕ is an open mapping.*

Therefore, if a linear map $\phi \in \mathscr{L}(E,F)$ is one-to-one and onto, it is an isomorphism (Banach isomorphism theorem).

When $E = \mathbb{R}^k$ and $F = \mathbb{R}^\ell$, the linear map $\phi: E \to F$ is of course represented by the matrix $\Phi = (a_{ij})$, $a_{ij} = \phi^j(e_i)$, where $e_i = (0\cdots1\cdots0)$ is the i-th coordinate vector, $i = 1\cdots k$ and we set $\phi(x) = (\phi^j(x))$, $j = 1\cdots\ell$. The transpose matrix $\Phi' = (b_{ij})$, $b_{ij} = a_{ji}$ is extended to the adjoint map.

Proposition E2. *Let E and F be Banach spaces. For each $\phi \in \mathscr{L}(E,F)$, there exists a unique $\phi^* \in \mathscr{L}(F^*,E^*)$ called the adjoint of ϕ satisfying $y^*(\phi(x)) = \phi^*(y^*)(x)$ for all $x \in E$ and all $y^* \in F^*$, and $\|\phi\| = \|\phi^*\|$.*

Let H be a Hilbert space with the inner product $\langle x,y\rangle$ for $x,y \in H$. Then, we have the Riesz' lemma,

Lemma E1 (Riesz). *For each $p \in H^*$, there exists $y \in H$ such that $\|p\| = \|y\|$ and $p(x) = \langle y,x\rangle$ for all $x \in H$.*

As a corollary, it follows that $H = H^*$, and the condition of the adjoint map of Proposition E2, is written as $\langle \boldsymbol{y}^*, \phi(\boldsymbol{x})\rangle = \langle \phi^*(\boldsymbol{y}^*), \boldsymbol{x}\rangle$, which could be of more transparent.

For $\phi \in \mathscr{L}(E,F)$, the null space and the range of ϕ are denoted by $\mathcal{N}(\phi) = \{\boldsymbol{x} \in E | \phi(\boldsymbol{x}) = \boldsymbol{0}\}$ and $\mathcal{R}(\phi) = \{\boldsymbol{y} \in F \mid \phi(\boldsymbol{x}) = \boldsymbol{y}$ for some $\boldsymbol{x} \in E\}$, respectively. Let M be a subspace of a Banach space E. The annihilator $M^{\perp}$ of M is defined by $M^{\perp} = \{\boldsymbol{x}^* \in E^* \mid \boldsymbol{x}^*(\boldsymbol{x}) = 0$ for all $\boldsymbol{x} \in M\}$. Similarly, for a subspace N of E^*, the annihilator $N^{\perp}$ of N is defined by $N^{\perp} = \{\boldsymbol{x} \in E \mid \boldsymbol{x}^*(\boldsymbol{x}) = 0$ for all $\boldsymbol{x}^* \in N\}$.

Proposition E3. *Let E and F be Banach spaces and $\phi \in \mathscr{L}(E,F)$. Then, $\mathcal{N}(\phi^*) = \mathcal{R}(\phi)^{\perp}$ and $\mathcal{N}(\phi) = \mathcal{R}(\phi^*)^{\perp}$.*

Let E be a Banach space and F a linear subspace of E. We can define an equivalence relation on E by $\boldsymbol{x} \equiv \boldsymbol{y}$ if and only if $\boldsymbol{x} - \boldsymbol{y} \in F$. Let $(\boldsymbol{x})$ be the equivalence class containing $\boldsymbol{x} \in E$. The set of all equivalent classes is called the quotient space and denoted by E/F. We can define the sum and the scalar multiplication on E/F by $(\boldsymbol{x}) + (\boldsymbol{y}) = (\boldsymbol{x} + \boldsymbol{y})$ and $\alpha(\boldsymbol{x}) = (\alpha\boldsymbol{x})$ for $\alpha \in \mathbb{R}$, and it can be easily shown that E/F is a vector space. When dimension $E/F = n$, we say that F is a subspace of E with the co-dimension n. The null space of $\boldsymbol{p} \in E^*$, $\mathcal{N}(\boldsymbol{p}) = \{\boldsymbol{x} \in E \mid \boldsymbol{p}(\boldsymbol{x}) = 0\}$ is an example of a subspace of E with the co-dimension 1.

As in the finite-dimensional spaces, we can approximate a general map between Banach spaces by a linear map (derivative). If this procedure is possible, the approximated map is called smooth, and one can get a lot of information from the approximating linear map which is generally simpler than the original map.

Definition E5. Let U be an open subset of a Banach space E. A map f from U to a Banach space F is called (Frećhet) differentiable at $x \in U$ if there is a continuous linear map $Df \in \mathscr{L}(E,F)$ such that:

$$\lim_{\boldsymbol{h}\to\boldsymbol{0}} \frac{\|f(\boldsymbol{x}+\boldsymbol{h}) - f(\boldsymbol{x}) - Df(\boldsymbol{x})\|}{\|\boldsymbol{h}\|} = 0.$$

The linear map $Df(\boldsymbol{x}) \in \mathscr{L}(E,F)$ is called the derivative of f at $\boldsymbol{x}$.

An advantage of this definition of the derivative is "coordinate free", hence it can be applied to the case of general Banach spaces. When we use the coordinates of $E = \mathbb{R}^k$ and $F = \mathbb{R}^{\ell}$, $f(\boldsymbol{x}) = (f^1(\boldsymbol{x}) \cdots f^{\ell}(\boldsymbol{x}))$, the derivative can be written in the standard matrix form:

$$Df(\boldsymbol{x}) = \begin{pmatrix} \partial_1 f^1(\boldsymbol{x}) \cdots \partial_k f^1(\boldsymbol{x}) \\ \cdots\cdots\cdots\cdots\cdots\cdots \\ \partial_1 f^{\ell}(\boldsymbol{x}) \cdots \partial_k f^{\ell}(\boldsymbol{x}) \end{pmatrix}.$$

This $k \times \ell$ matrix is often called the Jacobian matrix of f.

For every integer $r \geq 0$, the r-th derivative $D^r f(\boldsymbol{x})$ of f at $x \in U$ is defined inductively:

$$D^r f(x) \equiv D(D^{r-1}f)(x): U \to \mathscr{L}(E, \mathscr{L}^{r-1}(E,F)) \approx \mathscr{L}^r(E,F)$$

which maps $\boldsymbol{x}$ at U to an r-multilinear map of E to F. A map f is said to be of class C^r at $\boldsymbol{x} \in U$ if this map is continuous. When $E = \mathbb{R}^k$ and $F = \mathbb{R}^\ell$, it is equivalent with that every partial derivative (in the usual sense) $\frac{\partial^n f^j}{\partial x_{i1} \cdots \partial x_{in}}(\boldsymbol{x})$, $1 \leq i_1 \cdots i_n \leq k$, $1 \leq j \leq \ell$, $1 \leq n \leq r$ exists and continuous.

The fundamental property of the derivative is that it is linear.

Proposition E4. *Let E, F be Banach spaces, $U \subset E$ an open set. Let a be a real number. If $f, g: U \to F$ are of class C^r, then af and $f+g$ are also of class C^r and:*

$$D^r(f(\boldsymbol{x}) + g(\boldsymbol{x})) = D^r f(\boldsymbol{x}) + D^r g(\boldsymbol{x}) \text{ and } D^r(af)(\boldsymbol{x}) = aD^r f(\boldsymbol{x}).$$

Proposition E5 (Chain rule). *Let E, F, G be Banach spaces, $U \subset E$ and $V \subset F$ are open, and maps $f: U \to V$ and $g: V \to G$ be of Class C^1. Then, we have:*

$$D(g \circ f)(\boldsymbol{x}) = Dg(f(\boldsymbol{x})) \circ Df(\boldsymbol{x}).$$

Let σ be a permutation of $\{1 \cdots r\}$, or one-to-one and onto map of $\{1 \cdots r\}$ to itself. For $\psi \in \mathscr{L}^r(E,F)$, we define $\sigma\psi(\boldsymbol{x}_1 \cdots \boldsymbol{x}_r) = \psi(\sigma(\boldsymbol{x}_1) \cdots \sigma(\boldsymbol{x}_r))$. When $\sigma\psi = \psi$ for all σ, we say that the r multilinear map ψ is symmetric. Then, we have:

Proposition E6 (Euler). *Let E, F be Banach spaces, $U \subset E$ an open set. If $f: U \to F$ is of class C^r, then $D^r f \in \mathscr{L}^r(E,F)$ is symmetric.*

Proposition E7 (Leibniz rule). *Let E, F_1, F_2 be Banach spaces, and $U \subset E$ an open set. Suppose $f: U \to F_1$ and $g: U \to F_2$ are of class C^r, and $\psi \in \mathscr{L}(F_1, F_2, G)$. Let $f \times g(\boldsymbol{x}) = (f(\boldsymbol{x}), g(\boldsymbol{x}))$ and $\psi(f,g) = \psi \circ (f \times g)$. Then $\psi(f,g)$ is of class C^r by the Leibniz rule,*

$$D\psi(f,g)(\boldsymbol{x})\boldsymbol{h} = \psi(Df(\boldsymbol{x})\boldsymbol{h}, g(\boldsymbol{x})) + \psi(f(\boldsymbol{x}), Dg(\boldsymbol{x})\boldsymbol{h}).$$

Proposition E8 (Taylor's formula). *Let E, F be Banach spaces, and $U \subset E$ be an open set. If $f: U \to F$ is of class C^r, then we have:*

$$f(\boldsymbol{x} + \boldsymbol{h}) = \sum_{k=0}^{r} \frac{D^k f(\boldsymbol{x})\boldsymbol{h}^k}{k!} + R(\boldsymbol{x}, \boldsymbol{h})\boldsymbol{h}^r,$$

where $\boldsymbol{h}^k = (\boldsymbol{h} \cdots \boldsymbol{h}) \in E^k$ and $R(\boldsymbol{x}, \boldsymbol{h})$ is given by:

$$R(\boldsymbol{x}, \boldsymbol{h}) = \int_0^1 \frac{(1-t)^{r-1}}{(r-1)!}(D^r f(\boldsymbol{x} + t\boldsymbol{h}) - D^r f(\boldsymbol{x}))dt,$$

and $\boldsymbol{h}$ is taken so small that $\{\boldsymbol{x} + t\boldsymbol{h} \mid 0 \leq t \leq 1\} \subset U$.

Note that in the above formula, $R(\boldsymbol{x},\boldsymbol{h})$ is continuous and $R(\boldsymbol{x},\boldsymbol{0}) = 0$. A related concept of the derivative is:

Definition E6. Let U be an open subset of E and let F be a Banach space. We say that a map $f: U \to F$ has a derivative in the direction $\boldsymbol{h} \in E$ at $\boldsymbol{x}$ if,

$$\frac{d}{dt}f(\boldsymbol{x} + t\boldsymbol{h})_{t=0}$$

exists. We call this element of F the (Gateaux) derivative at $x \in U$, and if it exists everywhere in U, we say that the map f is (Gateaux) differentiable.

The Fréchet differentiability is stronger than the Gateaux differentiability according to:

Proposition E9. *If f is Fréchet differentiable at $\boldsymbol{x}$, then the Gateaux derivatives of f at $\boldsymbol{x}$ exist and they are given by:*

$$\frac{d}{dt}f(\boldsymbol{x} + t\boldsymbol{h})_{t=0} = Df(\boldsymbol{x})\boldsymbol{h}.$$

We can also give a "coordinate-free" definition of the partial derivative. Let $E = E_1 \oplus E_2$ (direct sum), where E_1 and E_2 are Banach spaces, and $U \subset E$ is open subset of E. For a map of class C^r $f: U \to F$, the partial derivative with respect to E_1 is defined by:

$$\partial_1 f(x) \equiv Df(x)(e_1, 0), \quad e_1 \in E_1.$$

The partial derivative with respect to E_2 is defined similarly. The next theorem is of central importance in the differential calculus on manifolds (Abraham *et al.*, (2nd ed) 1991, p. 121).

Theorem E2 (Implicit function theorem). *Let $U \subset E, V \subset F$ be open and $f: U \times V \to F$ be of class $C^r (r \geq 1)$. For some $\boldsymbol{x}_0 \in U$ and $\boldsymbol{y}_0 \in V$, assume that $\partial_2 f(\boldsymbol{x}_0, \boldsymbol{y}_0): F \to F$ is an isomorphism. Then, there are neighborhoods U_0 of $\boldsymbol{x}_0$ and $W \subset F$ of $f(\boldsymbol{x}_0, \boldsymbol{y}_0)$, and a unique C^r map $g: U_0 \times W \to V$ such that for all $(\boldsymbol{x}, \boldsymbol{z}) \in U_0 \times W$,*

$$f(\boldsymbol{x}, g(\boldsymbol{x}, \boldsymbol{z})) = \boldsymbol{z}.$$

The convex subsets of a normed space E, and the concavity of real-valued functions defined on a convex subset of E are defined as in the same way as in the case of $\mathbb{R}^\ell$, see Appendix A. The separation theorem (Theorem A1) and the Kuhn–Tucker theorem (Theorem A5) are extended to Banach spaces.

Theorem E3 (Hahn–Banach). *Let X and Y be non-empty convex subsets of a Banach space L, and interior $X \neq \emptyset$ or interior $Y \neq \emptyset$. If $X \cap Y = \emptyset$, then there exists a vector $\boldsymbol{p} \in L^*$ with $\boldsymbol{p} \neq \boldsymbol{0}$ such that $\boldsymbol{px} \leq \boldsymbol{py}$ for every $\boldsymbol{x} \in X$ and every $\boldsymbol{y} \in Y$.*

Theorem E4 (Kuhn–Tucker). *Let E be a Banach space and let f be a real-valued concave function defined on a convex set X of E, and $g_1 \dots g_m$ be concave real-valued functions defined on X. Suppose that the Slater's condition holds, namely that: there exists an $\boldsymbol{x}_0 \in X$ such that $g_j(\boldsymbol{x}_0) > 0, j = 1 \cdots m$.*

Then, a point $\hat{\boldsymbol{x}}$ achieves a maximum of $f(\boldsymbol{x})$ on X subject to $g_j(\boldsymbol{x}) \geq 0,\ j = 1 \cdots m$, if and only if there exists a non-negative vector $\hat{\boldsymbol{\lambda}} = (\hat{\lambda}_1 \cdots \hat{\lambda}_m) \geq \boldsymbol{0}$ such that:

$$\mathcal{L}(\boldsymbol{x}, \hat{\boldsymbol{\lambda}}) \leq \mathcal{L}(\hat{\boldsymbol{x}}, \hat{\boldsymbol{\lambda}}) \leq \mathcal{L}(\hat{\boldsymbol{x}}, \boldsymbol{\lambda})$$

for every $\boldsymbol{x} \in X$ and every non-negative $\boldsymbol{\lambda} \geq \boldsymbol{0}$, where $\mathcal{L}(\boldsymbol{x}, \boldsymbol{\lambda}) = f(\boldsymbol{x}) + \boldsymbol{\lambda} g(\boldsymbol{x}) = f(\boldsymbol{x}) + \sum_{j=1}^{m} \lambda_j g_j(\boldsymbol{x})$ is the Lagrangian.

We now give the definition of the differentiable manifold. In the following, we assume that $E_1 \cdots E_n = \mathbb{R}^k$ and $F = \mathbb{R}^\ell$. Therefore, every element ϕ of $\mathscr{L}^n(E, F)$ is continuous and we can identify the space of linear maps of $E = \mathbb{R}^k$ to $F = \mathbb{R}^\ell$ or $\mathscr{L}(E, F)$ with $\mathbb{R}^{k\ell}$, since a point ϕ in $\mathscr{L}(E, F)$ has $k\ell$ "coordinates" $\phi^j(\boldsymbol{e}^i)$, $i = 1 \cdots k$, $j = 1 \cdots \ell$, where $\boldsymbol{e}^i = (0 \cdots 1 \cdots 0)$ with 1 in i-th position. Similarly, we can identify $\mathscr{L}^n(E, F)$ with $\mathbb{R}^{k^n \ell}$.

Definition E7 (A Hausdor). Topological space M is a k-dimensional manifold if there exist an open cover $\{U_\alpha\}$ of M and local isomorphisms ϕ_α on U_α to $\mathbb{R}^k$ such that $\phi_\beta \circ \phi_\alpha^{-1} : \phi_\alpha(U_\alpha \cap U_\beta) \to \phi_\beta(U_\alpha \cap U_\beta)$ is bijective and of class C^r for each α and β. See the following diagram:

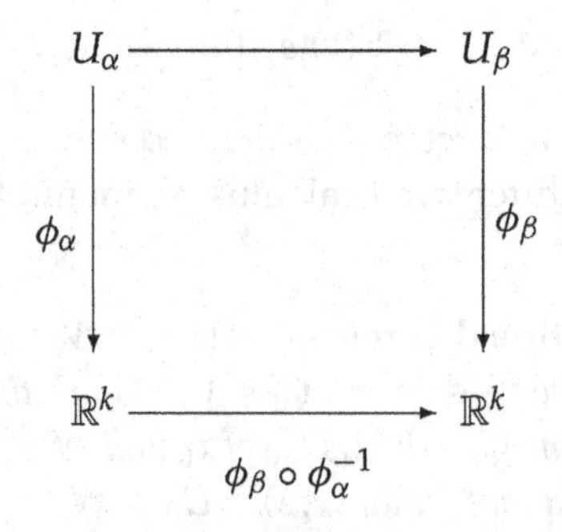

Since $(\phi_\beta \circ \phi_\alpha^{-1})^{-1} = \phi_\alpha \circ \phi_\beta^{-1} : \phi_\beta(U_\alpha \cap U_\beta) \to \phi_\alpha(U_\alpha \cap U_\beta)$, the inverse of $\phi_\beta \circ \phi_\alpha^{-1}$ is also of class C^r, so that it is C^r-diffeomorphism. (U_α, ϕ_α) is called the chart and the family of all charts is called an atlas.

Example 3. The ℓ-dimensional sphere:

$$S^\ell = \{\boldsymbol{x} = (x^0 \cdots x^\ell) \in \mathbb{R}^{\ell+1} \mid \sqrt{(x^0)^2 + \cdots + (x^\ell)^2} = 1\}$$

is a manifold. Indeed, we define $2(\ell + 1)$ open sets $U_\pm^t$, $t = 0, 1 \cdots \ell$ on S^ℓ by:

$$U_+^t = \{(x^0 \cdots x^\ell) \in S^\ell \mid x^t > 0\},$$

and

$$U_-^t = \{(x^0 \cdots x^\ell) \in S^\ell | x^t < 0\},$$

and $2(\ell + 1)$ maps $\phi_\pm^t$ on $U_\pm^t$ to the open disk $\{\boldsymbol{y} = (y^1 \cdots y^\ell) \in \mathbb{R}^\ell | \, \|\boldsymbol{y}\| = \sqrt{(y^1)^2 + \cdots + (y^\ell)^2} < 1\}$ by:

$$\phi_\pm^t(x^0 \cdots x^\ell) = (x^0 \cdots x^{t-1}, x^{t+1}, \ldots x^\ell).$$

Then, it is easy to see that:

$$\phi_\pm^s \circ (\phi_\pm^t)^{-1}(y^1 \cdots y^\ell) = \left(y^1 \cdots y^{s-1}, y^{s+1} \cdots y^{t-1}, \pm\sqrt{1 - \|\boldsymbol{y}\|^2}, y^t \cdots y^\ell\right),$$

which is obviously smooth.

Let $f\colon M \to N$ be a map from a manifold M to a manifold N. The map f is said to be at class C^r, if the map $\psi_\beta o f o \phi_\alpha^{-1}$ from $\phi_\alpha(U_\alpha)$ to $\psi_\beta(V_\beta)$ is of class C^r, where (U_α, ϕ_α) and (V_β, ψ_β) are charts of M and N, respectively. This definition does not depend on the choice of the co-ordinate charts. Indeed, let (U_γ, ϕ_γ) and (V_δ, ψ_δ) be another charts of M and N, respectively. Since $\psi_\delta o f o \phi_\gamma^{-1} = (\psi_\delta o \psi_\beta^{-1}) o (\psi_\beta o f o \phi_\alpha^{-1}) o (\phi_\alpha o \phi_\gamma^{-1})$ and $\psi_\delta \circ \psi_\beta{}^1$ and $\phi_\alpha \circ \phi_\gamma^{-1}$ are of class C^r by definition, $\psi_\delta \circ f \circ \phi_\gamma^{-1}$ is of class C^r.

As a special case, a C^r curve through $p \in M$ is a C^r map from $(-\epsilon, \epsilon)$ to M such that $c(0) = p$. Two C^r curves c and d are equivalent if and only if $\dot{c}(0)(\equiv \frac{d(\phi_\alpha o c)}{dt}(0)) = \dot{d}(0)$, where (U_α, ϕ_α) is a chart of M such that $p \in U_\alpha$. This definition of the equivalence relation is also independent of the choice of the coordinate chart. Let (U_β, ϕ_β) be another chart with $p \in U_\beta$. From $\phi_\beta \circ c = (\phi_\beta \circ \phi_\alpha^{-1}) \circ (\phi_\alpha \circ c)$ and $\phi_\beta \circ d = (\phi_\beta \circ \phi_\alpha^{-1}) \circ (\phi_\alpha \circ d)$, it follows that $d(\phi_\beta \circ c)/dt = D(\phi_\beta \circ \phi_\alpha^{-1})\frac{d(\phi_\alpha o c)}{dt}$ and $d(\phi_\beta \circ d)/dt = D(\phi_\beta \circ \phi_\alpha^{-1})\frac{d(\phi_\alpha o d)}{dt}$. Since $D(\phi_\beta \circ \phi_\alpha^{-1})$ is a linear isomorphism, $d(\phi_\beta \circ c)/dt = d(\phi_\beta \circ d)/dt$ if and only if $d(\phi_\alpha \circ c)/dt = d(\phi_\alpha \circ d)/dt$.

The equivalence class is denoted as $[c]_p$ and called a tangent vector at p. Let T_pM be the set of all tangent vectors at p and call the tangent space at p. It is easy to see that T_pM is a k-dimensional vector space. Let (U_α, ϕ_α) be a coordinate chart with $p \in U_\alpha$. Without loss of generality, we may assume that $\phi(p) = \mathbf{0} \in \mathbb{R}^k$. We define the smooth curves $c_i(t) = \phi^{-1}(0, \ldots, t, \ldots, 0)$, where t is at the i-th coordinate. Then the curves $[c_i]_p$, $i = 1 \cdots k$, make up with a basis of T_pM.

Let f be a C^r map from a manifold M to a manifold N. The tangent map (derivative) at $p \in M$ of a C^r map f is a linear map $Df(p)\colon T_pM \to T_{f(p)}N$ defined by:

$$Df(p)([c]_p) = [f \circ c]_{f(p)}.$$

We have to check that this definition is independent of the choice of curves representing the equivalence class. Let c_1 and c_2 be two curves such that $[c_1]_p = [c_2]_p$. This means that $d(\phi_\alpha \circ c_1)/dt = d(\phi_\alpha \circ c_2)/dt$, where (U_α, ϕ_α) is a chart on M with $p \in U_\alpha$. We want to show that $[f \circ c_1]_{f(p)} = [f \circ c_2]_{f(p)}$. Since $\psi_\gamma \circ f \circ c_i = (\psi_\gamma o f o \phi_\alpha^{-1}) o (\phi_\alpha o c_i)$, $i = 1, 2$, where (V_γ, ψ_γ) is a chart on N with $f(p) \in V_\gamma$, one has

$d(\psi_\gamma \circ f \circ c_i)/dt = D(\psi_\gamma \circ f \circ \phi_\alpha^{-1})\frac{d(\phi_\alpha \circ c_i)}{dt}, i = 1, 2$. Since $d(\phi_\alpha \circ c_1)/dt = d(\phi_\alpha \circ c_2)/dt$, we get $\frac{d(\psi_\gamma \circ f \circ c_1)}{dt}(0) = \frac{d(\psi_\gamma \circ f \circ c_2)}{dt}(0)$.

Let $f: M \to N$ be a smooth (C^r) map between manifolds M and N. A point $q \in N$ is a regular value of f (or f is transversal to $\{q\}$) if for every $p \in f^{-1}(q)$, $Df(p): T_pM \to T_{f(p)}N$ is surjective (onto). Note that when dimension $M <$ –dimension N, a point q is regular value only if $q \notin f(M)$. When dimension M = dimension N, $q \in N$ is regular value, if and only if $Df(p)$ is an isomorphism between T_pM and T_qN at every $p \in f^{-1}(q)$. A point of N which is not a regular value is called a critical value.

Theorem E5 (Regular value theorem). *Let $f: M \to N$ be a smooth (C^r) map between manifolds M and N such that dimension $M \geq$ dimension N and $q \in N$ a regular value of f. Then, $f^{-1}(q)$ is a submanifold of M such that* dimension $f^{-1}(q)$ = dimension M – dimension N.

A subset R of $\mathbb{R}^\ell$ is called a rectangular solid if it is of the form $R = \{(x^1 \cdots x^\ell) | a^h \leq x^h \leq b^h, h = 1 \cdots \ell\}$ for vectors $\boldsymbol{a} = (a^1 \cdots a^\ell)$ and $\boldsymbol{b} = (b^1 \cdots b^\ell)$ with $a^h \leq b^h$ for all $h = 1 \cdots \ell$. The volume of the rectangular solid R is defined by:

$$\operatorname{vol} R = \prod_{h=1}^{\ell} (b^h - a^h).$$

A subset $A \subset \mathbb{R}^\ell$ is said to have a measure zero if for every $\epsilon > 0$, there exist countably many rectangular solids $R_1, R_2, \ldots$ such that $A \subset \cup_{j=1}^\infty R_j$ and $\sum_{j=1}^\infty \operatorname{vol} R_j < \epsilon$.

The next theorem is a key for the analysis of Section 2.8.

Theorem E6 (Sard). *Let $f: M \to V$ be a C^r map where M is a manifold of dimension k and V is an open subset of $\mathbb{R}^\ell$ (hence a manifold of dimension ℓ). If $k \geq \ell$ and $r >$* $\max\{0, k - \ell\}$, *then the set of regular values of f has measure zero.*

Spaces of Closed Sets

Appendix

F

Let (X, d) be a metric space and let $\mathcal{K}(X)$ be the set of all non-empty compact subsets of (X, d). For every E and $F \in \mathcal{K}(X)$, we define the Hausdor distance δ by:

$$\delta(E, F) = \inf\{\epsilon \in [0, \infty) \mid E \subset clB(F, \epsilon) \text{ and } F \subset clB(E, \epsilon)\},$$

where $clB(E, \epsilon)$ is the (closed) ϵ-neighborhood of E, namely:

$$clB(E, \epsilon) = \{x \in X \mid \inf_{z \in E} d(x, z) \leq \epsilon\}.$$

Then, one can show that δ satisfies the condition (m-1), (m-2), and (m-3) of Appendix C, hence the space $(\mathcal{K}(X), \delta)$ is a metric space.

Proposition F1. *The metric space $(\mathcal{K}(X), \delta)$ has the following properties:*

(i) *the Hausdor distance topology on $\mathcal{K}(X)$ depends only on the topology of X and not on the particular metric on X;*
(ii) *if X is separable, then so is $(\mathcal{K}(X), \delta)$;*
(iii) *if X is separable and locally compact, then $(\mathcal{K}(X), \delta)$ is complete; and*
(iv) *if X is compact, then so is $(\mathcal{K}(X), \delta)$.*

Let $\{S_n\}$ be a sequence of subsets of (X, d). Recall the definition of the topological limes inferior and topological limes superior.

Definition D3 (re-stated). A topological limes inferior $Li(S_n)$ is a subset of X such that $x \in Li(S_n)$ if and only if for every neighborhood U of x, there is an integer N such that $U \cap S_n \neq \emptyset$ for all $n \geq N$.

A topological limes superior $Ls(S_n)$ is a subset of X such that $x \in Ls(S_n)$ if and only if for every neighborhood U of x, there are infinitely many n such that $U \cap S_n \neq \emptyset$.

We recall:

Proposition D7 (re-stated). *For every sequence $(F_n)_{n \in \mathbb{N}}$ of subsets of X, the following properties hold.*

(i) $L_i(F_n)$ *and* $L_s(F_n)$ *are closed (possibly empty) and* $Li(S_n) \subset Ls(S_n)$.
(ii) $x \in L_i(F_n)$ *if and only if there exists an integer* $N \in \mathbb{N}$ *and a sequence* $(x_n)_{n \in \mathbb{N}}$ *with* $x_n \in F_n$ *for all* $n \geq N$ *and* $x_n \to x$.
(iii) $x \in L_s(F_n)$ *if and only if there exists a subsequence* (F_{n_q}) *from which one can choose an element* $x_{n_q} \in F_{n_q}$ *for every* n_q *such that* $x_{n_q} \to x$.

A subset F of X is called the closed limit of a sequence $(F_n)_{n \in \mathbb{N}}$ if $L_i(F_n) = F = L_s(F_n)$.

Let (X, d) be a compact metric space and let $(\mathcal{F}_0, \delta)$ be the set of all non-empty closed subset of X with the topology of the Hausdor distance. Then, the metric space $(\mathcal{F}_0, \delta)$ is compact by Proposition F1. Moreover, we have:

Proposition F2. *A sequence* (F_n) *converges to F in* $(\mathcal{F}_0, \delta)$ *if and only if* $L_i(F_n) = F = L_s(F_n)$. *Every open set of* $(\mathcal{F}_0, \delta)$ *can be written as a union of the sets of the form:*

$$B(G, G_1 \cdots G_k) = \{F \in \mathcal{F}_0 | F \subset G \text{ and } F \cap G_i \neq \emptyset, \ i = 1 \dots k\},$$

where $G, G_1 \cdots G_k$ *are open sets of X.*

Let $\mathcal{K}_c$ be the family of all compact and convex subsets of $\mathbb{R}^\ell$. Then, by Proposition F1, $\mathcal{K}_c$ is metrizable by the Hausdor distance δ. Furthermore, Rådström (1952) proved:

Theorem F1 (Rådström). *The space* $(\mathcal{K}_c, \delta)$ *can be embedded as a convex cone in a real-normed space L in such a way that:*

(i) *the embedding is isometric;*
(ii) *the addition in L induces an addition in* $\mathcal{K}_c$; *and*
(iii) *the multiplication by non-negative scalars in L induces the same operation in* $\mathcal{K}_c$.

Let (X, d) be a metric space and let $\mathcal{F}(X)$ be the set of all closed subsets of X. Consider subsets of $\mathcal{F}(X)$ which are of the form:

$$[K, \mathcal{G}] \equiv \{F \in \mathcal{F} \mid F \cap K = \emptyset \text{ and } F \cap G \neq \emptyset \text{ for } G \in \mathcal{G}\},$$

where K is a compact subset of X and $\mathcal{G}$ is a finite family of non-empty open subsets of X.

Since a finite intersection of sets of this form is again in this form, the family of arbitrary unions of these sets satisfies the conditions (t-1), (t-2), and (t-3) of Appendix B, hence the family is a topology on the set $\mathcal{F}(X)$. This topology is called the topology of closed convergence and it is denoted by τ_c.

Then, one can prove that:

Theorem F2. *Let (X, d) be a locally compact and separable metric space. Then, the set $\mathcal{F}(X)$ of all closed subsets of X endowed with the topology of closed convergence, $(\mathcal{F}(X), \tau_c)$, is a compact metrizable space. A sequence $(F_n)_{n\in\mathbb{N}}$ converges to F if and only if $L_i(F_n) = F = L_s(F_n)$.*

Let (X, d) be a metric space. Then, we can endow the set $\mathcal{K}(X)$ with both the Hausdor distance topology and the topology of closed convergence. If X is not compact, then the topological space $(\mathcal{K}, \delta)$ is distinct from the topological space $(\mathcal{K}, \tau_c)$. The topology induced by the Hausdor distance δ is finer than the topology of closed convergence τ_c.

MEASURE AND INTEGRATION

Appendix

G

Let Ω be a set and $\mathcal{A}$ be a collection of subsets of Ω. Then, $\mathcal{A}$ is called a σ-field if it satisfies:

(f-1) $\Omega \in \mathcal{A}$;
(f-2) if $A \in \mathcal{A}$, then $\Omega \backslash A \in \mathcal{A}$; and
(f-3) if $A_1, A_2 \cdots \in \mathcal{A}$, then $\cup_{n=1}^{\infty} A_n \in \mathcal{A}$.

By (f-1) and (f-2), it is clear that $\emptyset \in \mathcal{A}$, and if $A_1, A_2 \cdots \in \mathcal{A}$, then $\cap_{n=1}^{\infty} A_n \in \mathcal{A}$. The pair $(\Omega, \mathcal{A})$ is called a measurable space, and each element A of $\mathcal{A}$ is called a measurable set. If $\mathcal{C}$ is a collection of subsets of Ω, the smallest σ-field containing $\mathcal{C}$ is called the σ-field generated by $\mathcal{C}$ and denoted by $\sigma(\mathcal{C})$.

Let $\mu: \mathcal{A} \to \mathbb{R} \cup \{+\infty\} \cup \{-\infty\}$ be a set function on a σ-field $\mathcal{A}$. We say that μ is (finitely) additive if and only if

$$\mu\left(\bigcup_{n=1}^{N} A_n\right) = \sum_{n=1}^{N} \mu(A_n)$$

for all disjoint measurable sets $A_1 \cdots A_N \in \mathcal{A}$. The set function μ is called countably additive if and only if

$$\mu\left(\bigcup_{n=1}^{\infty} A_n\right) = \sum_{n=1}^{\infty} \mu(A_n)$$

for all collection of countably many disjoint measurable sets $A_1, A_2 \cdots \in \mathcal{A}$.

A countably additive set function on a measurable space $(\Omega, \mathcal{A})$ is called a signed measure. It is called a measure if it is nonnegative or $\mu(A) \geq 0$ for every $A \in \mathcal{A}$. If in addition it satisfies $\mu(\Omega) = 1$, then we say that μ is a probability measure. The triple $(\Omega, \mathcal{A}, \mu)$ consisting of a set Ω, a σ-field $\mathcal{A}$, and a measure μ on $(\Omega, \mathcal{A})$ is called a measure space.

Example 1. Let $\Omega = \mathbb{N} = \{0, 1, \dots\}$ be the set of non-negative integers. Let $\mathcal{A}$ be the set of all subsets of $\mathbb{N}$ and we define the set function μ on $\mathcal{A}$ by $\mu(A) = \sharp A =$ the number of elements in A. Then, it is easy to see that $\mathcal{A}$ is a σ-field and μ is a measure. The measure μ is called the counting measure.

We say that a condition C is said to hold almost everywhere, and write it as C a.e, if there exists a measurable set of μ measure zero such that the condition C holds outside the set. A set of measure zero is often called null.

Let $(\Omega, \mathcal{A}, \mu)$ be a measure space. If the σ-field $\mathcal{A}$ contains all μ-null sets, we say that the measure space $(\Omega, \mathcal{A}, \mu)$ is complete. Let $\mathcal{N}$ be the set of all null sets of the measure space $(\Omega, \mathcal{A}, \mu)$, and let $\mathcal{A}_\mu$ be the σ-field generated by the family $\mathcal{A} \cup \mathcal{N}$. It can be shown that $\mathcal{A}_\mu = \{A \cup N \mid A \in \mathcal{A},\ N \in \mathcal{N}\}$. We can define the measure $\bar{\mu}$ on the measurable space $(\Omega, \mathcal{A}_\mu)$ by $\bar{\mu}(A \cup N) = \mu(A)$. Then, the measure space $(\Omega, \mathcal{A}_\mu, \bar{\mu})$ is called the completion of the measure space $(\Omega, \mathcal{A}, \mu)$.

Let $\Omega = X$ be a topological space. The Borel σ-field on X, denoted by $\mathcal{B}(X)$, is the σ-field generated by the open sets of X. Each element B of $\mathcal{B}(X)$ is called a Borel set.

Example 2. Let $\Omega = \mathbb{R}$ and $\mathcal{A}$ be the σ-field generated by the family of all intervals of the form $(a, b]$, $a \leq b$. It can be easily shown that $\mathcal{A} = \mathcal{B}(\mathbb{R})$. For an interval $(a, b] \in \mathcal{A}$, set $\lambda_0((a, b]) = b - a$. Then, λ_0 can be extended as a unique measure on $\mathcal{A}$. It is called the Borel–Lebesgue measure on $\mathbb{R}$.

Example 3. The measure λ_0 defined in the Example 2 is not complete. The completion λ of λ_0 is called the Lebesgue measure on $\Omega = \mathbb{R}$.

Let $(\Omega_1, \mathcal{A}_1)$ and $(\Omega_2, \mathcal{A}_2)$ be two measurable sets. The product σ-field $\mathcal{A}_1 \times \mathcal{A}_2$ is the σ-field on the set $\Omega_1 \times \Omega_2$ which is generated by all sets of the form $A_1 \times A_2$, $A_1 \in \mathcal{A}_1$ and $A_2 \in \mathcal{A}_2$. The measurable space $(\Omega_1 \times \Omega_2, \mathcal{A}_1 \times \mathcal{A}_2)$ is called the product measurable space. Similarly, the product measurable spaces $(\prod_{i=1}^m \Omega_i, \prod_{i=1}^m \mathcal{A}_i)$ for the measurable spaces $(\Omega_1, \mathcal{A}_1) \cdots (\Omega_m, \mathcal{A}_m)$ are defined.

Proposition G1. *Let Ω_1 and Ω_2 be separable metric spaces. Then, we have $\mathcal{B}(\Omega_1 \times \Omega_2) = \mathcal{B}(\Omega_1) \times \mathcal{B}_2(\Omega_2)$.*

Let $(\Omega, \mathcal{A}, \mu)$ be a measure space. A set $A \in \mathcal{A}$ with $\mu(A) > 0$ is called an atom if for all $B \subset A$, it follows that $\mu(B) = \mu(A)$ or $\mu(B) = 0$. A measure space $(\Omega, \mathcal{A}, \mu)$ (or sometimes the measure μ itself) is said to be atomless if it has no atom.

Theorem G1 (Liapunov). *Let $\mu_1 \cdots \mu_m$ be atomless measures on a measurable space $(\Omega, \mathcal{A})$. Then, the set $\{(\mu_1(A) \cdots \mu_m(A)) \in \mathbb{R}^m \mid A \in \mathcal{A}\}$ is a closed and convex subset in $\mathbb{R}^m$.*

Let $(\Omega_1, \mathcal{A}_1)$ and $(\Omega_2, \mathcal{A}_2)$ be measurable spaces. A mapping $f: \Omega_1 \to \Omega_2$ is said to be measurable if $f^{-1}(A) \in \mathcal{A}_1$ for every $A \in \mathcal{A}_2$. Then, it is sufficient for the mapping f to be measurable that $f^{-1}(A) \in \mathcal{A}_1$ for every $A \in \mathcal{C}$, where $\mathcal{C}$ is a family of subsets of Ω_2 which generates $\mathcal{A}_2$, that is to say, $\mathcal{A}_2 = \sigma(\mathcal{C})$.

Indeed, since the set $\{A \in \mathcal{A}_2 \mid f^{-1}(A) \in \mathcal{A}_1\}$ is a σ-field containing $\mathcal{C}$, it coincides with $\mathcal{A}_2$. In particular, if Ω_1 and Ω_2 are metric spaces, every continuous function is measurable with respect to $(\Omega_1, \mathcal{B}(\Omega_1))$ and $(\Omega_2, \mathcal{B}(\Omega_2))$. When $\Omega_2 = X$ is a metric space, a measurable function $f: (\Omega_1, \mathcal{A}_1) \to (X, \mathcal{B}(X))$ is called Borel measurable.

The next proposition is clear from the definition.

Proposition G2. *If two mappings $f: (\Omega_1, \mathcal{A}_1) \to (\Omega_2, \mathcal{A}_2)$ and $g: (\Omega_2, \mathcal{A}_2) \to (\Omega_3, \mathcal{A}_3)$ are both measurable, then the composition $g \circ f: (\Omega_1, \mathcal{A}_1) \to (\Omega_3, \mathcal{A}_3)$ is also measurable.*

Proposition G3. *Let f and g be Borel-measurable functions of $(\Omega, \mathcal{A})$ to $(\mathbb{R}, \mathcal{B}(\mathbb{R}))$ and $(f_n)_{n\in\mathbb{N}}$ a sequence of Borel-measurable mappings of $(\Omega, \mathcal{A})$ to $(\mathbb{R}, \mathcal{A})$.*

(i) *the functions $\omega \mapsto f(\omega) \cdot g(\omega)$ and $\omega \mapsto \sup\{f(\omega), g(\omega)\}$ are Borel measurable,*
(ii) *if $\lim_{n\to\infty} f_n(\omega)$ exists for every $\omega \in \Omega$, then the function $\omega \mapsto \lim_{n\to\infty} f_n(\omega)$ is Borel measurable.*

Proposition G4. *Let $f_1 \cdots f_m$ be measurable mappings of a measurable space $(\Omega, \mathcal{A})$ to measurable spaces $(\Omega_1, \mathcal{A}_1) \cdots (\Omega_m, \mathcal{A}_m)$, respectively. Then, the mapping $\omega \mapsto (f_1(\omega) \cdots f_m(\omega))$ of $(\Omega, \mathcal{A})$ to $(\prod_{i=1}^m \Omega_i, \prod_{i=1}^m \mathcal{A}_i)$ is measurable.*

A function $f: (\Omega, \mathcal{A}) \to (\mathbb{R}, \mathcal{B}(\mathbb{R}))$ is called simple if it can be written in the form: $f(\omega) = \sum_{i=1}^m a_i \mathbf{1}_{A_i}(\omega)$, where $a_i \in \mathbb{R}$, $A_i \in \mathcal{A}$, $i = 1 \cdots m$, and $\mathbf{1}_A$ is the indicator function of the set A, which is defined by:

$$\mathbf{1}_A(\omega) = \begin{cases} 1 & \text{for } \omega \in A, \\ 0 & \text{otherwise.} \end{cases}$$

It is clear that every simple function is measurable.

Proposition G5. *Every measurable function f is the pointwise limit of a sequence of simple functions, namely that there exists a sequence $(f_n)_{n\in\mathbb{N}}$ of measurable functions such that $f_n(\omega) \to f(\omega)$ for every $\omega \in \Omega$.*

We now define the integration of a Borel-measurable function f of $(\Omega, \mathcal{A}, \mu)$ to $\mathbb{R}$.

Step 1. When the function f is simple and written as $f(\omega) = \sum_{i=1}^m a_i \mathbf{1}_{A_i}(\omega)$, we define:

$$\int_\Omega f(\omega) d\mu = \sum_{i=1}^m a_i \mu(A_i),$$

as long as both $+\infty$ and $-\infty$ together do not appear at once in the sum of the right-hand side. When they do, we say that the integral does not exist.

Step 2. When the function f is nonnegative and Borel measurable, we define:

$$\int_{\Omega} f(\omega)d\mu = \sup\left\{\int_{\Omega} s(\omega)d\mu \mid s(\cdot) \text{ is simple and } 0 \leq s(\cdot) \leq f(\cdot)\right\}.$$

Note that this definition agrees with the definition of Step 1 when the function is simple.

Step 3. When f is an arbitrary measurable function, let $f^{+}(\omega) = \sup\{f(\omega), 0\}$ and $f^{-}(\omega) = \sup\{-f(\omega), 0\}$. Then we have $f(\omega) = f^{+}(\omega) - f^{-}(\omega)$, $|f(\omega)| = f^{+}(\omega) + f^{-}(\omega)$. By Proposition G2, the function $f^{+}(\cdot)$ and $f^{-}(\cdot)$ are both measurable. We define:

$$\int_{\Omega} f(\omega)d\mu = \int_{\Omega} f^{+}(\omega)d\mu - \int_{\Omega} f^{-}(\omega)d\mu,$$

if at least one of the two terms in the right-hand side is not ∞. Otherwise, we say that the integral does not exist. If both of the two terms are finite, the function f is said to be integrable (or μ-integrable).

For a measurable set $A \in \mathcal{A}$, we define the integral of f over A as:

$$\int_{A} f(\omega)d\mu = \int_{\Omega} f(\omega)\mathbf{1}_{A}(\omega)d\mu.$$

Since the integrand $f \cdot \mathbf{1}_{A}$ is measurable by Proposition G3, this definition is legitimate.

The next proposition follows easily from the definition of the integration.

Proposition G6. *Let f and g be integrable functions of a measure space $(\Omega, \mathcal{A}, \mu)$ to $\mathbb{R}$. If:*

(i) $f(\omega) \leq g(\omega)$ *a.e., then we have* $\int_{\Omega} f(\omega)d\mu \leq \int_{\Omega} g(\omega)d\mu$,
(ii) $\int_{\Omega} cf(\omega)d\mu = c\int_{\Omega} g(\omega)d\mu$ *for every* $c \in \mathbb{R}$, *and*
(iii) $\int_{\Omega} f(\omega) + g(\omega)d\mu = \int_{\Omega} f(\omega)d\mu + \int_{\Omega} g(\omega)d\mu$

and we can prove that:

Theorem G2 (Fatou's lemma). *Let f and $f_1, f_2 \cdots$ be Borel-measurable functions of a measure space $(\Omega, \mathcal{A}, \mu)$ to $\mathbb{R}$. Then, we have:*

(i) *if* $f \leq f_n$ *a.e. for all* n, *where* $-\infty < \int_{\Omega} f(\omega)d\mu$, *then* $\int_{\Omega} \liminf_{n\to\infty} f_n(\omega)d\mu \leq \liminf_{n\to\infty} \int_{\Omega} f_n(\omega)d\mu$,
(ii) *if* $f_n(\omega) \leq f(\omega)$ *a.e. for all* n, *where* $\int_{\Omega} f(\omega)d\mu < +\infty$, *then* $\limsup_{n\to\infty} \int_{\Omega} f_n(\omega)d\mu \leq \int_{\Omega} \limsup_{n\to\infty} f_n(\omega)d\mu$.

Let $(f_n)_{n\in\mathbb{N}}$ be a sequence of Borel-measurable functions of a measure space $(\Omega, \mathcal{A}, \mu)$ to $\mathbb{R}$. The sequence (f_n) is said to converge to a Borel-measurable function f almost everywhere (written $f_n \to f$ a.e.) if $\mu(\{\omega \in \Omega \mid \lim_{n\to\infty} f_n(\omega) \neq f(\omega)\}) = 0$.

From Theorem G2, we can deduce that:

Theorem G3 (Monotone convergence theorem). *Let f, g and $f_1, f_2 \cdots$ be Borel-measurable functions of a measure space $(\Omega, \mathcal{A}, \mu)$ to $\mathbb{R}$. Then, we have:*

(i) *if $g \le f_n(\omega)$ a.e. for all n, where $-\infty < \int_\Omega g(\omega)d\mu$, and $f_1 \ge f_2 \ge \cdots \ge f_n$ a.e., then $\int_\Omega f_n(\omega)d\mu \to \int_\Omega f(\omega)d\mu$,*

(ii) *if $f_n \le g$ for all n, where $\int_\Omega g(\omega)d\mu < +\infty$, and $f_1 \le f_2 \le \cdots \le f_n$ a.e, then $\int_\Omega f_n(\omega)d\mu \to \int_\Omega f(\omega)d\mu$.*

and we have:

Theorem G4 (Dominated convergence theorem). *Let f, g and $f_1, f_2 \cdots$ be Borel-measurable functions of a measure space $(\Omega, \mathcal{A}, \mu)$ to $\mathbb{R}$. If $|f_n(\omega)| \le g(\omega)$ a.e. for all n, where $\int_\Omega g(\omega)d\mu < +\infty$ and $f_n \to f$ a.e., then f is integrable and $\int_\Omega f_n(\omega)d\mu \to \int_\Omega f(\omega)d\mu$.*

A sequence $(f_n)_{n\in\mathbb{N}}$ of Borel-measurable functions of a measure space $(\Omega, \mathcal{A}, \mu)$ to $\mathbb{R}$ is said to converge to a measurable function f in measure if for every $\epsilon > 0$, $\mu(\{\omega \in \Omega \mid |f_n(\omega) - f(\omega)| \ge \epsilon\}) \to 0$ as $n \to \infty$.

Theorem G5. *Let the measure μ be finite, or $\mu(\Omega) < +\infty$. Then, a sequence of measurable functions on $(\Omega, \mathcal{A}, \mu)$ which converges almost everywhere to f converges in measure to f.*

Let f be a map from a measure space $(\Omega, \mathcal{A}, \mu)$ to a metric space X and assume that f is Borel-measurable. The distribution of f in the measure ν on the measurable space $(X, \mathcal{B}(X))$ is defined by $\nu = \mu \circ f^{-1}$, or $\nu(B) = \mu(\{\omega \in \Omega \mid f(\omega) \in B\})$ for every $B \in \mathcal{B}(X)$.

A sequence $(f_n)_{n\in\mathbb{N}}$ of Borel-measurable functions of a measure space $(\Omega, \mathcal{A}, \mu)$ to a metric space X is said to converge in distribution to a measurable mapping f if:

$$\int_\Omega g(\omega)d\nu_n \to \int_\Omega g(\omega)d\nu$$

for every bounded and continuous function g of X to $\mathbb{R}$, where $\nu = \mu \circ f^{-1}$ and $\nu_n = \mu \circ f_n^{-1}$ are distributions of f and f_n, $n = 0, 1, \ldots$ respectively.

Theorem G6. *Let f and $f_1, f_2 \cdots$ be Borel-measurable mappings of a measure space $(\Omega, \mathcal{A}, \mu)$ to a metric space (X, d). Then, the mapping $\omega \mapsto d(f_n(\omega), f(\omega))$ of $(\Omega, \mathcal{A})$ to $(\mathbb{R}, \mathcal{B}(\mathbb{R}))$ is Borel-measurable. Furthermore, if the sequence $\{d(f_n(\cdot), f(\cdot))\}_{n\in\mathbb{N}}$ converges in measure to zero, then the sequence $(f_n)_{n\in\mathbb{N}}$ converges in distribution to f.*

A measure μ on a measurable space $(\Omega, \mathcal{A})$ is said to be σ-finite if $\Omega = \cup_{n=1}^{\infty} A_n$, where $A_n \in \mathcal{A}$ for all n and $\mu(A_n) < \infty$. A countably additive set function λ on a measurable space $(\Omega, \mathcal{A})$ is said to be absolutely continuous with respect to a measure μ if $\mu(A) = 0$ implies that $\lambda(A) = 0$.

Theorem G7 (Radon–Nikodym). *Let μ be a σ-finite measure and λ a countably additive set function on a measurable space $(\Omega, \mathcal{A})$. Assume that λ is absolutely continuous with respect to the measure μ. Then, there exists a Borel-measurable function $g: \Omega \to \mathbb{R}$ such that,*

$$\lambda(A) = \int_A g(\omega) d\mu$$

for every $A \in \mathcal{A}$.

Let $(\Omega_1, \mathcal{A}_1, \mu_1)$ and $(\Omega_2, \mathcal{A}_2, \mu_2)$ be measurable spaces. Consider the product measurable space $(\Omega, \mathcal{A}) = (\Omega_1 \times \Omega_2, \mathcal{A}_1 \times \mathcal{A}_2)$. For each measurable set $A \in \mathcal{A} = \mathcal{A}_1 \times \mathcal{A}_2$, the section $A(\omega_1)$ of the set A at $\omega_1 \in \Omega_1$ is defined by $A(\omega_1) = \{\omega_2 \in A_2 \mid (\omega_1, \omega_2) \in A\}$. We define the set function μ on the product measurable space $(\Omega, \mathcal{A})$ by:

$$\mu(A) = \int_{\Omega_1} \mu_2(A(\omega_1)) d\mu.$$

Then, one can show that $\int_{\Omega_1} \mu_2(A(\omega_1)) d\mu = \int_{\Omega_2} \mu_1(A(\omega_2)) d\mu_2$ and μ is the unique measure on $(\Omega, \mathcal{A})$ such that $\mu(A \times B) = \mu(A)\mu(B)$ for all $A \in \mathcal{A}_1$ and $B \in \mathcal{A}_2$. The measure μ is called the product measure of μ_1 and μ_2 and written as $\mu = \mu_1 \times \mu_2$.

Then, we have:

Theorem G8 (Fubini). *Let f be an integrable function defined on the product measure space $(\Omega, \mathcal{A}, \mu) = (\Omega_1 \times \Omega_2, \mathcal{A}_1 \times \mathcal{A}_2, \mu_1 \times \mu_2)$. Then, the following equality holds.*

$$\int_\Omega f(\omega_1, \omega_2) d\mu = \int_{\Omega_1} \left(\int_{\Omega_2} f(\omega_1, \omega_2) d\mu_2 \right) d\mu_1 = \int_{\Omega_2} \left(\int_{\Omega_1} f(\omega_1, \omega_2) d\mu_1 \right) d\mu_2.$$

In the following paragraphs of this section, we shall consider the integration theory for the correspondences. In doing so, we start to consider the mapping f of $(\Omega, \mathcal{A}, \mu)$ to the ℓ-dimensional Euclidean space $\mathbb{R}^\ell$. Let $(\Omega, \mathcal{A}, \mu)$ be a measure space and $f : (\Omega, \mathcal{A}, \mu) \to (\mathbb{R}^\ell, \mathcal{B}(\mathbb{R}^\ell))$ be a Borel-measurable function. The function $\omega \mapsto (f^1(\omega) \cdots f^\ell(\omega))$ is said to be integrable if each coordinate function $f^t: \Omega \to \mathbb{R}$, $t = 1 \cdots \ell$ is integrable. The integral $\int_\Omega f(\omega) d\mu$ is then defined by:

$$\int_\Omega f(\omega) d\mu = \left(\int_\Omega f^1(\omega) d\mu \cdots \int_\Omega f^\ell(\omega) d\mu \right).$$

The set of integrable mappings of $(\Omega, \mathcal{A}, \mu)$ to $(\mathbb{R}^\ell, \mathcal{B}(\mathbb{R}^\ell))$ is denoted by $\mathcal{L}(\Omega, \mathcal{A}, \mu)$.

We now define the integration of a correspondence $\phi: \Omega \to \mathbb{R}^\ell$ as follows.

Definition G1. The set,

$$\left\{\int_{\Omega} f(\omega)d\mu \in \mathbb{R}^{\ell} \mid f \in \mathcal{L}(\Omega, \mathcal{A}, \mu), f(\omega) \in \phi(\omega) \text{ a.e.}\right\}$$

is called the integral of the correspondence ϕ and denoted by $\int_{\Omega} \phi(\omega)d\mu$.

The next theorem ensures that this definition makes sense for a large class of correspondences, namely that $\int_{\Omega} \phi(\omega)d\mu \neq \emptyset$ for many ϕs.

Theorem G9 (Measurable selection theorem). *Let ϕ be a correspondence of a measure space $(\Omega, \mathcal{A}, \mu)$ to a complete and separable metric space X with a measurable graph, that is to say,*

$$\text{Graph}\, \phi = \{(\omega, \xi) \in \Omega \times X \mid \xi \in \phi(\omega)\} \in \mathcal{A} \times \mathcal{B}(X).$$

Then, there exists a Borel-measurable function f of Ω to X such that $f(\omega) \in \phi(\omega)$ a.e.

The next related proposition will be also used in the text.

Proposition G7. *Let ϕ be a closed valued correspondence of a measurable space $(\Omega, \mathcal{A})$ to a complete and separable metric space X which is measurable in the sense that,*

$$\{\omega \in \Omega \mid \phi(\omega) \cap F \neq \emptyset\} \in \mathcal{A}$$

for every closed subset F of X. Then, there exists a countable family of measurable mappings $\{f_n\}_{n \in \mathbb{N}}$ of $(\Omega, \mathcal{A})$ to X such that $\phi(\omega) = cl\{f_n(\omega) \mid n \in \mathbb{N}\}$ for every $\omega \in \Omega$.

A correspondence ϕ of $(\Omega, \mathcal{A}, \mu)$ to $(\mathbb{R}^{\ell}, \mathcal{B}(\mathbb{R}^{\ell}))$ is said to be integrably bounded if there exists an integrable function g of $(\Omega, \mathcal{A}, \mu)$ to $\mathbb{R}^{\ell}_{+}$ such that $|\phi(\omega)| \leq g(\omega)$ a.e. Suppose that a correspondence ϕ is integrably bounded. Then, by the measurable selection theorem (Theorem G9) and Proposition G6, it is immediate that the integral of ϕ is nonempty, $\int_{\Omega} \phi(\omega)d\mu \neq \emptyset$.

Proposition G8. *Let ϕ be a correspondence of a measurable space $(\Omega, \mathcal{A})$ to $\mathbb{R}^{\ell}$ with a measurable graph. Then we have,*

(i) *if a function $h: (\Omega, \mathcal{A}) \to (\mathbb{R}^{\ell}, \mathcal{B}(\mathbb{R}^{\ell}))$ is Borel measurable, then the correspondence $\omega \mapsto \phi(\omega) + h(\omega)$ has a measurable graph,*
(ii) *if g is a measurable mapping of a measure space $(\Sigma, \mathcal{S})$ to $(\Omega, \mathcal{A})$, then the correspondence $\phi \circ g : \Sigma \to \Omega,\ \sigma \mapsto \phi(g(\sigma))$ has a measurable graph.*

Proposition G9. *Let ϕ be a correspondence of a measurable space $(\Omega, \mathcal{A}, \mu)$ to $\mathbb{R}^{\ell}$ with a measurable graph and $\int_{\Omega} \phi(\omega)d\mu \neq \emptyset$. Then, for every vector $\boldsymbol{p} \in \mathbb{R}^{\ell}$, it follows that,*

$$\sup\left\{\boldsymbol{px} \in \mathbb{R} \mid \boldsymbol{z} \in \int_{\Omega} \phi(\omega)d\mu\right\} = \int_{\Omega} \sup\{\boldsymbol{pz} \in \mathbb{R}^{\ell} \mid \boldsymbol{z} \in \phi(\omega)\}d\mu.$$

The next theorem is deduced from Liapunov's theorem (Theorem G1).

Theorem G10. *Let ϕ be a correspondence of an atomless measurable space $(\Omega, \mathcal{A})$ to $\mathbb{R}^\ell$. Then the integral $\int_\Omega \phi(\omega)d\mu$ is a convex subset of $\mathbb{R}^\ell$.*

The next theorem is known as the ℓ-dimensional version of the Fatou's lemma (Theorem G2).

Theorem G11 (Fatou's lemma in ℓ dimension). *Let $(\phi)_{n\in\mathbb{N}}$ be a sequence of integral functions of a measure space $(\Omega, \mathcal{A}, \mu)$ to $\mathbb{R}^\ell_+$. Suppose that $\lim \int_\Omega \phi_n d\mu$ exists. Then, there exists an integrable function $\phi: (\Omega, \mathcal{A}, \mu) \to \mathbb{R}^\ell_+$ such that:*

(i) $\phi(\omega) \in Ls(\phi_n(\omega))$ *a.e. in A,*
(ii) $\int_\Omega \phi(\omega)d\mu \leq \lim_{n\to\infty} \int_\Omega \phi_n(\omega)d\mu$.

Furthermore, if the sequence (ϕ_n) is uniformly bounded and if the set $\{\phi_n(\omega) \mid n \in \mathbb{N}\}$ is bounded a.e. in Ω, then there exists a measurable selection ϕ of $Ls(\phi_n)$ such that:

$$\int_\Omega \phi(\omega)d\mu = \lim_{n\to\infty} \int_\Omega \phi_n(\omega)d\mu.$$

If for every function ϕ of Ω to $\mathbb{R}^\ell$ with properties (i) and (ii), it follows that $\int_\Omega \phi(\omega)d\mu = \lim_{n\to\infty} \int_\Omega \phi_n(\omega)d\mu$, then the sequence (ϕ_n) is uniformly integrable, that is, there exists an integrable function $g: \Omega \to \mathbb{R}^\ell_+$ such that $|\phi_n| \leq g$ a.e. for all n. (For the definition of Ls, see Appendix D.)

From Theorem G11, we can deduce that,

Theorem G12. *Let $(\phi)_{n\in\mathbb{N}}$ be a sequence of correspondences of a measure space $(\Omega, \mathcal{A}, \mu)$ to $\mathbb{R}^\ell_+$ such that there exists an integrable function $g: (\Omega, \mathcal{A}, \mu) \to \mathbb{R}^\ell_+$ with $\phi_n(\omega) \leq g(\omega)$ a.e. for all n. Then, it follows that:*

$$Ls\left(\int_\Omega \phi_n(\omega)d\mu\right) \subset \int_\Omega Ls(\phi_n(\omega))d\mu.$$

From Theorem G12, we can easily obtain that:

Corollary G1. *If the correspondence ϕ of a measure space $(\Omega, \mathcal{A}, \mu)$ to $\mathbb{R}^\ell$ such that $b \leq \phi(\omega)$ a.e. for some $b \in \mathbb{R}^\ell$, is close-valued and integrably bounded, then the integral $\int_\Omega \phi(\omega)d\mu$ is compact.*

Proof. Since ϕ is bounded from below, we can assume without loss of generality that $\phi(\omega) \subset \mathbb{R}^\ell_+$ a.e. We define a sequence of measurable correspondences $(\phi_n)_{n\in\mathbb{N}}$ by $\phi_n = \phi,\ n = 1, 2, \ldots$ Then, $Ls(\phi_n(\omega)) = \phi(\omega)$ for all n, since $\phi(\omega)$ is a closed set. Hence, by Theorem G12, it follows that $Ls\left(\int_\Omega \phi(\omega)d\mu\right) \subset \int_\Omega \phi(\omega)d\mu$, in other words, every limit point of $\int_\Omega \phi(\omega)d\mu$ belongs to $\int_\Omega \phi(\omega)d\mu$. □

And we also have:

Corollary G2. *Let (X, d) be a metric space. Suppose that a correspondence $\phi: \Omega \times X \to \mathbb{R}^\ell_+$ satisfies that*

(i) *there exists an integrable function $g: (\Omega, \mathcal{A}, \mu) \to \mathbb{R}_+$ with $\|\phi(\omega, \mathbf{z})\| \leq g(\omega)$ a.e. for every $\mathbf{z} \in X$,*
(ii) *the correspondence $\phi(\omega, \cdot): X \to \mathbb{R}^\ell$ is closed at $\mathbf{z} \in X$ a.e.*

Then the relation $\mathbf{z} \mapsto \int_\Omega \phi(\cdot, \mathbf{z}) d\mu$ is closed at $\mathbf{z}$.

Proof. We need to show that for every sequence $(\mathbf{z}_n)_{n \in \mathbb{N}}$ converging to $\mathbf{z}$, one has $Ls\left(\int_\Omega \phi(\cdot, \mathbf{z}_n) d\mu\right) \subset \int_\Omega \phi(\cdot, \mathbf{z}) d\mu$. Since $\phi(\omega, \cdot)$ is closed at $\mathbf{z} \in X$, $Ls(\phi(\cdot, \mathbf{z}_n)) \subset \phi(\cdot, \mathbf{z})$. Therefore, by Theorem G12, $Ls\left(\int_\Omega \phi(\cdot, \mathbf{z}_n) d\mu\right) \subset \int_\Omega Ls(\phi(\cdot, \mathbf{z}_n)) d\mu \subset \int_\Omega \phi(\cdot, \mathbf{z}) d\mu$. □

Radon–Nikodym's theorem (Theorem G7) is extended to correspondences as follows.

Theorem G13 (Radon–Nikodym for correspondences). *Let $(\Omega, \mathcal{A}, \mu)$ be a measure space and Φ a correspondence of $\mathcal{A}$ to $\mathbb{R}^\ell$ with the following properties:*

(i) *$\Phi(\bigcup_{n=1}^\infty A_n) = \sum_{n=1}^\infty \Phi(A_n)$ for every pairwise disjoint sequence $(A_n)_{n=1}^\infty$ of $\mathcal{A}$;*
(ii) *$\Phi(A)$ is convex set for every $A \in \mathcal{A}$; and*
(iii) *$\Phi(A) = \{0\}$ if $\mu(A) = 0$.*

Then, there exists a convex-valued correspondence of ϕ of $(\Omega, \mathcal{A}, \mu)$ to $\mathbb{R}^\ell$ such that:

$$\int_A \phi(\omega) d\mu \subset \Phi(A) \quad \text{with } cl \int_A \phi(\omega) d\mu \subset cl\Phi(\omega) \quad \text{for every } A \in \mathcal{A},$$

where clS denotes the closure of the set S, and the correspondence ϕ is measurable in the sense that:

$$\{\omega \in \Omega \mid \phi(\omega) \cap F \neq \emptyset\} \in \mathcal{A} \quad \text{for every closed subset } F \text{ of } \mathbb{R}^\ell.$$

Let $(\Omega, \mathcal{A}, \mu)$ be a measure space, and $\phi(\omega)$ a mapping from Ω to a Banach space X (see Appendix E). ϕ is said to be weakly Borel measurable if for every $\mathbf{p} \in X^*$, the function $\mathbf{p} \circ \phi: \Omega \to \mathbb{R}$ is Borel-measurable. ϕ is said to be strongly measurable if there exists a sequence of simple functions $\{\phi_n\}$ with $\|\phi_n(\omega) - \phi(\omega)\| \to 0$ as $n \to \infty$ a.e.

If ϕ is strongly measurable, it is weakly measurable. To see this, take $\mathbf{p} \in X^*$. Since ϕ is strongly measurable, there exists a sequence $\{\phi_n\}$ of simple functions converging to ϕ in the norm topology for almost all $\omega \in \Omega$. Hence, $\mathbf{p} \circ \phi_n \to \mathbf{p} \circ \phi$ a.e. Since ϕ_ns are simple, they are Borel measurable for all n. Therefore, by Proposition G3, $\mathbf{p} \circ \phi$ is Borel measurable.

We say that ϕ is μ-almost separably valued if there exists a set N of μ-measure 0 such that $\{\phi(\omega) \in X | \ \omega \in \Omega \backslash N\}$ is separable. The next theorem is considered to be a generalization of Proposition G5 to infinite-dimensional spaces.

Theorem G14 (Pettis). *Suppose ϕ is weakly Borel-measurable and μ-almost separably valued, then ϕ is strongly Borel measurable. Furthermore, the approximating simple functions can be taken as:*

$$\{\phi_n(\omega) \mid \omega \in \Omega, n \in \mathbb{N}\} \subset \{\phi(\omega) \mid \omega \in \Omega\} \cup \{\mathbf{0}\}.$$

BANACH SPACES AND RELATED TOPICS

Appendix

H

Let X be a (real) vector space. Recall that a norm on X is a function denoted by $\|\cdot\|$ of X to $\mathbb{R}_+$ satisfying (n-1), (n-2), and (n-3) of Appendix E. The pair $(X, \|\cdot\|)$ is called a normed space. As usual, we often call X a normed space if the norm on it is understood. A normed space is a metric space defined by $d(\boldsymbol{x},\boldsymbol{y}) = \|\boldsymbol{x}-\boldsymbol{y}\|$. A normed space is called a Banach space if it is complete (that is, every Cauchy sequence in X is convergent to a point in X) with respect to this metric induced from thc norm.

Theorem H1 (Theorem of completion). *Let X be a normed space which is not complete. Then, X is isometrically isomorphic to a dense, linear subspace of a Banach space $\tilde{X}$, that is, there exists a one-to-one and onto mapping ξ of X to a dense, linear subspace of $\tilde{X}$ such that:*

$$\xi(a\boldsymbol{x}+b\boldsymbol{y}) = a\xi(\boldsymbol{x}) + b\xi(\boldsymbol{y}), \quad \|\xi(\boldsymbol{x})\|_{\tilde{X}} = \|\boldsymbol{x}\|_X$$

for all $\boldsymbol{x},\boldsymbol{y} \in X$ and all $a, b \in \mathbb{R}$.

A topological vector space is a vector space X with a topology such that the addition $+: X \times X \to X$ and the scalar multiplication $\cdot: \mathbb{R} \times X \to X$ are continuous. Then the normed vector space is a topological vector space with the metric topology induced by the norm. For if $\boldsymbol{x}_n \to \boldsymbol{x} \in X$, $\boldsymbol{y}_n \to \boldsymbol{y} \in X$, and $a_n \to a \in \mathbb{R}$, then $\|(\boldsymbol{x}_n+\boldsymbol{y}_n)-(\boldsymbol{x}+\boldsymbol{y})\| \le \|\boldsymbol{x}_n-\boldsymbol{x}\| + \|\boldsymbol{y}_n-\boldsymbol{y}\| \to 0$, and $\|a_n\boldsymbol{x}_n - a\boldsymbol{x}\| \le \|a_n\boldsymbol{x}_n - a_n\boldsymbol{x}\| + \|a_n\boldsymbol{x} - a\boldsymbol{x}\| \le |a_n|\|\boldsymbol{x}_n-\boldsymbol{x}\| + |a_n - a|\|\boldsymbol{x}\| \to 0$.

Recall that a linear functional is a real-valued linear function on a vector space. The set of continuous linear functionals on X is called the dual space of X and denoted by X^*. The dual space X^* has naturally the vector space structure. If X is a normed space, then X^* is also a normed space by the operator norm defined by $\|\boldsymbol{p}\| = \sup\{|\boldsymbol{p}\boldsymbol{x}|| \; \|\boldsymbol{x}\| \le 1\}$ for $\boldsymbol{p} \in X^*$. Obviously, this definition is equivalent to $\|\boldsymbol{p}\| = \sup\{|\boldsymbol{p}(\boldsymbol{x}/\|\boldsymbol{x}\|)| \; \boldsymbol{x} \ne \boldsymbol{0}\}$.

Proposition H1. *A linear functional $\boldsymbol{p}: X \to \mathbb{R}$ is continuous if and only if it is bounded, or $\|\boldsymbol{p}\| < +\infty$.*

A dual pairing is an ordered pair of vector spaces (X, L) and a real-valued bilinear function on $X \times L$ (for the definition of the multilinear map, see Appendix E). For $\boldsymbol{x} \in X$ and $\boldsymbol{p} \in L$, the value of the bilinear functional is denoted by $\boldsymbol{px}$, or $\boldsymbol{p}(\boldsymbol{x})$ and sometimes by $\langle \boldsymbol{p}, \boldsymbol{x} \rangle$. Given a dual pairing (X, L), there exists a weakest topology on X such that $X^* = L$, which is denoted by $\sigma(X, L)$. The $\sigma(X, L)$-topology is characterized by the net as follows.

Proposition H2. *A net $(\boldsymbol{x}_\lambda)_{\lambda \in \Lambda}$ in X converges to $\boldsymbol{x} \in X$ with respect to $\sigma(X, L)$-topology if and only if for every $\boldsymbol{p} \in L$, $\boldsymbol{px}_\lambda \to \boldsymbol{px}$.*

If X is already a topological vector space, then there exists naturally a dual pairing (X, X^*). Since by definition the $\sigma(X, L)$-topology is as weak or weaker than the original topology on X, the $\sigma(X, L)$-topology is often called the weak topology. The $\sigma(X^*, X)$-topology on X^* is called the weak* topology.

On the other hand, given a dual pairing (X, L), there is a strongest topology on X such that $X^* = L$. It is called the Mackey topology and denoted by $\tau(X, L)$. The next theorem characterizes the Mackey topology by nets.

Proposition H3. *A net $(\boldsymbol{x}_\lambda)_{\lambda \in \Lambda}$ in X converges to $\boldsymbol{x} \in X$ with respect to $\tau(X, L)$-topology if and only if for every $\sigma(L, X)$-compact, convex, and circled subset C of L, $\sup\{|\boldsymbol{px}_\lambda - \boldsymbol{px}|| \ \boldsymbol{p} \in C\} \to 0$, where a set C is circled if and only if $c \in C$ implies that $rc \in C$ for every $-1 \leq r \leq 1$.*

A fundamental result on the weak and the Mackey topology is the following.

Theorem H2 (Mackey). *If (X, L) is a dual pairing, every convex subset of X is closed in the $\sigma(X, L)$-toplogy if and only if it is closed in the $\tau(X, L)$-topology.*

The next result, which is known as Alaoglu's theorem, is also a fundamental theorem with respect to the weak* topology.

Theorem H3 (Alaoglu). *If X is a normed space, then the unit ball of X^*, $B = \{\boldsymbol{p} \in X^* \,|\, \|\boldsymbol{p}\| \leq 1\}$ is compact in the $\sigma(X^*, X)$-topology.*

We now give a few examples. In Appendix E, we define for every integer p with $1 \leq p < +\infty$, the spaces ℓ^p by:

$$\ell^p = \left\{ \boldsymbol{x} = (x^t) \,\middle|\, \sum_{t=0}^{\infty} |x^t|^p < +\infty \right\}$$

with the norm,

$$\|x\|_p = \left(\sum_{t=0}^{\infty} |x^t|^p \right)^{1/p}.$$

The conditions (n-1) and (n-2) can be easily verified. The condition (n-3) follows from the Minkowski's inequality:

$$\left(\sum_{t=0}^{\infty}|x^t+y^t|^p\right)^{1/p}\leq\left(\sum_{t=0}^{\infty}|x^t|^p\right)^{1/p}+\left(\sum_{t=0}^{\infty}|y^t|^p\right)^{1/p}$$

which can be proved for every integer p with $1 \leq p < +\infty$. Moreover, we can show that the space ℓ^p is complete with respect to this norm, hence it is a Banach space.

One more important example of a Banach space of sequences is the space ℓ^∞ defined by:

$$\ell^\infty = \{\boldsymbol{x} = (x^t) \mid \sup_{t\geq 0}|x^t| < +\infty\}.$$

The norm of the vector $\boldsymbol{x} \in \ell^\infty$ is given by $\|\boldsymbol{x}\| = \sup_{t\geq 0}|x^t|$. The conditions (n-1), (n-2), and (n-3) can be easily verified and one can prove that the space ℓ^∞ is complete with respect to this norm. Hence, the space ℓ^∞ is also a Banach space.

Proposition H4. *For an integer with $1 \leq p < +\infty$, the dual space ℓ^{p*} of ℓ^p is isomorphic to the space ℓ^q, where q is an integer such that $\frac{1}{p}+\frac{1}{q}=1$ when $p > 1$, and we set $q = +\infty$ when $p = 1$.*

Then, we can consider $\sigma(\ell^\infty,\ell^1)$ and $\tau(\ell^\infty,\ell^1)$ topologies on the space ℓ^∞, and $\sigma(\ell^1,\ell^\infty)$ and $\tau(\ell^1,\ell^\infty)$ topologies on the space ℓ^1, respectively. The next propositions are particular to the case of ℓ^∞ (and ℓ^1), but useful.

Proposition H5. *Let X be a (norm) bounded subset of ℓ^∞. Then on the set X, the Mackey topology $\tau(\ell^\infty,\ell^1)$ coincides with the product topology, or the topology of the coordinate-wise convergence.*

Since the product toplogy is weaker than the $\sigma(\ell^\infty,\ell^1)$-topology, the product topology, $\sigma(\ell^\infty,\ell^1)$, and $\tau(\ell^\infty,\ell^1)$ topologies are all equal on bounded subsets of ℓ^∞.

Proposition H6. *Let Π be a $\sigma(\ell^1,\ell^\infty)$ compact subset of ℓ^1. Then, the paring map from $\Pi \times \ell^\infty$ to $\mathbb{R}$, $(\boldsymbol{p},\boldsymbol{x}) \mapsto \boldsymbol{px}$ is jointly continuous with respect to $\sigma(\ell^1,\ell^\infty)\times\tau(\ell^\infty,\ell^1)$ topology on $\Pi \times \ell^\infty$.*

Consider the measurable space $(\mathbb{N}, 2^{\mathbb{N}})$, or the set of non-negative integers in which every subset is a measurable set. Let *ba* be the set of finitely additive set functions with bounded variation, namely that:

$$ba = \{\pi\colon 2^{\mathbb{N}} \to \mathbb{R} \mid \sup_{E\subset\mathbb{N}}|\pi(E)| < +\infty,\ \ \pi(E\cup F) = \pi(E)+\pi(F) \text{ whenever } E\cap F = \emptyset\}.$$

Then, we can show that the space *ba* is a Banach space with the norm:

$$\|\pi\| = \sup\left\{\sum_{i=1}^{n}|\pi(E_i)| \,\middle|\, E_i\cap E_j=\emptyset \text{ for } i\neq j, n\in\mathbb{N}\right\}.$$

Proposition H7. *The dual space $\ell^{\infty *}$ of ℓ^{∞} is isomorphic to the space ba, or $\ell^{\infty *} = ba$.*

The set of all countably additive set functions on $\mathbb{N}$ which is a subspace of *ba* and denoted by *ca* can be identified with the space ℓ^1.

We can generalize these spaces as follows. Let $\rho = (\rho^0, \rho^1, \dots)$ be a sequence of strictly positive real numbers; $\rho^t > 0$ for all $t \in \mathbb{N}$. For an integer p with $1 \le p < +\infty$, we define the "weighted" ℓ^p space by:

$$\ell^p_\rho = \left\{ \boldsymbol{x} = (x^t) \middle| \sum_{t=0}^{\infty} |\rho^t x^t|^p < +\infty \right\}$$

with the norm,

$$\|x\|_p = \left(\sum_{t=0}^{\infty} |\rho^t x^t|^p \right)^{1/p}.$$

For $p = \infty$, the "weighted"-ℓ^∞ space, ℓ^∞_ρ, is defined analogously by:

$$\ell^\infty_\rho = \{\boldsymbol{x} = (x^t) | \sup_{t \ge 0} |\rho^t x^t| < +\infty\}.$$

The norm of the vector $\boldsymbol{x} \in \ell^\infty_{\rho^t}$ is given by $\|\boldsymbol{x}\| = \sup_{t \ge 0} |\rho^t x^t|$.

Finally, we define the 'weighted' *ba* space by:

$$ba_\rho = \left\{ \pi : 2^{\mathbb{N}} \to \mathbb{R} \middle| \sup_{E \subset \mathbb{N}} \int_E \rho d|\pi| < +\infty, \ \pi(E \cup F) = \pi(E) + \pi(F) \right.$$
$$\left. \text{whenever } E \cap F = \emptyset \right\},$$

where the integral $\int_E \rho d|\pi|$ is defined in a similar way as the integral of the countably additive measure.

The space ba_ρ is endowed with the norm:

$$\|\pi\| = \sup \left\{ \sum_{i=1}^{n} \left| \int_{E_i} \rho d|\pi| \right| \middle| E_i \cap E_j = \emptyset \text{ for } i \ne j, n \in \mathbb{N} \right\}.$$

These weighted spaces are Banach spaces. Indeed, it can be easily verified that the space ℓ^p_ρ is isometrically isomorphic to the space ℓ^p for $1 \le p \le +\infty$, and the space ba_ρ is also isometrically isomorphic to the space *ba*.

Proposition H8. *For an integer with $1 \le p < +\infty$, the dual space ℓ^{p*}_ρ of ℓ^p_ρ is isomorphic to the space $\ell^q_{\rho^{-1}}$, where $\rho^{-1} = (1/\rho^0, 1/\rho^1 \cdots)$ and q is an integer such that $\frac{1}{p} + \frac{1}{q} = 1$ when $p > 1$, and we set $q = +\infty$ when $p = 1$.*

Proposition H9. *The dual space $\ell^{\infty *}_\rho$ of ℓ^∞_ρ is isomorphic to the space $ba_{\rho^{-1}}$.*

The space ℓ^p and *ba* can be extended to the spaces of functions on a measure space $(\Omega, \mathcal{A}, \mu)$. The space $L^p(\Omega, \mathcal{A}, \mu)$ for $1 \le p < +\infty$, is defined as the set of all Borel-measurable functions f such that $\int_\Omega |f(\omega)|^p d\mu < +\infty$.

For a Borel-measurable mapping $g: \Omega \to \mathbb{R}$, we define the essential supremum of g as:

$$\text{esssup } g = \inf\{c \ge 0 | \mu(\{\omega \in \Omega | g(\omega) > c\}) = 0\}.$$

The space $L^\infty(\Omega, \mathcal{A}, \mu)$ is the set of all Borel-measurable functions with finite essential supremum.

Unfortunately, the "norm" of f, $\|f\| = \left(\int_\Omega |f(\omega)|^p d\mu\right)^{1/p}$ for $f \in L^p(\Omega, \mathcal{A}, \mu)$, and $\|f\| = \text{esssup}|f|$ for $f \in L^\infty(\Omega, \mathcal{A}, \mu)$ are not exactly norms, since even if $\|f - g\| = 0$, it may be the case that $f(\omega) \ne g(\omega)$ on a set of measure 0. We can get the true norm by considering $f \in L^p(\Omega, \mathcal{A}, \mu)$ for $1 \le p \le +\infty$ as an equivalence class determined by the equivalence relation $f \sim g$ if and only if $f = g$ a.e. In this case, we denote by $\mathcal{L}^p(\Omega, \mathcal{A}, \mu)$ for $1 \le p \le +\infty$ the set of the equivalence classes.

The space $ba(\Omega, \mathcal{A}, \mu)$ is defined as the collection of finitely additive set functions which are absolutely continuous with respect to μ and have bounded (finite) total variation norm. Here, the total variation norm of $\pi: \mathcal{A} \to \mathbb{R}$ is given by:

$$\|\pi\| = \sup\left\{\sum_{i=1}^{n} |\pi(E)||E_i \cap E_j = \emptyset \text{ for } i \ne j, E_i \in \mathcal{A}, i = 1 \cdots n, n \in \mathbb{N}\right\}.$$

The Radon Nikodym's theorem (Theorem G7) says that the natural embedding from $L_1(\Omega, \mathcal{A}, \mu)$ to $ba(\Omega, \mathcal{A}, \mu)$ identifies $L_1(\Omega, \mathcal{A}, \mu)$ with the set of countably additive set functions in $ba(\Omega, \mathcal{A}, \mu)$. The set of all finitely additive set functions on $(\Omega, \mathcal{A})$ which is denoted by $ba(\Omega, \mathcal{A})$ endowed with the total variation norm given above is defined similarly.

The set function $\pi \in ba(\Omega, \mathcal{A})$ is said to be purely finitely additive if $\rho = 0$ whenever $\rho \in ba(\Omega, \mathcal{A}$ is countably additive and $0 \le \rho \le \pi$.

Theorem H4 (Yosida–Hewitt). *If $\pi \in ba(\Omega, \mathcal{A})$ and $\pi \ge 0$, then there exist set functions $\pi_c \ge 0$ and $\pi_p \ge 0$ in $ba(\Omega, \mathcal{A}$ such that π_c is countably additive and π_p is purely finitely additive and satisfy $\pi = \pi_c + \pi_p$. This decomposition is unique.*

Let K be a compact metric space. Recall that $\mathcal{B}(K)$ is the set of all Borel-measurable subsets of K (see Appendix G). The countably additive set function on $(K, \mathcal{B}(K))$ is called the signed measure. The set of all bounded signed Borel measures on K is denoted by $ca(K, \mathcal{B}(K))$. The positive orthant $ca_+(K, \mathcal{B}(K))$ is nothing but the set of all measures on $(K, \mathcal{B}(K))$, which is denoted by $\mathcal{M}(K, \mathcal{B}(K))$, or $ca_+(K, \mathcal{B}(K)) = \mathcal{M}(K, \mathcal{B}(K))$.

Proposition H10. *Let $C(K)$ be the set of all continuous functions on K. Then the dual space of $C(K)$ is $ca(K, \mathcal{B}(K))$, or $C^*(K) = ca(K, \mathcal{B}(K))$ by the pairing defined by:*

$$\boldsymbol{px} = \int_K \boldsymbol{p}d\boldsymbol{x} \quad \text{for } \boldsymbol{p} \in C(K) \text{ and } \boldsymbol{x} \in ca(K, \mathcal{B}(K)).$$

Therefore, by definition, the weak* topology on $ca(K, \mathcal{B}(K))$ is the topology of pointwise convergence on $C(K)$, or a net $(\boldsymbol{x}_\alpha)$ on $ca(K, \mathcal{B}(K))$ converges to $\boldsymbol{x} \in ca(K, \mathcal{B}(K))$, if and only if $\boldsymbol{px}_\alpha \equiv \int_K \boldsymbol{p}(t)d\boldsymbol{x}_\alpha \to \int_K \boldsymbol{p}(t)d\boldsymbol{x}$ for every $\boldsymbol{p} \in C(K)$.

Proposition H11. *The weak* topology on $ca(K, \mathcal{B}(K))$ is separable. Moreover, norm-bounded subsets of $ca(K, \mathcal{B}(K))$ are compact and metrizable.*

By Proposition C3, the compact metric space K is separable. Hence, there exists a countable dense subset of K. For $t \in K$, the Dirac measure δ_t is defined by:

$$\delta_t(B) = \begin{cases} 1 & \text{if } t \in B, \\ 0 & \text{otherwise.} \end{cases}$$

The following proposition is a basic tool for the proof of Theorem 3.1, in which we approximate an infinite-dimensional economy by a sequence of finite-dimensional sub-economies.

Proposition H12. *Let $\{t_n\}$ be a countable dense subset of K. Then, $ca(K, \mathcal{B}(K))$ is the closure of the set of finite linear combinations of the Dirac measure δ_{t_n}.*

Let (K, d) be a compact metric space and consider a sequence of pairs of closed subsets K_n of K and continuous and non-negative functions $\boldsymbol{p}_n(t) \in C_+(K)$ on K, $(K_n, \boldsymbol{p}_n)$. If $\boldsymbol{p} \in C_+(K)$ is nonnegative, we write $(K_n, \boldsymbol{p}_n) \to (K, \boldsymbol{p})$ if and only if $K_n \to K$ in the topology of closed convergence (see Appendix F) and for all sequences n_k, t_{n_k} with $t_{n_k} \in K_{n_k}$ and $t_{n_k} \to t$, we have $p(t_{n_k}) \to p(t)$. Then, we have:

Proposition H13. *Suppose $(K_n, \boldsymbol{p}_n) \to (K, \boldsymbol{p})$ and for a bounded sequence $\boldsymbol{x}_n$ with support $(\boldsymbol{x}_n) \subset K_n$, $\boldsymbol{x}_n \to \boldsymbol{x}$. Then, $\boldsymbol{p}_n\boldsymbol{x}_n \to \boldsymbol{px}$.*

The sequence $(K_n, \boldsymbol{p}_n)$ is said to be equi-continuous if and only if for all $\epsilon > 0$, there exists a $\delta > 0$ such that for all $n \in \mathbb{N}$ and all $t, s \in K$ with $d(t, s) < \delta$, it follows that $|p_n(t) - p_n(s)| < \epsilon$.

Proposition H14. *Let K_n, $\boldsymbol{p}_n$ be as above with $K_n \subset K_{n+1}$ and $K_n \to K$ in the topology of closed convergence. If $(K_n, \boldsymbol{p}_n)$ is equi-continuous and $\boldsymbol{p}_n$ are uniformly bounded, there exists a subsequence n_k and a continuous function $\boldsymbol{p} \in C(K)$ such that $(K_{n_k}, \boldsymbol{p}_{n_k}) \to (K, \boldsymbol{p})$.*

The examples of Banach spaces which have been given so far are equipped with the order relation which is compatible with the vector space structure and they are called the ordered vector spaces. That is, they have a reflexive, transitive, and anti-symmetric relation $\leq$ on them which satisfies,

(i) if $\boldsymbol{x} \leq \boldsymbol{y}$ and $\alpha \in \mathbb{R}$, then $\alpha\boldsymbol{x} \leq \alpha\boldsymbol{y}$,
(ii) if $\boldsymbol{x} \leq \boldsymbol{y}$ and $\mathbf{0} \leq \boldsymbol{z}$, then $\boldsymbol{x} + \boldsymbol{z} \leq \boldsymbol{y} + \boldsymbol{z}$.

Let L be an ordered vector space. We define the positive cone L_+ by $L_+ = \{\boldsymbol{x} \in L | \mathbf{0} \leq \boldsymbol{x}\}$. Obviously, the positive cone L_+ is convex and the proper cone, or if $\boldsymbol{x} \in L_+ \cap -L_+$, then $\boldsymbol{x} = \mathbf{0}$. If a vector space L has a proper convex cone C, then we can make L an ordered vector space by giving the order structure defined by $\boldsymbol{x} \leq \boldsymbol{y}$ if and only if $\boldsymbol{y} - \boldsymbol{x} \in C$. An ordered vector space L is called an ordered topological vector space if it is a topological vector space and the positive cone L_+ is closed. Note that if L is an ordered topological vector space, then the dual space L^* is also an ordered vector space, with positive cone $L^*_+ = \{\boldsymbol{p} \in L^* | \mathbf{0} \leq \boldsymbol{p}\boldsymbol{x} \text{ for every } \boldsymbol{x} \in L\}$. A linear functional $\boldsymbol{p} \in L^*_+$ is called positive. Moreover, the positive cone L^*_+ is evidently $\sigma(L^*, L)$- closed.

For $\boldsymbol{x}, \boldsymbol{y} \in L$, define the order interval $[\boldsymbol{x}, \boldsymbol{y}] = \{\mathbf{z} \in L | \boldsymbol{x} \leq \mathbf{z} \leq \boldsymbol{y}\}$. We say that a subset $X \subset L$ is solid if $[\boldsymbol{x}, \boldsymbol{y}] \subset X$ whenever $\boldsymbol{x}, \boldsymbol{y} \in X$.

A subset $X \subset L$ has a supremum (least upper bound) if there exists an element $\sup X \in L$ such that $\boldsymbol{x} \leq \sup X$ for every $\boldsymbol{x} \in X$ and $\sup X \leq \boldsymbol{y}$ for every $\boldsymbol{y} \in L$ which satisfies that $\boldsymbol{x} \leq \boldsymbol{y}$ for every $\boldsymbol{x} \in X$. Similarly, a subset $X \subset L$ has an infimum (greatest lower bound) if there exists an element $\inf X \in L$ such that $\inf X \leq \boldsymbol{x}$ for every $\boldsymbol{x} \in X$ and $\boldsymbol{y} \leq \inf X$ for every $\boldsymbol{y} \in L$ which satisfies that $\boldsymbol{y} \leq \boldsymbol{x}$ for every $\boldsymbol{x} \in X$. We usually write $\boldsymbol{x} \vee \boldsymbol{y}$ rather than $\sup\{\boldsymbol{x}, \boldsymbol{y}\}$ and $\boldsymbol{x} \wedge \boldsymbol{y}$ rather than $\inf\{\boldsymbol{x}, \boldsymbol{y}\}$. If every pair $\boldsymbol{x}, \boldsymbol{y}$ of an ordered vector space L has the supremum $\boldsymbol{x} \vee \boldsymbol{y}$ and the infimum $\boldsymbol{x} \wedge \boldsymbol{y}$, then we call L a vector lattice or Riesz space. We write $\boldsymbol{x}_+ = \boldsymbol{x} \vee \mathbf{0}$ and $\boldsymbol{x}_- = (-\boldsymbol{x}) \vee \mathbf{0}$ and call the positive part and the negative part of $\boldsymbol{x}$, respectively. Then, $\boldsymbol{x} = \boldsymbol{x}_+ - \boldsymbol{x}_-$ and we write $|\boldsymbol{x}| = \boldsymbol{x}_+ + \boldsymbol{x}_-$ and call the absolute value of $\boldsymbol{x}$.

A fundamental property of the vector lattices is the Riesz decomposition property.

Theorem H5 (Riesz decomposition property). *Let L be a vector lattice and let* $\boldsymbol{x}_1 \cdots \boldsymbol{x}_n$, $\mathbf{z}$ *be the positive elements of L such that* $\mathbf{z} \leq \sum_{i=1}^{n} \boldsymbol{x}_i$. *Then, there exist positive elements* $\mathbf{z}_1 \cdots \mathbf{z}_n$ *of L such that* $\mathbf{z} = \sum_{i=1}^{n} \mathbf{z}_i$ *and* $\mathbf{z}_i \leq \boldsymbol{x}_i$ *for each i.*

A topological vector lattice L is called a Banach lattice if it is a Banach space and satisfies that $\|\boldsymbol{x}\| \leq \|\boldsymbol{y}\|$ whenever $\mathbf{0} \leq \boldsymbol{x} \leq \boldsymbol{y}$. The previous spaces given above are all examples of Banach lattices.

NOTES

Appendix I

Appendix A: Most of the materials in this section can be found in standard text books on point-set topology and linear algebra, for example, see Nikaido (1968), Chap. 1 or Takayama (1986), Chap. 0. For the separation hyper-plane theorem (Theorem A1) and K–K–M lemmas (Theorems A2 and A3), we recommend Ichiishi (1983). For Theorem A4, see Rockafellar (1970), p. 84. For the Kuhn–Tucker theorem (Theorem A5), we refer Takayama (1985), Chap. 1.

Appendix B: The basic reference for this section is Kelly (1975). In the text, we derive the notion of nets and its limit from the topology which is a family of open sets satisfying the conditions (t-1) to (t-3). On the contrary, we can derive the concept of topology from the nets which is initially given as follows. Therefore, the concepts of topology and nets are equivalent.

Let $(D_\omega, \succsim_\omega)_{\omega\in\Omega}$ be a family of directed sets. The product order $\succsim$ on the product set $\prod_{\omega\in\Omega} D_\omega$ is defined by:

$$x \succsim y \quad \text{if and only if} \quad x(\omega) \succsim_\omega y(\omega) \quad \text{for all } \omega \in \Omega.$$

The pair $(\prod_{\omega\in\Omega} D_\omega, \succsim)$ is called the product-directed set. It is easy to verify that the product-directed set is, in fact, directed set.

Let X be a set on which the net x_λ is defined and $\mathcal{C}$ be a family consisting of pairs (x_λ, x) of the net and a point of X. We say that a net x_λ converges $(\mathcal{C})$ to x if and only if $(x_\lambda, x) \in \mathcal{C}$ and denote it by $\lim_\lambda x_\lambda = x(\mathcal{C})$. The family $\mathcal{C}$ is called a convergence class for X if and only if it satisfies the following conditions:

(C-1) if $x_\lambda = x$ for each λ, then $(x_\lambda, x) \in \mathcal{C}$,

(C-2) if $(x_\lambda, x) \in \mathcal{C}$, then $(x_{\lambda_p}, x) \in \mathcal{C}$ for every subnet (x_{λ_p}, x) of x_λ,

(C-3) if $(x_\lambda, x) \notin \mathcal{C}$, then there exists a subnet of x_λ such that for any subnet (x_{λ_p}, x) of it, $(x_{\lambda_p}, x) \notin \mathcal{C}$, and

(C-4) let Λ be a directed set, Δ_λ be also a directed set for each $\lambda \in \Lambda$, and $x(\lambda, \delta)$ be a point of X for each $\lambda \in \Lambda$ and $\delta \in \Delta_\lambda$, denoting $z(\lambda, D) = (\lambda, D(\lambda))$ for $(\lambda, D) \in \Lambda \times \prod_{\lambda\in\Lambda} \Delta_\lambda$, if $\lim_\lambda \lim_\delta x(\lambda, \delta) = x\ (\mathcal{C})$, then $(x \circ z, x) \in \mathcal{C}$.

The next theorem shows that there exists a one-to-one correspondence between the topologies on a set X and the convergence classes on it.

Theorem. *Let C be a convergence class for a set X and for each subset S of X, let clS be the set of all points x of X such that for some net x_λ in S, x_λ converges (C) to x. Then, there exists a unique topology on X such that clS defined above is indeed the closure of S, and $(x_\lambda, x) \in C$ if and only if x_λ converges to x with respect to this topology.*

The correspondence between the topologies and the convergence classes is the order inverting in the sense that $C_1 \subset C_2$ if and only if $\tau_2 \subset \tau_1$, where C_1 and C_2 are convergence classes on a set X, and τ_1 and τ_2 are the corresponding topologies.

Appendix C: All materials in Appendix C are standard. See for example, Royden (1988).

Appendix D: A classical text book for the correspondences is Berge (1963). For much of the exposition in this appendix, we owe to Hildenbrand (1974). For the Kakutani's fixed-point theorem and related propositions, see Ichiishi (1983), Chap. 3. A readable proof of the Brower's fixed-point theorem is available in Milnor (1965). Theorem D4 is proposed by Hildenbrand (1974).

Appendix E: For the coordinate-free definition of the differentiable manifolds, see Abraham *et al.* (1988) or Lang (1972). For Banach space theory including Hilbert spaces and the linear mappings on them, see Rudin (1991) or Yosida (1968). The infinite-dimensional version of the Kuhn–Tucker theorem, see Luenberger (1969). A clear and readable text books of manifold theory, including in particular, regular-value theorem and Sard's theorem, are Milnor (1965) and Guillmin and Pollack (1974).

Appendix F: We owe most of the expositions to Hildenbrand (1974). Theorem F1 is proposed by Rådström (1952).

Appendix G: Basic references are Halmos (1974) and Royden (1988). Integration theory for the correspondences, see Hildenbrand (1974). For Pettis' theorem (Theorem G14), see, for example, Yosida (1980, 6th edition, p. 131). The latter part on the range of functions is not explicitly stated in his book, but a careful observation of his proof will validate it.

Appendix H: Basic reference is Dunford and Schwartz (1958). See also Royden (1988), Rudin (1991) and Yosida (1968). Propositions H5 and H6 can be found in Bewley (1991). Propositions H13 and H14 are proposed by Mas-Colell (1975). For the theory of Banach lattices including Theorem H5, see Shaefer (1974).

REFERENCES

Abraham, R, JE Marsden and T Ratiu (1988). *Manifolds, Tensor Analysis, and Applications*, 2nd Ed. (1991) Berlin and New York: Springer.

Aliprantis, CD, D Brown and O Burkinshaw (1987). Edgeworth equilibria. *Econometrica*, 55, 1109–1137.

Annalen des Deutchen Reiches (Annals of German Empire), 1911, Zahn, pp. 165–169.

Araujo, A (1987). The non-existence of smooth demand in general Banach spaces. *Journal of Mathematical Economics*, 17, 1–11.

Arrow, KJ and G Debreu (1954). Existence of an equilibrium for a competitive economy. *Econometrica*, 22, 265–290.

Arrow, KJ and F Hahn (1971). *General Equilibrium Analysis*. Amsterdam: North-Holland.

Arrow, KJ and L Hurwicz (1958). On the stability of the competitive equilibrium. *Econometrica*, 26, 522–552.

Arrow, KJ, HD Block and L Hurwicz (1959). On the stability of the competitive equilibrium II. *Econometrica*, 27, 82–109.

Aumann, RJ (1964). Markets with a continuum of traders. *Econometrica*, 32, 39–50.

Aumann, RJ (1966). Existence of competitive equilibria in markets with a continuum of traders. *Econometrica*, 34, 1–17.

Balasko, Y (1997a). Pareto optima, welfare weights, and smooth equilibrium analysis. *Journal of Economic Dynamics and Control*, 21, 473–503.

Balasko, Y (1997b). Equilibrium analysis of the infinite horizon model with smooth discounted utility functions. *Journal of Economic Dynamics and Control*, 21, 783–829.

Balasko, Y (1997c). The natural projection approach to the infinite horizon model. *Journal of Mathematical Economics*, 27, 251–265.

Beato, P (1982). The existence of equilibria of marginal cost pricing equilibria with increasing returns. *Quarterly Journal of Economics*, 389, 669–688.

Berge, C (1963). *Topological Spaces*. Edinburgh: Oliver and Boyde (Reprinted (1997), NY: Dover).

Bewley, TF (1970). Existence of equilibria with infinitely many commodities. *Journal of Economic Theory*, 4, 514–540.

Bewley, TF (1991). A very weak theorem on the existence of equilibria in atomless economies with infinitely many commodities. In *Equilibrium Theory in Infinite Dimensional Spaces*, M Ali Khan and N Yannelis (eds.), Berlin and New York: Springer-Verlag.

Brown, DJ (1991). Equilibrium analysis with non-convex technologies. In *Handbook of Mathematical Economics*, H Hildenbrand and H Sonnenschein (eds.), pp. 1963–1995 Amsterdam and New York: North-Holland.

Chamberlin, EH (1933). *The Theory of Monopolistic Competition*. Cambridge, MA: Harvard University Press.

Chipman, JS (1965). A survey of the theory of international trade: Part II, the neoclassical theory. *Econometrica*, 33, 685–760.

Chipman, JS (1970). External economies of scale and competitive equilibrium. *Quarterly Journal of Economics*, 84, 347–385.

Cournot, A (1838). *Recherches sur les Principes Mathématiques de la Théorie des Richesses*. In 1929 translated as *Researches into the Mathematical Principles of the Theory of Wealth*. New York: Macmillan.

Debreu, G (1952). A social equilibrium existence theorem. *Proceedings of the National Academy of Sciences of the U.S.A.*, 38, 886–893.

Debreu, G (1954), Valuation equilibrium and pareto optimal. *Proceedings of the National Academy of Science*, 40, 588–592.

Debreu, G (1959). *Theory of Value*. New York: John Wiley and Sons.

Debreu, G (1970). Economies with a finite set of equilibria. *Econometrica*, 38, 387–392.

Debreu, G (1974). Excess demand functions. *Journal of Mathematical Economics*, 1, 15–21.

Debreu, G and H Scarf (1963). A limit theorem on the core of an economy. *International Economic Review*, 4, 235–246.

Dehez, P and JH Dreze (1988a). Competitive equilibria with quantity taking producers and increasing returns to scales. *Journal of Mathematical Economics*, 17, 209–230.

Dehez, P and JH Dreze (1988b). Distributive production sets and equilibria with increasing returns. *Journal of Mathematical Economics*, 17, 231–248.

Dehez, P, JH Dreze and T Suzuki (2003). Imperfect competition à la negishi, also with fixed costs, *Journal of Mathematical Economics*, 39, 219–237.

Dierker, E (1972). Two remarks on the number of equilibria of an economy. *Econometrica*, 50, 867–881.

Dierker, E (1974). Topological methods in Walrasian Economics, Lecture notes in economics and mathematical systems, Vol 92, Berlin: Springer.

Dunford, N and J Schwartz (1958). *Linear Operators, Part I*. New York: Wiley.

Edgeworth, FY (1881). *Mathematical Psychics*, London: Routledge and Kegan Paul.

Edgeworth, FY (1905). Review of *A Geometrical Political Economy* by Henry Cunynghame. *Economic Journal*, 15, 62–71.

Edgeworth, FY (1925). *Papers Relating to Political Economies*, I–III, London: Macmillan and Co Ltd. 2nd printing, New York: Burt Franklin.

Gabszewicz, JJ and JP Vial (1972). Oligopoly à la Cournot in a general equilibrium analysis. *Journal of Economic Theory*, 4, 381–400.

Gale, D and A Mas-Colell (1975). An equilibrium existence theorem for a general model without ordered preferences. *Journal of Mathematical Economics*, 2, 9–16.

Gale, D and A Mas-Colell (1979). Corrections to an equilibrium existence theorem for a general model without ordered preferences. *Journal of Mathematical Economics*, 6, 297–298.

Gale, D and H Nikaido (1965). The Jacobian matrix and global univalence of mappings. *Mathematische Annalen*, 159, 81–93.

Graham, FD (1923). Some aspects of production further considered. *Quarterly Journal of Economics*, 37, 199–227.

Graham, FD (1925). Some fallacies in the interpretation of social cost. A reply. *Quarterly Journal of Economics*, 39, 324–330.

Guillemin, V and A Pollack (1974). *Differential Topology*. Englewood Cliffs, New Jersy: Prentice-Hall.

Halmos, P (1974). *Measure Theory*. Reprint, Berlin and New York: Springer-Verlag.

Harrod, RH (1967). Increasing returns. In *Monopolistic Competition Theory: Studies in Impact; Essays in Honor of Edward H. Chamberlin*, RE Kuenne (ed.), New York: John Wiley and Sons, Inc.

Hicks, JR (1939). *Value and Capital*. Oxford: Oxford University Press.

Hildenbrand, W (1974). *Core and Equilibria of a Large Economy*. Princeton, New Jersy: Princeton University Press.

Hildenbrand, W (1983). On the "law of demand". *Econometrica*, 51, 997–1020.

Hildenbrand, W and A Kirman (1986). *Introduction to Equilibrium Analysis*, Amsterdam and New York: North-Holland.

Hotelling, H (1929). Stability in Competition. *Economic Journal*, 39, 41–57.

Ichiishi, T (1983). *Game Theory for Economic Analysis*. New York and London: Academic Press.

Jevons, WS (1871). *The Theory of Political Economy*. London and New York: Macmillan.

Jones, L (1984). A competitive model of commodity differentiation. *Econometrica*, 52, 507–530.

Kelly, JL (1975). *General Topology*. Berlin and New York: Springer-Verlag.

Knight, F (1924). Some fallacies in the interpretation of social cost. *Quarterly Journal of Economics*, 38, 582–606.

Knight, F (1925) On decreasing cost and comparative cost. A rejoinder. *Quarterly Journal of Economics*, 39, 331–333.

Lancaster, K (1971). Consumer demand: A new approach. New York: Columbia University Press.

Lang, S (1972). *Differential Manifolds*. Reading, Massachassets: Addison-Wesley.

Lenin, VI (1917). *Imperialism, the Highest Stage of Capitalism*. English translation (1947). Moscow: Foreign Language Publishing House.

Lipsey, RG (1960). The theory of customs unions: A general survey. *Economic Journal*, 70, 496–513 (reprinted in A.E.A., *Readings in International Economics*, Homewood, Illinois: Richard D. Irwin, Inc., 1968).

Luenberger, DG (1969). *Optimization by Vector Space Method*. New York: Wiley.

Mantel, R (1979). Equilibrio con rendimiento crecientes a escala. *Anales de la Asociation Argentine de Economia Politica*, 1, 271–283.

Mantel, R (1974). On the characterization of aggregate excess demand. *Journal of Economic Theory*, 7, 348–353.

Marshall, A (1890). *Principles of Economics*. London and New York: Macmillan.

Mas-Colell, A (1974). An equilibrium existence theorem without complete or transitive preferences. *Journal of Mathematical Economics*, 1, 237–246.

Mas-Colell, A (1975). A model of equilibrium with differentiated commodities. *Journal of Mathematical Economics*, 2, 263–296.

Mas-Colell, A (1977). Indivisible commodities and general equilibrium theory. *Journal of Economic Theory*, 16, 443–456.

Mas-Colell, A (1983). Walrasian equilibria as limits of noncooperative equilibria, Part I: Mixed strategies. *Journal of Economic Theory*, 30, 153–170.

Mas-Colell, A (1985). *The Theory of General Economic Equilibrium: A Differentiable Approach.* Cambridge, England: Cambridge University Press.

Mas-Colell, A (1986). The price equilibrium existence problem in topological vector lattices. *Econometrica*, 54, 1039–1054.

Mas-Colell, A, MD Whinston and J Green (1995). Microeconomic theory. New York and Oxford: Oxford University Press.

Mas-Colell, A and WR Zame (1991). Equilibrium theory in infinite dimensional spaces. In *Handbook of Mathematical Economics*, H Hildenbrand and H Sonnenschein (eds.), Chap. 34, pp. 1835–1898. Amsterdam and New York: North-Holland.

McKenzie, L (1954). On equilibrium in Graham's model of world trade and other competitive systems. *Econometrica*, 22, 147–161.

McKenzie, L (1956–1957). Demand theory without utility index. *Review of Economic Studies*, 24, 185–189.

McKenzie, L (1960). Stability of equilibrium and the value of positive excess demand. *Econometrica*, 28, 606–617.

McKenzie, L (1981). The classical theorem of competitive equilibrium. *Econometrica*, 49, 819–841.

McKenzie, L (2002). *Classical General Equilibrium Theory*. Cambridge, MA: MIT Press.

Mead, JE (1952). *A Geometry of International Trade*. London: George Allen and Unwin Ltd.

Menger, C (1871). *Grundsätze der Volkswirtschaftslehre*, Wilhelm Braumüller: Wien.

Milnor, JW (1965). *Topology from the Differentiable Viewpoint*. Charlottesville: Virginia University Press.

Moore, JC (1975). The existence of "compensated equilibrium" and the structure of the Pareto efficiency frontier. *International Economic Review*, 16, 267–300.

Nash, J (1950). Equilibrium points in N-person games. *Proceedings of the National Academy of Sciences of the U.S.A.*, 36, 48–49.

Negishi, T (1960). Welfare economics and existence of an equilibrium for a competitive economy. *Metroeconomica*, 12, 92–97.

Negishi, T (1961). Monopolistic competition and general equilibrium", *Review of Economic Studies*, 28, 196–201.

von Neumann, J (1937). Über ein Ökonomisches Gleichungssystem und eine Verallgemeinerung des Browerschen Fixpunktsatzes. *Ergebnisse eines Mathematischen Kolloquiums*, 8, 73–83.

von Neumann, J (1945). A model of general economic equilibrium. *Review of Economic Studies*, 13, 1–9. (English translation of Neumann (1937).

Newman, P (1960). The erosion of Marshall's theory of value. *Quarterly Journal of Economics*, 74, 587–600.

Nikaido, H (1956). On the classical multilateral exchange problem. *Metroeconomica*, 8, 135–145.

Nikaido, H (1968). *Convex Structures and Economic Analysis*, New York: Academic Press.

Nishino, H (1971). On the occurence and existence of competitive equilibria. *Keio Economic Studies*, 8, 33–67.

Noguchi, M (1997). Economies with a continuum of consumers, a continuum of suppliers and an infinite dimensional commodity space. *Journal of Mathematical Economics*, 27, 1–21.

Novshek, W (1980). Cournot equilibrium with free entry. *Review of Economic Studies*, 47, 473–486.

Novshek, W and H Sonnenschein (1983). Walrasian equilibria as limits of noncooperative equilibria, Part II: Pure strategies. *Journal of Economic Theory*, 30, 171–187.

Pareto, V (1909). *Manuel d'Economie Politique*. Paris: Giard.

Peleg, B and M Yaari (1969). Markets with countably many commodities. *International Economic Review*, 11, 369–370.

Pigou, AC (1920). Principles 8th ed., pp. 388–390.

Pigou, AC (1920). *The Economics of Welfare*. London: Macmillan and Co. Ltd., 2nd edition (1924), 3rd edition (1929) and 4th edition (1932)).

Rådström, H (1952). An embedding theorem for spaces of convex sets. *Proceedings of American Mathematical Society*, 3, 165–169.

Ramsey, FP (1928). A mathematical theory of savings. *Economic Journal*, 38, 543–559.

Roberts, J and H Sonnenschein (1977). On the foundations of the theory of monopolistic competition. *Econometrica*, 45, 101–113.

Robertson, DH (1924). Those empty boxes. *Quarterly Journal of Economics*, 34, 16–30.

Robertson, DH (1924). The trees of the forest. *Economic Journal*, **40** (March 1930), pp. 80–89.

Robertson, DH (1957). *Lectures on Economic Principles*. London: Staples Press.

Robinson, J (1933). *Economics of Imperfect Competition*, London: Macmillan.

Rockafellar, RT (1970). *Convex Analysis*. Princeton, NJ: Princeton University Press.

Romer, P (1986). Increasing returns and economic growth. *Journal of Political Economy*, 94, 1002–1037.

Rosen, S (1974). Hednic prices and implicit markets: Product differenciation in pure competition. *Journal of Political Economy*, 82, 34–55.

Royden, HL (1988). *Real Analysis*. New York: Macmillan.

Rudin, W (1991). *Functional Analysis*. New York: McGraw-Hill.

Samuelson, PA (1947). *Foundations of Economic Analysis*. Cambridge: Harvard University Press.

Samuelson, PA (1948). International trade and the equalization of factor prices. *Economic Journal*, 58, 163–184.

Samuelson, PA (1953–1954). Prices of factors and goods in general equilibrium. *Review of Economic Studies*, 21, 1–20.

Scarf, H (1986). Notes on the core of a productive economy. In *Contributions to Mathematical Economics: In Honor of Gerald Debreu*, W Hildenbrand and A Mas-Colell (eds.), Chap. 21, Amsterdam and New York: North-Holland.

Schmeidler, D (1969). Competitive equilibria in markets with a continuum of traders and incomplete preferences. *Econometrica*, 37, 578–585.

Shaefer, HH (1974). *Banach Lattices and Positive Operators*. Berlin and New York: Springer-Verlag.

Shafer, W and H Sonnenschein (1975). Equilibrium in abstract economies without ordered preferences. *Journal of Mathematical Economics*, 2, 345–348.

Shapley, L (1973). On balanced games without side-paymets. In *Mathematical Programming*, TC Hu and SM Robinson (eds.), New York: Academic Press.

Shannon, C and WR Zame (2002). Quadratic concavity and determinacy of equilibrium. *Econometrica*, 70, 631–662.

Sonnenschein, H (1971). Demand theory without transitive preference with applications to the theory of competitive equilibrium. In *Preference, Utility and Demand*, J Chipman, L Hurwicz, M Richter and H Sonnenschein (eds.), Harcourt, Brace, Jovanovich, New York.

Sonnenschein, H (1972). Market excess demand functions. *Econometrica*, 40, 549–563.

Sonnenschein, H (1973). Do Walras' identity and continuity characterize the class of community excess demand functions? *Journal of Economic Theory*, 6, 345–354.

Sraffa, P (1926). The laws of return under competitive conditions. *Economic Journal*, 36, 535–550.

Statistical abstract of the United States, 1912, p. 202

Suzuki, T (1992). Nonconvexities, externalities, and increasing returns. Unpublished PhD Thesis, University of Rochester.

Suzuki, T (1995). Nonconvex production economies. *Journal of Economic Theory*, 66, 158–177.

Suzuki, T (1996). Intertemporal general equilibrium model with external increasing returns. *Journal of Economic Theory*, 69, 117–133.

Takayama, A (1986). *Mathematical Economics*. Cambridge, England: Cambridge University Press.

Wald, A (1933–1934). Über die Eindeutige Positive Löbarkeit der Neuen Produktions-Gleichungen. *Ergebnisse eines Mathematischen Kolloquiums*, 6, 12–20.

Wald, A (1934–1935). Über die Produktionsgleichungen der ökonomische Wertlehre. *Ergebnisse eines Mathematischen Kolloquiums*, 7, 1–6.

Wald, A (1936). Über einige Gleichungssysteme der Mathematschen Ökonomie. *Zeitschrift für Nationalökonomie*, 7, 637–670.

Wald, A (1951). On some systems of equations of mathematical economics. *Econometrica*, 19, 368–403. (English translation of Wald (1936).)

Walras, L (1874, 1877). *Elements d'Economie Politique Pure*, Lausanne: L Corbaz. In 1954 translated by W Jaffé as *Elements of Pure Economics*. Homewood, Illinios: Richard D. Irwin.

Wealth and Welfare, p. 178

Yamazaki, A (1978). An equilibrium existence theorem without convexity assumptions. *Econometrica*, 46, 541–555.

Yamazaki, A (1981). Diversified consumption characteristics and conditionally dispersed endowment distribution: regularizing effect and existence of equilibria. *Econometrica*, 49, 639–654.

Yamazaki, A (1986). *Foundations of Mathematical Economics*. Tokyo: Keiso-Shobo (in Japanese).

Yosida, K (1968). *Functional Analysis*. Berlin and New York: Springer-Verlag, 6th edition (1980).

Young, A (1913). Pigou's wealth and welfare. *Quarterly Journal of Economics*, 27, 672–686.

Young, A (1928). Increasing returns and economic progress. *Economic Journal*, 38, 527–542.

Zame, WR (1987). Competitive equilibria in production economies with an infinite dimensional commodity space. *Econometrica*, 55, 1075–1108.

Index